Shang Gao • Sui Pheng Low

Lean Construction Management

The Toyota Way

Springer

Shang Gao
School of Architecture and the Built Environment
The University of Newcastle, Australia (Singapore campus)
Singapore, Singapore

Sui Pheng Low
Department of Building
National University of Singapore
Singapore, Singapore

ISBN 978-981-10-1238-9 ISBN 978-981-287-014-8 (eBook)
DOI 10.1007/978-981-287-014-8
Springer Singapore Heidelberg New York Dordrecht London

Printed on acid-free paper

Springer is part of Springer Science+Business Media (www.springer.com)

Lean Construction Management

Preface

The Chinese construction industry has witnessed many remarkable achievements over the past years. Yet it still suffers from many problems, including low product quality, low working efficiency, projects finishing over budget, huge construction wastes and others. As very little past studies have shed light on the Chinese construction industry in the context of implementing lean or Toyota Way practices, this study addresses this research gap with the aim of generating useful insights that may better guide large Chinese construction firms in embarking on a lean transformation exercise by means of deploying the Toyota Way principles.

The Toyota Way was historically the first domain, where the practices and principles of lean production or lean construction were formulated and developed. It can be easily seen that lean construction has already borrowed some principles and techniques of the lean concept or Toyota Way and has become an established theme in the construction domain. The aim of this study is to establish the implementation framework of the Toyota Way model for large construction firms in China. It begins with an extensive literature review of the lean concept, the Toyota Way and the relevant frameworks of lean construction. A theoretical framework for the Toyota Way model within the construction context has been developed and is accompanied by a list of Toyota Way-styled attributes, which fit into the construction context. It is worth highlighting that the focus has been put on the Toyota Way model, over other existing frameworks of lean construction, because of the comprehensiveness of the Toyota Way model, which contains four layers—the philosophy model, the process model, the people and partners model, and the problem-solving model. Most importantly, it has addressed the technical and social aspects of the lean concept.

In order to assess Toyota Way practices within large Chinese construction firms, a mixed research method was adopted at different stages of the study. For a start, a structured questionnaire based on the identified Toyota Way-styled attributes was developed, and data was collected from building professionals with large construction firms in China. The quantitative data outlines the status quo of the Toyota Way-styled practices implemented in the Chinese construction industry, as well as the extent to which these attributes were perceived. The results showed that all the actionable attributes derived from the Toyota Way model were appreciated by the respondents, but some attributes fall short of implementation. To further investigate why implementation was uneven, and also to understand how these Toyota Way practices could be implemented in real-life projects, interviews and case studies were carried out as part of the investigation. At this stage, from the interview findings, the evaluations of the case study projects and the comparisons with the theoretical model of the Toyota Way, the findings have enhanced the understanding of Toyota Way practices in the Chinese construction context. Furthermore, the results highlighted that the gap between actual practice and Toyota Way-styled practices is enormous, and implementation faces considerable challenges. Based on all the findings, this study then employs the SWOT analysis to present a picture that addresses the strengths, weaknesses, opportunities and threats to the implementation of the Toyota Way in China. It is also confirmed that the Toyota Way model presented in this book is considered appropriate for use in Chinese construction firms and may additionally be used as a holistic assessment tool for measuring the maturity of firms in terms of their Toyota Way implementation. Management would then be in a better position to develop plans for Toyota Way implementation by focusing on weak areas, and thus increasing the likelihood of success in the implementation of the Toyota Way.

Keywords: Toyota Way model; Lean construction; Large Chinese construction firms; SWOT

Singapore, Singapore Shang Gao
Singapore, Singapore Sui Pheng Low

Abbreviations

ACE	Architectural, Engineering and Construction
BIQ	Built-in Quality
BIM	Building Information Modeling
CCA	Chinese Construction Association
CCETB	China Construction Engineering Third Bureau
CCTV	Closed-circuit Television
CI	Corporate Image
CII	Construction Industry Institute
CICA	China International Contractors Association
CICC	Chinese International Construction Company
COE	Collective-owned Enterprises
CSCEC	China State Construction Engineering Corporation
DI	Design Institute
GDP	Gross Domestic Product
GM	General Motors
HRM	Human Resource Management
JIT	Just-in-Time
LCCF	Large Chinese Construction Firm
LCI	Lean Construction Institute
LPS	Last Planner System
MHURD	Minister of Housing Urban-Rural Development
MOC	Ministry of Construction
NAOC	National Audit Office of China
NUMMI	New United Motor Manufacturing, Inc
NBSC	National Bureau of Statistics of China
NIE	National Inspection Exemption
PDCA	Plan-Do-Check-Act
POM	Production and Operations Management
PE	Project Engineer
PM	Project Manager
PMBOK	Project Management Body of Knowledge
PPC	Percent Plan Complete
PPT	Project Planning Taskforce

QCs	Quality Circles
Rebar	Reinforcement bar
RMB	Renminbi
SME	Small and Medium Enterprises
SOE	State-Owned Enterprises
SOP	Standardized Operating Procedure
SQC	Statistical Quality Control
SWOT	Strengths, Weaknesses, Opportunities and Threats
TFV	Transformation-Flow-Value
TPS	Toyota Production System
TQC	Total Quality Control
TQM	Total Quality Management
WIP	Work-in-process
WTO	World Trade Organization

List of Non-English Terms Used

5-S	5-S is the acronym for Sort (Seiri), Simplify (Seiton), Sweep (Seiso), Standardize (Seiketsu), and Self-discipline (Shitsuke): a visually oriented system for organizing the workplace to minimize the waste of time
Andon	A visual control device to notify management, maintenance staff and operators of quality or process problems
Gemba	The actual place where the real added-value work is done
Genchi Genbutsu	Go and see for yourself
Guanxi	Relationships among various parties that cooperate and support one another
Hansei	Relentless reflection
Heijunka	Production smoothing or leveling the production schedule
Hoshin kanri	Also called policy deployment. It is a step-by-step planning, implementation and review process to manage change
Jianli	This is the supervision firm in the Chinese construction industry. The *jianli's* main role is to ensure that a project is constructed safely and to the quality standards as required under the law
Jidoka	Providing machineries and operators with the ability to detect when an abnormal condition has occurred and to immediately stop work
Kaizen	Continuous, incremental improvement of an activity to create more value with less waste
Kanban	A signal, often a card attached to suppliers or equipment that regulates pull by signaling upstream production or delivery
Muda	Waste or non-value-adding activities
Muri	Overburden—when workers or machines are pushed beyond their capacity
Mura	Unevenness—when some workers and machines work below their capacities for some of the time, while others may overproduce
Nemawashi	A commonly used Japanese consensus building technique

Poka-yoke	Mistake-proofing by employing visual signals that prevent mistakes or defects
Takt time	A German word for cycle time which is calculated as the available production time divided by the rate of customer demand

Contents

List of Figures

List of Tables

Introduction

1

1.1 Background

The construction industry in China is in a period of rapid expansion, witnessing a steady growth rate of 10 % in recent years. The latest statistics released by the National Bureau of Statistics of China (NBSC 2011) indicate that the construction sector contributes 6.7 % to the Chinese Gross Domestic Product (GDP) (RMB 40.12 billion) at the end of 2010 and will continue to play a significant role in expanding the Chinese economy. However, within China, the construction industry is still perceived as one of the less developed sectors and has a public image of producing low-quality products (Chen 1998; Lam and Cheng 2004; China Daily 2010a) with low productivity (Xue et al. 2008; Xu et al. 2005), low profit margins (China Daily 2010a; Cheah et al. 2007), and poor on-site working conditions (Lu and Fox 2001). Hence, there is a great need to help the industry improve its competitiveness, change its image, and contribute to the national economy.

Taiichi Ohno, a Toyota Motor Corporation engineer, revolutionized the thinking on process inefficiency or "waste" in the early 1950s, leading to the creation of the Toyota Production System (hereinafter referred to as TPS). This system helped propel Toyota Motor Corporation from a small truck-maker in the early 1950s to the world's largest auto-maker by 2007. Over the years, Toyota has been able to sustain a strategic competitive advantage by applying TPS as a process of innovation, as measured by quality, reliability, productivity, cost reduction, sales and market share growth, and market capitalization. The principles underlying the TPS are embodied in Liker's (2004) book, *The Toyota Way*, where he outlined the strategic organization-level principles that guide behaviour in Toyota. More recently, many organizations have tried to replicate Toyota's success by adopting TPS or derivative philosophies, such as lean production and lean thinking principles, into their respective business environments. Its implementation beyond manufacturing firms has also been reported in the literature, such as in the health care sector (Collins and Muthusamy 2007) and the construction industry (Koskela 1992; Ballard 2000).

S. Gao and S.P. Low, *Lean Construction Management*,
DOI 10.1007/978-981-287-014-8_1,

The Toyota Way philosophy thus has the potential to help solve the problems that plague the Chinese construction industry with a view to changing its poor image for the better. This study examined how the Toyota Way principles can be used as guidelines for the large Chinese construction firms (LCCFs) to improve their performance and enhance competitiveness.

1.2 Problem Statement

China's construction industry, as one of the pillars of China's economy, has been developing very rapidly in recent years (Han and Ofori 2001; Low and Jiang 2003a). Membership of the World Trade Organization (WTO) in 2001, as well as the successful hosting of the Olympic Games in 2008 and the Shanghai World Expo in 2010 helped to draw much attention to the industry. While many of the remarkably successful construction projects were highly appreciated by their western counterparts, China's construction industry nonetheless still suffers from many problems: poor product quality (Chen 1998; Lam and Cheng 2004); low work efficiency (Xu et al. 2005); over budget (Cheah et al. 2007; China Daily 2010a; Liao 2004); high frequency of accidents at construction sites (Fang et al. 2008); and huge construction waste (China Daily 2010b). Although these problems have been highlighted repeatedly by the government, hardly any progress has been seen. Moreover, the Ministry of Housing Urban-Rural Development (MOHURD, 2008) has begun to recognize the gaps between China's construction firms and their counterparts in the developed nations in terms of management capabilities. MOHURD (2008) has outlined a government agenda for improving the management level of firms and has highlighted that the business and project management of Chinese construction firms need to adapt to standardization, normalization, and fine-tuning at all stages of management processes, in order to develop a unique management method for them. The highlighting of these problems within China's construction industry has encouraged people to think about how construction project methods can be changed. This is especially the case with China's becoming a WTO member and with the world's economy being in a recession; the competition faced by the construction industry is no longer regional, but is global, and therefore, there is an urgent need for new management ideas and methods to improve the industry's performance.

Inspired by the manufacturing sector, a significant number of studies have been undertaken in construction aimed at reaping similar benefits (e.g. Egan 1998). In line with these, it is important to acknowledge the lessons learnt in the manufacturing industry, where lean implementation took off. For example, many applications of lean practices and principles have focused more narrowly on the technical or physical aspect of the lean system (see Liker 2004; Paez et al. 2004), while ignoring the implications for human resource management (Green 2002). This study therefore builds upon Liker's (2004) Toyota Way model, which is well considered with the social and technical aspects of the lean system.

Moreover, the challenge of dealing with the issue of poor quality and high cost of construction, while speeding up the value-delivery process of construction projects has therefore led to the need to explore production management systems (such as TPS or lean philosophy) which optimize value delivery to customers while minimizing waste. If the application of Toyota Way principles to the Chinese construction industry is feasible, it seems that this new production management philosophy would allow the industry to have the opportunity to improve its performance. Although the authors of the book *The Machine that Changed the World* claimed that these lean techniques could apply universally in all sectors (Womack et al. 1990), it is necessary for the Chinese construction industry to develop its own implementation framework based on lean principles if it is to solve the above problems.

1.3 Research Aim, Objectives and Hypotheses

1.3.1 Aim and Objectives

Based on the above discussion, this study aims to establish the implementation framework of the Toyota Way model that can guide and enable LCCFs to embark on lean implementation in the future. To achieve this, the research pursues the following key objectives:

Objective 1

Break the Toyota Way principles down into measurable or quantifiable parameters that are appropriate to the construction context.

To establish understanding of the subject matter, literature with relevance to Toyota Way principles is reviewed in the areas of lean production, TPS, and lean construction. The sub-objectives are:

1. To review the state of art in lean approaches and the Toyota Way.
2. To identify the actionable attributes of Toyota Way-styled practices in the construction context.
3. To develop a conceptual framework incorporating lean, Toyota Way, as well as relevant theories from the domain of production and management

Objective 2

Investigate the status quo of Toyota Way principles that have been implemented by LCCFs.

This investigates the current practices of LCCFs and specific problem areas. An industry-wide questionnaire survey is conducted to identify

1. *The extent to which* the attributes derived from Toyota Way principles have been implemented by LCCFs,
2. *The extent to which* the attributes derived from Toyota Way principles are perceived as important factors in managing the projects, and

3. The relationship between the implementation level of Toyota Way practices and project performance.

Objective 3

Understand how these Toyota Way-styled practices can be better implemented within LCCFs.

This focuses on understanding the gaps that exist between the current practices of LCCFs and the Toyota Way standards. Constraints that would hinder the implementation of the Toyota Way principles—specially of those rated poorly in the questionnaire survey—are reviewed and investigated in real-life projects, to achieve the following:

1. Understanding the current practices of LCCFs in relation to the Toyota Way-styled practices.
2. Investigate the constraints in real-life projects that hinder the successful implementation of Toyota Way principles.

Objective 4

Establish the Toyota Way implementation guidelines for LCCFs.

This focuses on establishing the Toyota Way guidelines as a holistic approach for LCCFs. The sub-objectives are:

1. To use SWOT analysis to summarize the strengths, weaknesses, opportunities, and threats from the collective findings on the basis of all the fieldwork carried out.
2. To propose strategies to mitigate the threats and minimize the weaknesses of LCCFs in terms of Toyota Way implementation.
3. To refine some of the Toyota Way implementation guidelines to better suit the Chinese context.

In fulfilling the first objective of the study, a detailed review of the 14 Toyota Way principles is required. Efforts are also needed to operationalize the underlying principles and behaviours of the Toyota Way into measurable parameters. Moreover, comparing the status quo in LCCFs with each Toyota Way principle would outline the gaps and challenges that might hinder the process of implementation of the Toyota Way model within China's construction industry. Eventually, based on Liker's (2004) 4P model, as well as findings arising from the fieldwork, a set of implementation guidelines are proposed for LCCFs to commence their implementation with philosophy, process, people, and problem-solving, all of which are intertwined in various ways.

1.3.2 Research Hypotheses

Based on the research objectives stated above, this study sets out to test a number of hypotheses: first, that LCCFs have implemented Toyota Way principles (H_1); and second, that Toyota Way principles and attributes are perceived as important factors in firm performance by Chinese building professionals (H_2). The first two hypotheses are derived from the first objective of this study, which is related to the approach of LCCFs in conducting business, i.e. in the area of philosophy, site operations (process), people management, and problem-solving practices. The third hypothesis is formulated to test the differences between the extent to which Chinese building professionals perceived Toyota Way attributes to be important and the extent to which they have implemented Toyota Way attributes (H_3). Moreover, in examining the extent to which potential hindrances may affect the implementation of Toyota Way principles in China's construction industry, two further hypotheses are developed:

1. There are hindrances when Chinese construction firms implement Toyota Way principles ($H_{4.1}$).
2. There are no significant differences in the perceptions of the barriers to Toyota Way implementation between premier and first-grade Chinese construction firms ($H_{4.2}$).

The final hypothesis tests the relationship between Toyota Way implementation level and the performance measurements. It states that there is a positive relationship between the level of Toyota Way implementation and performance measurements (H_5). A correlation analysis was performed to test this hypothesis.

1.4 Research Scope

This research is driven by the rising recognition of the constant quality problems and other issues reported in the Chinese construction industry. The study focuses on examining the operations of LCCFs to establish how Toyota Way principles can be incorporated in their business operations. It is necessary to define several specific boundaries, including:

1.4.1 Research Focus

The Toyota Way is the real-life model from which all understanding of lean production originates. Since the 1990s, there are certain terms, such as lean construction and lean thinking principles, that have been applied in the construction industry; these all refer to the same model inspired and derived from the Toyota Production System (Koskela 2004). In contrast to the various models or frameworks of the lean approach or lean construction that narrowly focus on the technical aspects of lean, this study will be expanded to give a more holistic approach to lean construction. In the context of this research view, lean in construction is based on the Toyota Way model (Liker 2004) in order to explain its adoption across four key areas:

1. The Toyota Way Philosophy model: the cornerstone of the Toyota Way, which encourages managers to base their decisions on a well-articulated long-term vision.
2. The Toyota Way Process model: this contains the "tactical" or "operational" aspects of the Toyota Way.
3. The Toyota Way People and Partners model: this concerns how Toyota's strategy relates to its people and partners.
4. The Toyota Way Problem-solving model: this has been seen to be critical in solving problems and sustaining improvements in performance

The four models contain 14 principles in total and are explained in detail in Chap. 4, which also explains why the Toyota Way model was chosen from among the various models of lean production and lean construction.

1.4.2 Measurable Factors

The Toyota Way model was formulated by Liker (2004). The Toyota Way principles are explained using cases that show how these are implemented in the development of the Lexus and Prius. It is necessary to operationalize the 14 management principles into quantifiable attributes, in order to fit into the construction environment. This is because the construction processes and their peculiarities are the two main reasons that prevent the industry from adopting lean principles (Koskela 1992). Thus, some Toyota Way principles may need modification in order to test their applicability in the Chinese construction industry.

1.4.3 The Targeted Firms

Construction firms in China are divided into four categories in terms of ownership: state-owned enterprises (SOEs), collective-owned enterprises (COEs), enterprises with shares, and private enterprises (Wang et al. 2006). More recently, a group of leading Chinese construction firms has emerged with large-scale operations, solid construction capabilities, and strong initiative in China's construction industry. This study aims to shed light on how the Toyota Way principles can be implemented in the Chinese construction industry, by focusing on leading construction firms whose qualification falls into the "premier" and "first" categories. There are several reasons for choosing LCCFs for this study:

(1) These firms are relatively large in terms of size, and who employ a large number of managers (such as project managers, project directors, and quality managers). This factor can assist in securing the cooperation of a relatively large number of managers to participate in this study.

(2) It is generally accepted that they represent a typical business model in terms of management style, site management practice, and human resource management in China.

(3) It is more likely that LCCFs will be able to absorb the management philosophy than SMEs. Efforts have been made by many scholars in exploring the opportunities of implementing information management (Love and Irani 2004), knowledge management (Hari, et al. 2005), innovation (Barrett and Sexton 2006), and other approaches in SMEs in the construction industry. All these studies have indicated that SMEs tend to focus more on securing the next project, rather than on implementing contemporary management approaches, due to their limited resources. Similarly, implementing the Toyota Way—an approach that originated outside of construction—would inevitably initiate changes and would require commitment and resources. Given the constraints faced by SMEs, it appears to be more challenging for them to adopt the Toyota Way in the first instance.

Moreover, following Long's (2006) work on differentiating units of analysis and units of observation, this study has taken LCCFs as the unit of analysis. For example, the survey data were aggregated at the firm level by computing the mean scores on various Toyota Way style attributes measuring the implementation level of the Toyota Way. On the other hand, the units of observation in this study are the building professionals working for the LCCFs, as these are the individuals who participated in this study.

1.5 Research Methodology

To achieve the aim and objectives, this research involves the following key phases:

1. Literature review: The available literature on lean production in general, lean construction, and the Toyota Way was reviewed. Possible theories from the domains of management and production were also reviewed and linked to the Toyota Way model. One of the outcomes of the literature review is to identify a basket of actionable attributes underpinning the Toyota Way principles.
2. Two-phased research: The mixed-methods research is adopted. Data were collected from Chinese building professionals working at LCCFs using a questionnaire survey and interviews, in order to investigate both quantitative and qualitative elements of the research objectives. Additionally, case studies of three ongoing construction projects in China enhanced understanding of the application of Toyota Way principles in real-life projects.
3. Validation: The findings arising from the fieldwork, as well as strategies proposed to better implement the Toyota Way within LCCFs, were further validated by six Chinese building professionals who are currently working in Singapore (see Chap. 11).

1.6 Significance of this Research

This study makes the following contributions to knowledge and practice, particularly to the improved understanding of lean management and Toyota Way-styled practices in construction. Further details can be found in the final chapter of this study. With respect to its practical significance, the contribution includes the following results: (1) the research has produced a checklist of Toyota Way-styled attributes that can be used easily to evaluate gaps (if any) in the implementation of such management philosophy, and practices for firms intending to embark on lean transformation or Toyota Way implementation; (2) a report on the status quo of lean or Toyota Way principles in China's construction context, where the relevant literature is found lacking; and (3) a series of implementation guidelines for Toyota Way model, with the necessary modifications to facilitate better implementation in the Chinese construction industry. The theoretical contributions consist of: (1) the proposition of an alternative model to the Toyota Way for the lean construction community, possessing better conceptual underpinnings from theories of management and production; (2) addressing human resource issues in the context of lean construction while also acknowledging the importance of the technical aspects of lean system; and (3) integrating the current two major schools of thoughts of lean construction into the Toyota Way model.

1.7 Organization of Chapters

This book is structured as follows:

Chapter 1 introduces the study by outlining the background and the research problem. Chapter 1 also sets out the research aim and objectives of this study.

Chapter 2 reviews the production management related literature as the Toyota Production System falls within this domain. In this chapter, Koskela's (2000) Transformation-Flow-Value (TFV) model of production is introduced to complement the need to further enhance the theory of production. Chapter 2 also presents a brief history of management thoughts for each management school to find its practice in the relevant production template.

Chapter 3 presents a summary of the literature in the body of knowledge relating to production paradigms. In this chapter, three production templates will be reviewed. These production paradigms include the craft-production paradigm, the mass-production paradigm, and the lean-production paradigm. Chapter 3 also provides a detailed explanation of the lean construction philosophy. It presents various issues of lean production and analyses the origins of and advances in lean production. In addition, Chap. 3 reviews the current frameworks of lean construction and their application in various countries.

Chapter 4 presents a review of the 14 management principles of the Toyota Way and clarifies why the Toyota Way model is chosen as the basis for this study. Each principle, grouped into the four categories of the Toyota Way model, is reviewed. The advantages and disadvantages of the Toyota Way model are also discussed, as are the barriers to the effective implementation of the Toyota Way. Chapter 4 also presents a review of how Toyota Way principles can be implemented in construction.

Chapter 5 develops a conceptual framework based on the findings of the literature review. It ties the various theories together for coherence, which then forms the basis for developing the survey questionnaire.

Chapter 6 argues for the need to introduce the Toyota Way model as a framework to the large Chinese contractors and reviews the status quo of the Chinese construction industry in the areas of quality, productivity, profitability, and others, which this research focuses on.

Chapter 7 presents the research methods adopted in this research. The discussion therein answers three questions: firstly, why the two-phased methods are used in this research; secondly, how data is to be collected; and thirdly, what operationalized measurements will be used in this research.

Chapter 8 presents the first part of the empirical results. This chapter deals with descriptive statistics and statistical data analysis. It presents statistical data on the current state of Toyota Way-styled practices with LCCFs, as well as the perceived importance of the Toyota Way attributes. It highlights the implementation and importance gap. The relationships between levels of implementation with project performance are also investigated. The results provide an insightful overview of what the strengths and weaknesses are in terms of the implementation capacities of the Toyota Way.

Chapter 9 presents the interview results, which focuses on understanding the gaps that exist between the current practices of LCCFs and the Toyota Way standard. Moreover, constraints that hinder the implementation of the Toyota Way principles are also discussed during the interviews and presented in this chapter.

Chapter 10 presents the three case studies conducted. The case studies further test the proposed model of the Toyota Way in the Chinese construction context.

Chapter 11 employs the SWOT analysis to summarize the results. The implications of this analysis are discussed. The chapter concludes with a set of strategies proposed to better implement Toyota Way-styled practices in the Chinese construction industry.

Chapter 12 concludes the book. In this chapter, the theoretical and practical contributions of the research for academics, practitioners and policy-makers are discussed. Finally, the chapter highlights the limitations of the research and recommends possible future research directions.

Production Management 2

2.1 Introduction

Production and operations management (POM) is the management of the production process by which goods and services are made. Research on production management can be found in a large and growing volume of literatures. However, in most POM textbooks (see Gaither and Frazier 1999), it appears to have been repeated on a few topics, the application of methods and frameworks. In recent years, POM research embraces a number of concepts derived from Japanese automobile industry. According to Filippini (1997), Just in time (JIT) and Quality Control (QC)—two building blocks of the Toyota Production System (Ohno 1988)—are becoming two key areas of production and operations management discipline. Moreover, some non-manufacturing industries such as construction are encouraged to emulate the managerial practices proved in manufacturing with the hope of gaining similar benefits (Egan 1998). In this respect, it must further be understood the manufacturing industry and its production management. In this chapter, the first part of the literature review is presented with an effort to cover a number of things. Firstly, this chapter starts with reviewing production management from a systematic perspective, mainly regarding its definition and elements. Secondly, this chapter adopts two approaches to study theoretical aspect of production. One is to search a "theory" of production through economics lens (e.g. Coombs et al. 1987; Perloff 2001), the other way is to review production in production management discipline, in which this study largely draws on Koskela's (1992, 2000) study. Lastly, as Chase and Aquilano (1992) emphasized the need to put management back into production management, it infers that reviewing production management cannot be isolated with its managerial aspect. Moreover, Toyota Way is a management philosophy used by Toyota (Liker 2004). Hence the most important schools of thought within managerial theory in general are reviewed.

S. Gao and S.P. Low, *Lean Construction Management*,
DOI 10.1007/978-981-287-014-8_2, © Springer Science+Business Media Singapore 2014

2.2 Overview of Production Management

POM emerged from World War II and entered the 1950s as a manufacturing oriented subject, which had its basis on concepts and techniques from the scientific management era (Andrew and Johnson 1982). The management of manufacturing of products is referred to as production management (Chase and Aquilano 1992; Gaither and Frazier 1999). While, the functions dealing with the operation of services as well as manufacturing and organizations are covered under operations management (Hopp and Spearman 2000), which is broader than the scope of this study that only concerns management of production. Production management deals with the direct production resources of the firms. These resources may be thought of as an amalgam of five aspects of work including People, Plants, Product, Process, and Planning and control (Lockyer 1984; Chase and Aquilano 1992). The people are the direct and indirect work force; the plants include the factories where production is conducted; the processes include the equipment and the steps by which production is accomplished; planning and control are the procedures and information used by management to operate the system.

Furthermore, production management, as defined by most scholars (e.g. Abramowitz 1967, p.8; Neely 1991; Ogawa 1984), consists of two main functions. First, there is production, which is the act of manufacturing goods for which a consumer is willing to pay. The underlying principles of production are outlined by O'Connor (1994, p.136) as given below:

1. The first principle of production is to convert designs into products, at the lowest cost. A production system takes inputs—raw materials, capital, machinery, labour, information, time and other resources—and transforms them into outputs in the form of products and services of higher value than the inputs. It may also be reviewed as a value-adding process.
2. The second principle of production is that all processes are operated or influenced by people, even though the automation has been increasingly adopted to replace human efforts.
3. The third principle of production is that, as far as practicable, nothing should be made that cannot be billed immediately as it leaves the factory.

Second, there are production managers (also called managers), managing the production system. Management was the process of planning, scheduling, commanding, coordinating and controlling business activities (Wren and Bedeian 2009; Ogawa 1984), and their primary concern is with the activities of the conversion process or production (Gaither and Frazier 1999). Drucker (1986) pointed out that production is not the application of tools to materials; it is the application of logic to work. Management needs to understand the logic behind each system of production and applies these principles consistently and thoroughly (Drucker 1986).

Ogawa (1984) pointed out that production management, originally focusing on managing the production line, has evolved into a means that is directly related to corporate strategy such as to cope with systematization, computerization, automation, respect for human, ecological control, safety and welfare. This change is likely to be related to the birth of industrial giants having complex production systems, such as the Toyota Production System (Ogawa 1984). In other words, production management should encompass not only quality, time and cost as three traditional goals (Hopp and Spearman 1996), but also flexibility, corporate strategy, and the changing business environment (Ogawa 1984).

2.3 Theory of Production: Search in Economics

Economists describe a production process either (1) as an arrangement of productive operations (or tasks) or (2) as a mapping of input quantities into output quantities (Scazzieri 1993). The former approach was favoured by a number of classical economists (Smith and Marx in particular). The latter approach is common to "neoclassical" theory of production, which occupies a rather central place in economics (Coombs et al. 1987). The neoclassical theory of production concerns the first aspect of production management, which focuses on the relationships between quantities of input (factors of production) and outputs in the productive unit. The function that describes the amount of output obtained for specified amounts of the inputs is called the *production function*, and mathematically it takes the form:

$$Q = q(X_1, X_2, X_3, X_4, \ldots, X_n) \text{ or } Q = f(\mathrm{K}, \mathrm{L})$$

where Q denotes the quantity of output and X_i is the ith input. The inputs encompass all things required for production, including raw material, machines, employees, managers, utilities and so on. However, most of these inputs can be grouped into three broad categories, namely capital (K), labor (L) and material (M) (Coombs et al. 1987; Perloff 2001). Under the neoclassical theory of production, the firm is built upon several important assumptions (Perloff 2001). One of the most important is the presumption that firms maximize profits and reduce cost (Coombs et al. 1987; Perloff 2001). This implies that the firm will attempt to exploit all opportunities to make more money and avoid any project that is not expected to make the firm richer (McCormick 1993). Because at any time there is a given level of technology which determines the techniques available for production, therefore among the available techniques the firm will choose the one which, given existing levels of production factors, minimizes total production costs (Coombs et al. 1987). In Toyota, or any other manufacturing firms, profit can be obtained only by reducing costs. According to Ohno (1988), Toyota Production System was born at the age of slow economic growth worldwide with a focus to develop human ability to their fullest capacity, to utilize facilities and machines well, and to eliminate all waste to achieve the cost minimization goal of the company.

2.4 New Production Philosophy: An Integrated View

2.4.1 Overview

The economic explanation of production only captures one aspect of production theory, which focused on the relationship between input and output. There is a general agreement of the formulation $Q = f(\mathrm{K}, \mathrm{L})$ by which the production function involves the transformation (conversion) of inputs into useful products and services. In Koskela's (1992, 2000) view, this conventional view of production was in line with Walrasian production model, which depicts the transformation process of production factors into finished product. Shingo (1988), however, highlighted that this conventional model of production confuses the difference between "operation" and "process", by which they all refer to a worker works on different products. Shingo (1988, p.5) added that there is distinction between process and operation:

1. Process: it refers to the flow of products from one worker to another, that is, the stages through which raw materials gradually move to become finished products.
2. Operation: it refers to the discrete stages at which a worker may work on different products and spatial flow that consistently centres around the worker.

In the meantime, the Japanese owe their leadership in manufacturing quality as a result from the guidance of quality gurus such as W. Edwards Deming and Joseph Juran in the 1950s and 1960s (Drucker 1990). The quality concepts such as Statistical Quality Control (SQC) were developed from statistical theory in 1930s. With SQC's rigorous methodology, Japanese assembly line could deliver built-in process control. Greatly influenced by Shingo's (1988) work which focused on the flow of material as well as quality control concept of production, the term "new production philosophy" has been coined by Koskela (1992). Furthermore, Koskela (1992) outlined that the genesis of new production philosophy was in the Japanese Just in time (JIT) and Total quality control (TQC) efforts in automobile manufacturing and the most prominent application was the Toyota Production System. Additionally, Shingo (1988) outlined that the Toyota Production System represents a pioneering attempt at a new production philosophy over the conventional preoccupation with operations. After that, attempts have been made by Koskela (2000) to develop a model of production that synthesizes all important features of production, especially those that are lacking in the conversion model. Koskela (2000) integrated three different views on the production process, namely the transformation concept, the flow concept and the value generation concept and termed it as a new production model (Koskela 2000).

2.4.2 Production as a Transformation Activity

The transformation concept has deep roots in the present Western thinkings about production (Ogawa 1984; Frisch 1965) and it has been commonly conceptualized in POM textbook as: "*A production system receives inputs in the form of material,*

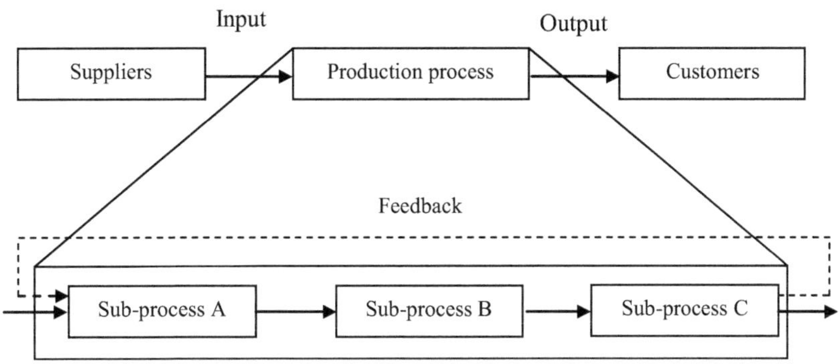

Fig. 2.1 A transformation process of a production process

personnel, capital, utilities, and information. These inputs are changed in a conversion subsystem into the desired products and services, which are called outputs" (Gaither and Frazier 1999, p.154).

According to Koskela (2000), there are three principles in the *transformation* model:

1. Production can be divided into smaller and more manageable sub-processes, finally into tasks, in which all inputs are available and assign these tasks to operatives or workstations [see Fig. 2.1 as adapted from Koskela (2000)].
2. Cost can be minimized by reducing the cost of each sub-process.
3. The output value of a process is associated with the costs (or value) of its input. In practice, the value of the output can be raised by utilizing better materials and more skilled labour.

The transformation concept not only appears in the production management domain, but also can be found in the microeconomics theory of production which employs *production function* to discuss the relationship between input and output. Koskela (1992, 2000) outlined, this transformation concept is predominantly applied in construction industry, where management efforts are centred on task management. This brings to a high chance of causing extra variability that if task management is poorly implemented (Koskela and Vrijhoef 2001).

2.4.3 Production as a Flow Activity

According to Koskela (2000), the transformation model of production had not been challenged until the 1980s when Shingo's (1988) invention on the theoretical rationale of the JIT movement, that highlighted two core points; one is the introduction of time as an input in production, and the second is in the observation that time is consumed by two types of activities: transformation activities and non-transformation activities. The flow view of production was further developed in Japan, especially in the automobile manufacturing at Toyota (Koskela et al. 2002), which was later embodied in "Lean production", a term to characterize the Toyota Production System (Womack et al. 1990). Koskela (2000) explained that the basic thrust of the flow concept of production is to eliminate waste from the flow processes, along with its three types of principles:

1. *Reducing the share of non-value-adding activities (waste)* is the first principle that also serves part of the theoretical and conceptual foundation.
2. There are principles of "*reduce the lead time*" and "*reduce variability*" derived from the flow model.
3. A set of core heuristic principles includes "*simplicity*", "*increase flexibility" and "increase transparency*" are derived based on their usefulness in practice but less direct connections with theory.

2.4.3.1 Reduce the Share of Non-value-Adding Activities (Waste)

Koskela (2000) outlined that the three root causes of non-value-adding activities: (1) the structure of the production system, (2) the way production is controlled and (3) the inherent nature of production attributed the non-value-adding activities in the different time frame of the process (i.e. design, control and improvement of production). With respect to all these root causes, Koskela (2000) proposed the following principles to reduce waste.

2.4.3.2 Reduce the Lead Time

Lead time refers to the time required for a particular piece of material to traverse the flow and can be interpreted in the given formula (Koskela 2000; Monden 1998, p.106).

$$\begin{aligned}\text{Lead time} = {} & \text{queue time before processing} + \text{processing time} + \text{waiting time}\\ & + \text{moving time}\end{aligned}$$

Table 2.1 identified a set of strategies to compress the lead time by elimination of queuing, processing, waiting and moving.

2.4.3.3 Reduce Variability

The principle of reducing variability is to deal with two types of variability, namely process-time variability and flow variability (Hopp and Spearman 1996). Process-time variability refers to the time required to process a task at one workstation, which consists of natural variability such as set-ups, operator availability and rework. The flow variability refers to the variability of the arrival of jobs to a single work station (Koskela 2000).

2.4.3.4 Simplicity

According to Koskela (2000), simplification is the result of the reduction of the number of components or steps that link in a material/information flow. Practical approaches can include shortening the flows by consolidating activities, standardizing parts and minimizing the amount of control information needed. Moreover, organizational changes can also bring about simplification, such as multi-skilled and autonomous teams.

Table 2.1 Strategies to compress the lead time

	Explanation	Strategies to gain reduction
Lead time	• Time required making products. Simply speaking, it is a sum of the following items	• Reducing the time required for the following four elements
Queue time before processing	• The time prior to the commencement of operations	• Establishing one-piece flow through set-up reduction along with the pull method to reduce the lot delay • Standardizing the work in order to reduce the process delay
Processing time	• Process time comprises set-up time and run time	• Small sized lot production • Using conveyor system
Waiting time	• Waiting time after process is the time that inventory must wait before being conveyed to the next process	• Small sized lot production
Moving time	• The duration required to move between employee/machines operations	• A process-based layout so that transport distances are eliminated

Source: Koskela (2000) and Monden (1998)

2.4.3.5 Increase Flexibility

The thrust of JIT production was based on mix flexibility (numbers of different products produced). The practical means to increase flexibility comprise: (1) minimize lots sizes to closely match demand, (2) reduce the difficulty of set-ups and changeovers, (3) training a multi-skilled workforce and (4) training the workforce in operational flexibility and so on.

2.4.3.6 Increase Transparency

Transparency can be used as an instrument to increase the motivation of workers for improvement, reduce the propensity of errors and increase the visibility of errors (Koskela 2000). Koskela (2000) further listed a number of practical approaches for increasing the level of transparency that can include the adoption of 5-S, standardization, using visual controls to enable anyone to capture the difference between the standards and deviation, reduce the interdependence of production units and so on.

2.4.4 Production as a Value Generation Activity

Value creation is the major concern in many modern theories of production management. In the same timeline when the critique originating from the flow concept moved against the transformation concept, the value generation concept was also employed as another approach to evaluate the foundation of production (Koskela 2000). This is a contrast to the transformation concept, which focuses on internal production matters rather than the customers' needs. The value of a product emphasizes more on the customer side, and the goal of production is to satisfy customers' needs. The quality-based movement and marketing-oriented value-based method are two diffusion and practice means of value generation concept

(Koskela 2000). The quality movement originated and disseminated in Japan, under the guidance of Deming, Juran and other quality management techniques (i.e. Quality Control, Total Quality Control, etc.). The value-based approach was fulfilled when a growing number of companies adopted various value generation models including value-based management, customer-driven company, customer orientation and mass customization (Koskela 2000). Overall, the value generation concept of production can be structured into the following five principles according to Koskela (2000, p.79–81):

1. Ensure that all customer requirements, both explicit and latent, have been captured.
2. Ensure that relevant customer requirements are available in all phases of production, and that these are not lost when progressively transformed into design solutions, production plans and products.
3. Ensure that customer requirements have a bearing on all deliverables for all roles of the customer.
4. Ensure the capability of the production system to produce products as required.
5. Ensure by measurements that value is generated for the customer.

2.4.5 TFV Model of Production

Koskela (2000, p.88) highlighted that each concept of production focuses on certain aspects of the production phenomenon and has its own methods and practices, but they are complementary. For integration purpose, Koskela (2000) proposed the "Transformation-flow-value generation" or TFV model of production by conceptualizing the above three complementary ways as shown in Table 2.2.

A closer examination of TFV model of production revealed that each of the three production concepts is closely related to one of the traditional objectives manufacturing firms that strive for, namely cost, time and quality.

1. Cost: cost reduction can be achieved by minimizing the cost of sub-process which transformation concept supports.
2. Time: time can be pressed through eliminating the non-value-adding activities in the flow concept.
3. Quality: the value generation view was started by and later on refined in the total quality movement framework. Ensuring customers' requirement are met in good manner enables the quality of product should be further underscored.

Koskela's (2000) TFV production model, however, has received criticisms. For example, Winch (2006) highlighted the following conceptual weakness that all three conceptual pillars share:

1. Focus on the production as material process. Winch (2006) argued that it ignored the factors that some phases in the production process involve non-transformation activities, for example, supplier service to client in the context of construction.
2. Absence of a concept of organization in the analysis.
3. Lack of any analysis of the implications of risk and uncertainty.

Table 2.2 Integrated TFV view on production

	Transformation view	Flow view	Value generation view
Conceptualization of production	As a transformation of inputs into outputs	As a flow of material, composed of transformation, inspect, moving and waiting	As a process where value for the customer is created through fulfilment of his requirements
Main principles	To make production efficiently	Elimination of waste (non-value-adding activities)	Elimination of value loss (achieve value in relation to best possible value)
Methods and practices (examples)	Work breakdown structure, MRP, organizational responsibility chart	Continuous flow, pull production control, continuous improvement	Methods for requirement capture, Quality function deployment (QFD)
Practical contribution	Ensure what has to be done	Ensure what is unnecessary is done as little as possible	Ensure customer requirements are met in the best possible manner
Suggested name for practical application of the view	Task management	Flow management	Value management

Source: Koskela (2000)

4. The unitary concept of value derived from quality management is inadequate for the value generation concept applied through the construction process.

Nevertheless, the development of TFV production model heavily draws on production management literature (Valence 2010) and has addressed how the three aspects of production, namely tasks, flow and value (quality) can be managed. This is important to an understanding of production management.

2.5 Historical Milestones of Management Thoughts: Search in Production Management

In today's competitive world, it is necessary to understand how production systems should be designed and put into operation in order to support competitive industrial production. Santos et al. (2002b) pointed out that production has been one of the critical laboratories for developing management theories throughout history. The evolution of management thoughts had direct influence on the way how production system was designed and management which reflect its connection with the second aspect of production management. Santos et al. (2002b) revisited the recent history of production management theory and outlined a roadmap for its evolution in Fig. 2.2.

Similar efforts have been made by Mullins (2006), who suggested four main approaches with different focus could identify main trends in the development of organizational behaviour and management theory; the earliest emphasizing production efficiency (classical approach), the second emphasizing human behaviour, the

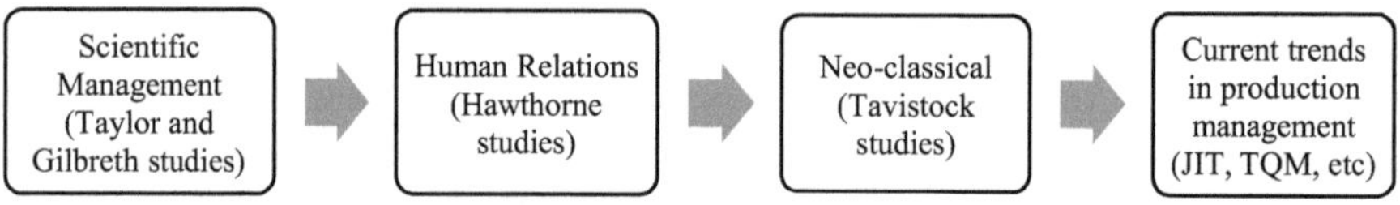

Fig. 2.2 Key contribution of production to the evolution of management

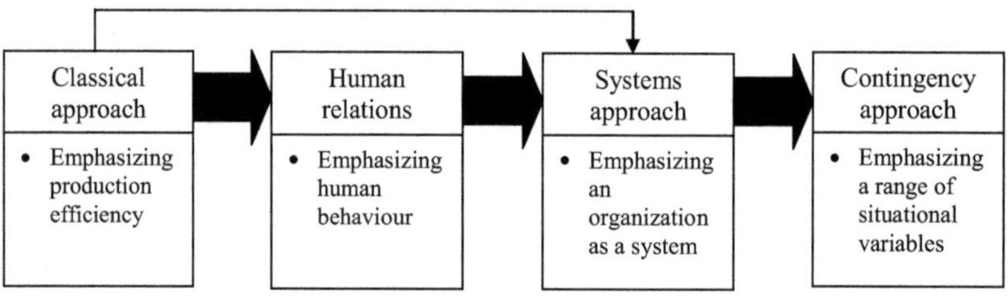

Fig. 2.3 Main approaches to organization, structure and management

third emphasizing organizations as systems and the fourth emphasizing a range of situational variables that determine the success of the organization [see Fig. 2.3 as adapted from Mullins (2006)]. It reflects that the management not only concerns the production process, plants, programmes, but also deals with the people, organization and others. Following Fig. 2.3, the next section reviews these mainstream management thoughts in the domain of production management.

2.5.1 The "Classical" Approach: Scientific Management

The field of production management is generally considered to be an outgrowth of the scientific management movement fostered by Frederick W. Taylor. Notable co-workers of Taylor were Frank Gilbreth (motion study) and Henry Gantt (Scheduling, Gantt chart). Each of these individuals offered great contribution to the scientific management movement and pioneered the evolving methods (Abramowitz 1967). The ideas of scientific management developed by them have had a huge influence on the discipline of production management in the twentieth century (Koskela 2000).

2.5.1.1 The Taylor System of Scientific Management

According to the theory of scientific management, each supervisor and manager is expected to have a total view of the process, define its objectives and steer daily work so that the targets are met. Taylor summarized his method in four principles (Taylor 1934):

1. The proper design of the work tasks such that the absolute maximum amount of work can be extracted from a given labour (using time and motion studies).
2. Scientific selection of the proper workers (finding workers who are highly motivated and controllable).

3. Cooperate with the workers so as to ensure that all of the work is being done in accordance with the principles of the science which has been developed.
4. An almost equal division of the work and the responsibility between the management and the workers.

Time study, used by Taylor to discover "what was possible" in improving job performance, became the foundation of Taylor's work. With a stopwatch, weight scale, and tape, Taylor literally measured the distances that workers and materials covered. As Wren and Bedeian (2009) outlined, Taylor's time study had two phases: analysis and synthesis. In the analysis phase, each job was broken into its elementary movements. Non-essential movements were discarded and the remainder carefully examined to determine the quickest and least wasteful means of performing a job. In the synthesis stage, the elementary movements were combined in the correct sequence to determine the time and the exact method for performing a job. This phase also led to improvements in tools, machines, materials, methods, and the ultimate standardization of all elements surrounding and accompanying a job.

2.5.1.2 The Contributions of Frank Gilbreth

Another important pioneer of the scientific management movement was Frank Bunker Gilbreth. His concept of scientific management can best be described as the search for the one best way to do work (Abramowitz 1967). His early work focused on motion study, which aimed to eliminate those variables that affect motion, develop, standardize and determine the best practice. In doing so, they paved the way for modern work simplification by cataloguing 17 different hand motions, such as "grasp" and "hold" (Kreitner 2007). Rather than Taylor's endeavours on the quality of the operative, Gilbreth offered the view that each worker can be trained in the correct way to sustain those best practices. He sought to improve operator's performance through reducing unnecessary motions (e.g. unnecessary motions can be eliminated through better design of the workplace) and limiting fatigue by placing far greater emphasis on the total working environment (Shelderake 1996). The motion study had generated a great influence on the later concepts such as waste elimination (Ohno 1988), which became the cornerstone of the Toyota Production System.

2.5.1.3 The Contributions of Gantt

Gantt is perhaps best known for his development of the graphic methods of describing plans and making possible better managerial control. He emphasized the importance of time, as well as cost, in planning and controlling work. This led eventually to the famous Gantt chart which is still in wide use today. Due to its simplicity, ease of preparation and graphical format, the Gantt chart is widely used as a construction-scheduling tool (Shelderake 1996).

2.5.1.4 Lessons from the "Classic" Approach

Taylor's "one best way" method became the standard for managerial work, and has been both celebrated and criticized over the years. Kreitner (2007, p.40) comments:

"*within the context of haphazard, turn-of-the-twentieth-century industrial practices, scientific management was indeed revolutionary with its emphasis on promoting production efficiency and waste elimination*". Nevertheless, much of the criticisms being directed towards scientific management were concerned that this management approach and techniques have dehumanized people by making them act like machines (Kreitner 2007). In the 1920s, these aspects were given more attention, which eventually led to *the Human Relations movement.*

2.5.2 Human Relations Approach

In the late 1920s and early 1930s, observers of business management began to develop the human relations school of managerial thought. During that time, workers gradually realized the weaknesses in the scientific management system and started to exploit them. The dehumanization of work on the shop floor, where the imperatives of working with machines had tended to dominate the work of people, had become more evident as mechanization and automation proceeded, threatening jobs which depended on continually expanding markets (Pearson 2009). The human relations school exclusively focused on management's relationship with people at work. Mayo and his colleagues' observations at Western Electric's Hawthorne Works were the first thorough experimental social science study of industrial work, and commonly viewed as having generated great influence on this school. Pearson (2009, p.138) noted that "*Understanding in the field of human relations…is of first importance to the executive; for human relations are the essence of managerial, employee, public and political relations*".

2.5.2.1 Hawthorne Studies

A series of studies, now known as the Hawthorne studies, was conducted from 1924 to 1932 at the Hawthorne Works of the Western Electric Company, as an attempt to investigate how characteristics of the work setting (specifically the level of lighting or illumination) affected worker fatigue and performance. During the experiment, it was found that production output increased when lighting was improved. When lighting was subsequently decreased, however, production again increased. The result suggested that people were strongly affected not only by physical conditions, but also by mental factors. The so-called Hawthorne effect seemed to suggest that workers' attitudes toward their managers affect the level of workers' performance (Wren and Bedeian 2009). This experiment also emphasized the importance of social and psychological factors in the work environment and the recognition of informal organization structures at work, in contrast to the assumptions of scientific management that motivation was simply a matter of payment by results.

2.5.2.2 The Influence of Psychology: Neo-human Relations

According to Mullins (2006), the Hawthorne studies did not address the link between "satisfaction" and work productivity. This is because the link between the two was not always correlated clearly and positively. A group of notable writers such as Abraham Maslow (1954), Frederick Herzberg (1959) and McGregor (1960) made their attempts to understand the forces which motivated people at work and the way in which individual adjustment, group relations and leadership styles impacted on worker motivation (Mullins 2006).

2.5.2.3 Lessons from the Human Behaviour Approach

The human behaviour approach strove for a better understanding of people's psychological and social needs at work as well as improving the process of management (Mullins 2006). According to Kreitner (2007), the human behaviour approach makes it clear to present and future managers that people are the key to productivity and technology, and that work rules, and standards do not necessarily guarantee good job performance. In contrast, success depends on motivated and skilled individuals who are committed to organizational objectives.

2.5.3 System Approach

Whereas classical approaches focused the technical requirements of the organization without the people, and the human relations approaches emphasized the psychological and social aspects of work, excluding the organization, the system approach attempts to reconcile these two earlier approaches by addressing the interrelationships of structure and behaviour, and the range of variables within the organization (Mullins 2006). Lugwig Von Bertalanffy (1973) was the first to use the term "system theory", and who was often cited as the founder of this school. From his perspective as a biologist, an organization is seen as a combination of interdependent parts or subsystems which collectively make up the whole (Mullins 2006). The value of system theory to the study of organizations is its ability to simplify complex situations by considering its subcomponents (subsystems) as well as with the relationship and interdependencies between these subsystems (Mullins 2006). In the system theory the socio-technical system will be discussed as it pertains to production management.

2.5.3.1 The Socio-technical System

The concept of the organization as a "socio-technical" system is concerned with the interactions between the psychological and social factors and the needs and demands of the human part of the organization, and its structural and technological requirements (Mullins 2006). Broadly speaking, the social system is viewed as anything having to do with the selection, development, and characteristics of an organization's people and the culture that emerges through the interaction of those people. The technical system includes not only machines but also the policies and standard operating procedures of an organization. Recognition of the socio-technical approach is of particular important today because people must be considered as at least an equal priority along with investments in technology (Mullins 2006). Morgan and Liker (2006) employed this model to describe Toyota's product development system with three primary subsystems: (1) process, (2) people, and

(3) tool and technology. These three subsystems are interrelated and interdependent and affect an organization's ability to achieve its external purpose.

2.5.4 Contingency Approach

The contingency approach can be seen as an extension of the system approach that highlights possible means of differentiating among alternative forms of organization structures and systems of management (Mullins 2006). According to Kreitner (2007), the contingency approach is an effort to determine which managerial practices and techniques are appropriate in specific situations. This approach to management also acknowledges that there is no one single best way to manage people or work in every situation (Dubrin 2008). It is true that in real-life management, the success of any given technique is dictated by the situation. Given the nature of this management approach, caution should therefore be exercised in this current research study that the so-called best practice of lean or Toyota Way is contingent upon the circumstances and projected outcomes of each unique organization. Simply because it has generally worked well in Japanese manufacturing plants, or because it has now become internationally accepted, it does not necessarily mean that such management practices would work as well in the Chinese construction industry. It is more important to select and/or tweak the principles or combinations of principles to achieve the targeted performance, or to adjust where necessary to better suit the Chinese context.

2.5.5 Discussion

The above literature review has clearly shown that the evolution of management theory from different schools of management thoughts mirrored the changes in the surrounding economic and social environment in the production management discipline. It confirms production as one of the critical laboratories for developing management theories. Moreover, the evolution of the management thoughts had a direct influence on the way how the production system was designed. For example, time and motion study inspired Toyota to eliminate wastes in seven different forms. In addition, the human relations approach laid significant emphasis on people and acknowledged that people are the key to productivity and technology success, while the system approach equally treats the technical aspects and human resource aspects of the organization. The Toyota Way was also developed on the socio-technical system thinking in that the Toyota Way model is incorporated with the people and process parts.

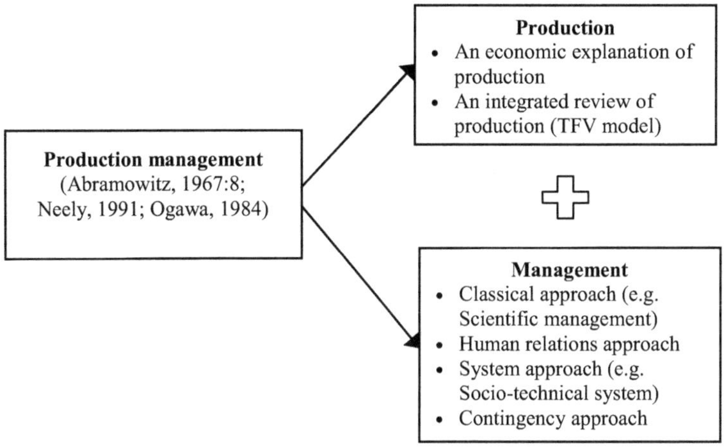

Fig. 2.4 Conceptualization of production management

2.6 Summary

Figure 2.4 outlines contemporary developments in the production management domain with two different emphases, namely on production and management. It parallels Adam's (1983) production management typology that contains two dimensions: the technical transformation axis included the design and operations activities for products and services. The managerial axis separated the classical, behavioural, system, and contingency approaches often used when responding to production and operations problems.

In the course of reviewing production, a brief economic explanation of production has been presented. The development of the neoclassical theory of production was reviewed in particular, which was based on a cost minimizing, profit maximizing firm, with a given level of technology. The economic explanation of production acknowledges all conceivable transformations that can be achieved with given inputs. In order to widen this unitary perspective of production, Koskela (1992, 2000) collectively reviewed three different views of production and integrated them into a new production model, which also laid the foundation for the development of lean construction (see Sect. 3.4). Furthermore, following a chronological order of theory development, this chapter reviewed the evolution of management thoughts from the classical approach to the human relations approach to the system and contingency approaches. Changes in the economy and society worldwide have resulted in a workforce that no longer accept it as being treated like another piece of machinery. In the human relations school, management theorists placed emphasis on motivation, leadership, etc. In the system approach to management, people are treated as equal to technology. The review of conceptualizations of production as well as the various approaches to management theories provides this research with a general theoretical background to review the Toyota Way model, which not only focuses on the manufacturing process, but is also a management philosophy per se.

From Lean Production to Lean Construction 3

3.1 Overview

The last few decades have witnessed three major phases or paradigm shifts of industrial production in the modern world (Womack et al. 1990; Smith 1992). These phases are generally concluded as craft production, mass production and lean production. Shook (1998) highlighted that the concepts such as mass production and lean production reflect ways of thinking about production within particular cultures and eras rather than simply focus on production systems. Moreover, these production paradigms can be related easily to the automobile manufacturing factories, where they are created, exercised and eventually evaluated to the next phase. The first part of this chapter provides a history of evolution of production practices and philosophies through the automobile manufacturing industry from the craft production era to current lean production. The differences in production philosophies, their limits, and their impacts in the construction industry are also reviewed. The second part of this chapter attempts to answer the following question: Can the success of lean principles be replicated in the construction industry? A number of scholars hold positive attitudes towards this proposition (including Koskela 1992, 2000; Ballard 2000). To address this issue, it starts with brief discussion of the peculiarities of construction that differentiate that industry from manufacturing industry. The following discussion covers the terminology of lean construction, along with its principles, tools, and implementation frameworks, as well as criticisms of it. It also sheds light on governments' roles in promoting lean construction in Western countries and outlines a few cases of lean practices that have emerged in developing countries.

S. Gao and S.P. Low, *Lean Construction Management*,
DOI 10.1007/978-981-287-014-8_3, © Springer Science+Business Media Singapore 2014

3.2 From Craft Production to Mass Production

Craft production employs skilled workers to make non-standard products. Womack et al. (1990) summarized the following characteristics of the age of craft production:

1. A highly skilled work force who transformed the inputs with their own hands, using tools to facilitate their work on a job-by-job basis.
2. Organizations were decentralized, although concentrated within small geographic locations.
3. Only producing a very low production volume of unique products.

Hormozi (2001) recognized the benefits that craft production can bring about that revolutionized the creative workers from the burdens of difficult manual labour and allowed them the time to utilize their creativity to increase their income and their standard of living. Many other benefits were reported by Nesan and Holt (1999) within the craft-based approach in the context of construction that include first, the integration of design and construction eliminated buildability problems; second, work processes were decentralized which led craftsmen to fully enjoy authority and control over the construction process. But craft production could not compete with the next phase in manufacturing, namely mass production in terms of its low work efficiency and costly but yet high quality final products.

Mass production was developed to increase production capacity and to reduce unit cost. Henry Ford revolutionized the automobile manufacturing process through the use of scientific management methods, when developing his assembly line for producing the famous Model T, in 1913. The interchangeability of parts and the simplicity of assembly were the manufacturing innovations that made the assembly line possible (Womack et al. 1990). The advantage of the mass production paradigm is that it brings about economies of scale as demonstrated in the moderate quality but low cost products as production volume increase. This gave Ford a substantial advantage over his craft-based competition. Moreover, mass production had a big impact on the organization of work. This was because Taylor's job specialization concept nicely fit with the concept of mass production which employed relatively unskilled labour. However, the disadvantage of mass production is that the job satisfaction is generally low and the cost of switching the models to vary the production is high (Womack et al. 1990). Furthermore, Henry Ford failed to appreciate the potential for producing a variety of end products from a common set of standardized parts, which led to Ford's declining market share. These problems forced American companies to look at lean production techniques as an alternative to mass production.

3.3 The Development of Lean Production

The term "lean production" was first brought to attention through the book—*The Machine that Changed the World* (Womack et al. 1990), in which the authors critically contrasted the differences between Toyota plants and three U.S. Motor giants. The authors claimed that the production philosophy and system of Toyota were superior to all the others, because it used less human effort, less manufacturing space, less investment in tools, less time spent on new product development but generated high quality, less inventory, and a greater variety of products (Womack et al. 1990). Since then, a large volume of publications on lean production emerged, which usually considered the Toyota Production System (TPS), or JIT production, or lean production as synonymous and equal (see Shah and Ward 2007; Womack et al. 1990). In order to understand lean production more precisely, it requires first of all an understanding of its historical evolution which was discussed in the previous sections as well as its mother platform—the TPS.

3.3.1 Toyota Production System

Toyota first drew the world's attention back in the 1980s, when it became clear that there was something special in terms of Japanese quality and efficiency (Liker 2004). It began with Eiji Toyoda's determination to implement American manufacturing methods (mass production techniques) when he studied in the USA in the 1950s. It then took considerable time for Toyota to find ways to maintain economies of scale in manufacturing and procurement with small-lot production. One principle that was considered as worthwhile to adapt from Ford's production system was the continuous flow. On the basis of continuous flow, Toyota created a one-piece flow with a unique flexibility to satisfy the customers' demands. Toyota learnt from the essence of, but did not follow the mass production approach blindly (Ohno 1988). Under the leadership of Eiji Toyoda and Taiichi Ohno, Toyota led the way in developing what is now called the TPS. Ohno (1988) in his book—*Toyota Production System*—credited Ford's mass production and the American supermarket was behind his JIT thinking. Moreover, Toyota adapted quality thinking from the American pioneers within quality engineering, including Edwards Deming and Joseph Juran. Deming encouraged the Japanese to adopt a more systematic approach towards problem-solving. Later this approach became known as the Deming cycle or the Plan-Do-Check-Act cycle (PDCA) which is a pillar for continuous improvement (*Kaizen*) (Imai 1986). These techniques evolved into what is now described as lean production.

TPS's goal is to reduce cost without increasing production volume. The basis to achieve is the elimination of waste and this idea marked the start of the present Toyota Production System (Ohno 1988). It has been widely acknowledged that the two pillars of the Toyota Production System are Just-in-Time and *Jidoka* (Ohno 1988; Monden 1998; Liker 2004; Sugimori et al. 1977) (see Fig. 3.1, as adapted from Liker (2004) for the TPS house).

Just in Time Just-in-time manufacturing prescribes the required units needed to produce the required quantities at the required time, wasting neither raw material nor time. A manufacturing company establishing this flow throughout can ideally approach zero inventories. Just in time is hardly an easy task, as it requires the

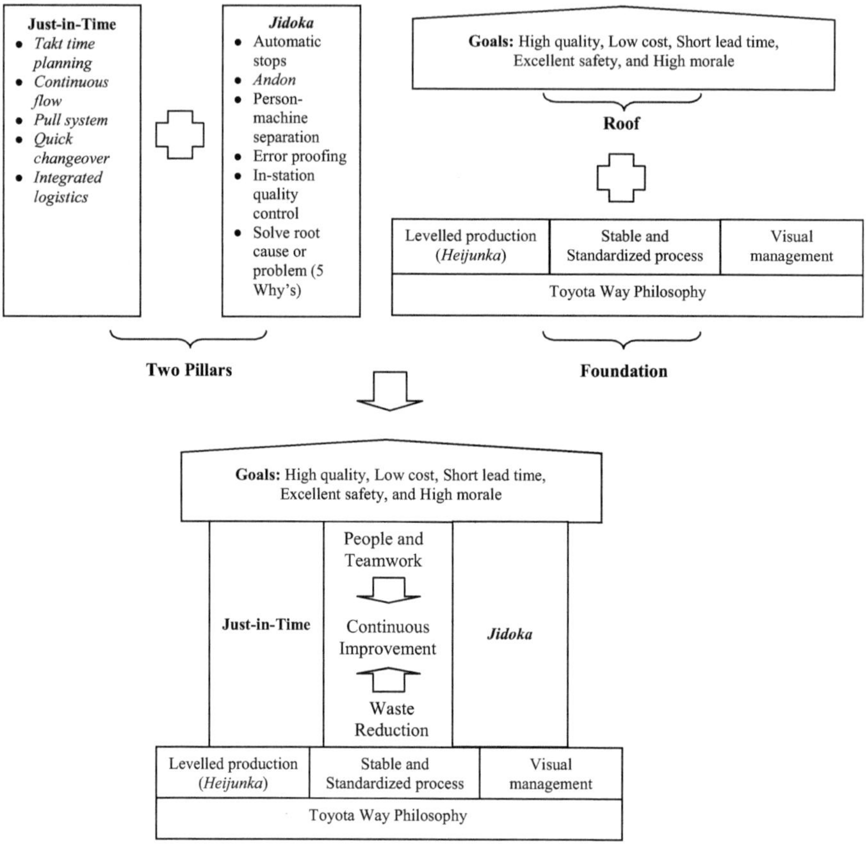

Fig. 3.1 The Toyota production system

coordination of potentially thousands of components/parts arriving where and when needed in just the right quantities, with all parts meeting the quality parameters. Ohno (1988) commented that JIT operates from a different paradigm than mass production and requires a different managerial and leadership mindset than mass production. Trying to achieve the JIT system led Ohno (1988) and others within Toyota to develop TPS tools such as continuous flow, pull system, quick change-over and integrated logistics (see Fig. 3.1—the TPS house).

Autonomation It is often known as its Japanese abbreviation *jidoka*, meaning "never let a defect pass into the next station and freeing people from machines" (Liker 2004). In all Toyota manufacturing plants, most machines, whatever old or new, are equipped with such devices as well as various safety devices to prevent defective products. The idea is to build quality in the process by distinguishing between normal and abnormal conditions, stopping production line once there is a problem being detected. It calls attention to the abnormal to ensure that its root cause is found and eliminated. In addition, Toyota also uses tools such as *andon*, error proofing, and visual management to build quality into its processes.

In the TPS house, Liker (2004) places "people and teamwork" in the centre of the system because only through continuous improvements can the operation ever attain the needed stability. People must be trained to identify waste and solve problems at the root causes. Finally, there are various foundational elements, which include the need for standardized, stable, and levelled processes. All of these are elaborated in the succeeding chapter.

3.3.2 Lean Production and Lean Principles

At the beginning, industry practitioners observed Toyota facilities and saw many tools and methods that were very different from what they practised. Believing this was the source of Toyota's competitive edge, many companies set out to emulate them. According to Koskela (1992), 11 important principles are essential to the lean philosophy, including:

1. Reduce the share of non-value-adding activities (also called waste)
2. Increase output value through systematic consideration of customer requirements
3. Reduce variability
4. Reduce cycle time
5. Simplify by minimizing the number of steps, parts and linkages
6. Increase output flexibility
7. Increase process transparency
8. Focus control on the complete process
9. Building continuous improvement into the process
10. Balance flow improvement with conversion improvement
11. Benchmark

These principles of lean production as reflected in the early days, suggested that lean principles were process focused. Of the principles identified by Koskela (1992), none was relevant to human resource or social aspect of lean. Implementing a few lean tools could result in some improvements, but it would never come close to the benefits that were possible from implementing the whole system (Liker 2004). Shah and Ward (2007, p. 791) conceptualized lean production as: "*an integrated socio-technical system, whose main objective is to eliminate waste by concurrently reducing or minimizing supplier, customer, and internal variability*". This is echoed by Paez et al. (2004) that a lean enterprise should be viewed as a socio-technical system that looks to maximizing production performance with minimal resources.

Waste Elimination Identifying and eliminating waste is fundamental to a lean organization such as Toyota. Liker (2004) highlighted that the heart of the Toyota

Production System is eliminating waste. Waste is anything that absorbs resources but creates no value (Womack and Jones 1996). Ohno (1988) identified the following seven wastes or "*muda*", namely (1) overproduction, (2) waiting, (3) transportation, (4) over processing, (5) inventory, (6) movement and (7) defect products and highlighted that the preliminary step towards application of the Toyota Production System is to identify wastes completely. Liker (2004) added one more waste: (8) waste of unused employee creativity, which resulted in losing time, ideas, skills, improvements, and learning opportunities by not engaging or listening to employees. This is applicable to employees in any industries. For example, construction companies were used to employ their employees for manual work, but have appeared to forget that their employees are able to think (Druker et al. 1996). The fact is that by capitalizing on employees' creativity, companies can eliminate the other seven wastes to continuously improve their performance. The first five "*muda*", namely overproduction, waiting, transportation, over processing, and inventory refer to the flow of materials. The last two "*muda*", namely movement and defective products, together with the waste of unused employee creativity are related to work of labour. Womack and Jones (1996) highlighted that lean is a powerful antidote to *muda* (waste). Likewise, Koskela (1992, 2000) proposed that eliminating non-value-adding activities from production is the fundamental principle based on the flow concept of production.

Customer Value Hine et al. (2004) argued that a misunderstanding of lean leads to value creation being viewed as equal to cost reduction. In the shop floor, all unnecessary production has been categorized as waste or non-value-adding activities, while what is necessary for creating a "perfect" product is called value-adding activities (Koskela 1992, 2000). By eliminating all the non-value-adding activities, cost can be reduced, which ultimately helps in creating value. Since value was defined as the first principle of lean thinking (Womack and Jones 1996), Hines et al. (2004) outlined that lean had moved away from a merely "shop-floor-focus" on waste and cost reduction, to an approach that contingently sought to enhance value (or perceived value) to customers by adding product or service features and/or removing waste. According to Hines et al. (2004), this shift from a mere waste reduction focus to a customer value focus essentially provides an alternative perspective on value creation:

1. Value is created if internal waste is reduced, as the wasteful activities and the associated costs are reduced, increasing the overall value proposition for the customer.
2. Value is also increased, if additional features or services are offered, which are valued by the customer. This could entail a shorter delivery cycle or smaller delivery batches, which might not add additional cost, yet enhance customer value.

In the context of lean construction, Jørgensen and Emmitt (2008) highlighted that value is either unaddressed, or it is largely discussed in the construction process, not the resultant building (the product). A number of challenges were identified by Jørgensen and Emmitt (2008) when the concept of customer value is applied to construction:

1. Construction is a long-term investment and is designed to function for 100 years or more.
2. A number of different owners and users, who may have different interests in projects or have different perceptions on value. Winch (2010) highlighted that even within the client organization, there exists different interest groups, who may have different functional requirements. In this case, project definition is likely to become a compromise, which may unravel as more information becomes available to those groups throughout the project life cycle relating to what the facility will be like (Winch 2011).
3. The level of complexity increases when the concept of value is discussed in the field of architectural design, in which more micro level issues should be taken into account.

Lean Thinking Lean thinking was elaborated in Womack and Jones's (1996) book—*Lean Thinking*—as additional theoretical framework and principles associated with lean production. The five main guiding principles of lean thinking are summarized by Womack and Jones (1996) as:

1. *Value:* value can only be defined by the ultimate customer (Womack and Jones 1996), where the customer can be considered as all downstream operations. It can be applied in the construction context in which end customers are multiple and the construction client can rarely be considered as the single ultimate customer (Jørgensen and Emmitt 2008).
2. *Value stream:* value stream analyses three types of actions along with the value stream: first, activity creating value; second, activity creating no value but is unavoidable with current technologies and production assets; third, activities creating no value and are determined to be avoidable.
3. *Flow:* once a company has reduced or eliminated waste and variation from a single process and streamlined the value stream, the next step is to make the remaining process steps flow. It is the opposite of batch and queue. The goal of this principle is to have a product move from concept to customer without interruption or delay.
4. *Pull:* this principle is closely related to the "pull" system which the TPS firstly created. The end users pull the production such that it is only produced to suit their requirements.
5. *Pursue perfection:* this principle indicates "*the complete elimination of muda so that all activities along a value-stream create value*" (Womack and Jones 1996, p. 350). The lean concept associated with perfection is *kaizen*, a Japanese word which is interpreted as continuous improvement in the West. By applying the previous four principles each time, the organization gains more and finds more hidden wastes that can be eliminated.

These principles have been successfully implemented in the manufacturing industry as well as in construction. In the context of construction, the Egan's

(1998) report adopted the essence of lean thinking and recommended that the industry should work toward the following aims:

1. Elimination of non-value activities which can represent up to 95 % of time and effort.
2. Removal of waste from all activities involved in delivering the end product.
3. Establishment of relationships with all members of the supply chain.
4. Removal of delays in the design and production process using just-in-time management.

However, the principles of lean thinking were challenged by Koskela (2000) that these failed to provide a proper theory of lean production due to the discussion of lean thinking being practically confined to the flow conceptualization of production without incorporating the transformation and value generation concepts. It lacks an adequate conceptualization of production, which has led to imprecise concepts, such as the term "value" (Koskela 2004).

In summary, lean production exists at both strategic and operational levels (Hines et al. 2004): at the strategic level, the concept helps one to understand customer value and identify the value stream; while at the operational level, it is a bundle of practices and tools that lead to the elimination of waste and encourage continuous improvement. In a similar vein, according to Shah and Ward (2007), lean production is generally viewed either philosophically or practically. The first point of view is related to guiding principles and overarching goals (see Womack and Jones 1996 and Spear and Bowen 1999); the second point of view relates a set of management practices, tools, or techniques that can be observed directly.

3.3.3 Implementation Frameworks of Lean

Empirically, based on the description of Womack et al. (1990) for lean production, Karlsson and Åhlström (1996) developed an implementation model based on conceptualizing lean production as consisting of a number of principles characterizing different functional areas and the overall strategy of the lean company. These functional areas consist of lean development, lean procurement, lean manufacturing and lean distribution as well as the factors are given below:

1. Lean development: supplier involvement, cross-functional teams, simultaneous engineering, integration instead of coordination, strategic management, and black-box engineering.
2. Lean procurement: supplier hierarchies and larger subsystems from fewer suppliers.
3. Lean manufacturing: elimination of waste, continuous improvement, multifunctional teams, vertical information systems, decentralized responsibilities, and pull system.
4. Lean distribution: lean buffers, customer involvement, and aggressive marketing.

This indicates that the utilization of lean production can affect the whole enterprise. Furthermore, Karlsson and Åhlström (1996) enumerated the following fundamental principles of lean production: elimination of waste, continuous improvement, zero defects/JIT, pull instead of push, and multifunctional teams. By taking the next step, Karlsson and Åhlström (1996) identified a number of measurable determinants within each principle, which are able to reflect changes in an effort to become lean. Karlsson and Åhlström's (1996) study is significant to this research because based on this, it offers management an opportunity to follow a checklist for what to aim at when trying to implement lean production. Following the work of Karlsson and Åhlström (1996), Sanchez and Perez (2001) introduced a checklist model with 36 indicators and tested the sample empirically in their study. The importance of various indicators can be analysed to develop manufacturing strategies for the manufacturing companies. Moreover, it offers a practical framework for its introduction in manufacturing companies and provides a good reference for lean implementation in the construction firms. In a similar vein, Paez et al. (2005) presented an integrated framework of lean enterprise based on the combination of human and technological subsystems. The framework contains four steps of guidelines:

1. It starts from the management commits to lean underpinned by three goals: waste elimination, flow, and pull.
2. The scope of activities is identified to conform the value stream, namely design, supply, and manufacturing.
3. This step aims to develop the workforce capabilities: problem-solving focus, teamwork, and creative thinking.
4. The final step is the implementation of lean techniques. It comprises of kanban system, production smoothing and autonomation. Additionally, it welcomes unique or individual elements that fit a specific company.

Shah and Ward (2007) proposed ten factors which constituted the operational complements from the contents and objectives of the historical roots in the TPS to the philosophy of lean production. This includes supplier feedback, JIT delivery, developing suppliers, involved customer, flow, pull, low set-up, involved employees, productive maintenance, and controlled processes. Of these ten factors, three factors measured supplier involvement issues, one factor measured customer involvement, and the remaining six factors addressed issues that are internal to the firm.

One missing link in the implementation framework of lean production is that, as Olivella et al. (2008) argued, the above-mentioned frameworks of lean have barely dealt with work organizations. Olivella et al. (2008) pointed out that there are indeed work organization practices characteristic of lean production and summarized their central concepts as given below: (1) standardization, discipline and control: that aim to obtain uniformity of the work, (2) continuing training and learning, (3) team-based organization: i.e. an organization where work is assigned to and done by teams, (4) participation and empowerment: the fact that the functions assumed by direct workers are more extended, (5) multi-skilling and adaptability, (6) common values, and (7) compensation and rewards to support lean production.

Forza (1996) highlighted that there is a priority order between the various work organization practices: some are indispensable for a lean system, while others adapt well to it but require more time and can be introduced later, for example, aspects of work organization which involve hierarchy: supervisors favour workers to act as a team, and decentralization of authority was found less of a differentiating factor in a lean plant.

3.4 Lean Construction

3.4.1 What is Lean Construction?

The success of lean principles in manufacturing and the benefits arising from its use is one of the main motivations for adopting lean principles in construction (Egan 1998). Lean first emerged in the construction industry a few years after it had gained full acceptance in Western manufacturing industries. Several authors have attempted to provide an account of the lean construction perspectives. These include Koskela's (1992) early discussion on the potential of what he termed "the new production philosophy" in the construction industry in his seminal Stanford report. Koskela (2000) later synthesized three different perspectives on the construction process (discussed earlier in Chap. 2), which formed the foundation for what has now become known as lean construction. A simple definition of lean construction was given by Koskela et al. (2002, p. 211):

> lean construction is a way to design production systems to minimize waste of materials, time, and effort in order to generate the maximum possible amount of value.

The Lean Construction Institute (LCI) defines lean construction as a management-based production approach to project delivery that is particularly useful on complex, uncertain, and quick projects (see http://www.leanconstruction.org). The definition of lean construction due to Koskela et al. (2002) indicates that lean construction strives for the same goals as lean production, namely to eliminate waste and to maximize value. LCI's definition, on the other hand, implies that industrial approaches in manufacturing are directly applicable to construction. Apart from Koskela's (1992, 2000) work, an alternative interpretation of the concepts of lean construction is illustrated in Fig. 3.2 (Koskela et al. 2002; Winch 2006). This school of thought discusses the application of lean production methods to construction. The best known of these is the Last Planner approach to the planning and management of the construction process (Ballard and Howell 1998a; Ballard 2000). Its goal is to create a reliable work-flow by having the project team, including all affected firms, collaboratively create a phase plan for a segment of the work (such as the foundations). This is a social process involving discussion with site staff and planning to ensure that work is not waiting on workers, and that workers are not waiting on work.

The coexistence of different interpretations of lean construction has also been observed by Green and May (2005, p. 503), who have pointed out that "*lean*

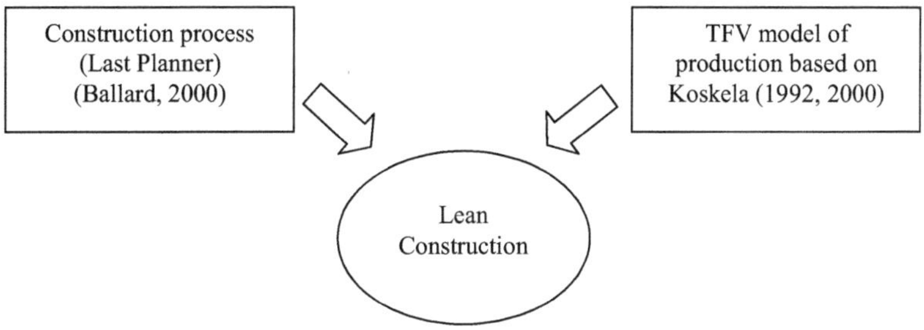

Fig. 3.2 Two core interpretations of lean construction

construction can be interpreted as a set of techniques, a discourse, a 'socio-technical paradigm' or even a cultural commodity". The Egan report (Egan 1998) has been responsible for popularizing the "lean" label amongst construction professionals in the UK (Green and May 2005), who see lean thinking (Value, Value stream, Flow, Pull, and Perfection) primarily as a set of techniques that can be directly applied to construction. As interest in lean construction has steadily grown, the research has covered almost all construction stages employing lean concepts. As Emmitt (2007, p. 135) noted: *"the term lean construction tends to be interpreted quite widely, ranging from a term to include design and construction activities to very narrow interpretations related to specific production functions and/or application of tools by contractors"*.

In addition, the term covers a range of project types, such as industrialized housing (Höök and Stehn 2008; Dentz et al. 2009; Yu et al. 2009), high-rise buildings (Sacks and Goldin 2007), prefabrication projects (Low and Chan 1997), refurbishment project (Bryde and Schulmeister 2012), and others. It has also extended to project areas such as project definition, the design process (Ballard and Zabelle 2000), supply-chain mapping and simulation (Tommelein 1998), and construction site practices (Picchi and Granja 2004; Salem et al. 2006). It has been reported that lean construction can bring about benefits to the construction sector (see Sect. 3.4.5). Although its achievements in construction so far have not been as extraordinary as in the case of Toyota, the outcome is indeed encouraging.

3.4.2 Peculiarities of the Construction Industry

The construction industry differs in many ways from the manufacturing industry. Thus, this study cannot suggest simply applying the Toyota Way model to the construction industry without modification. Previous studies have listed a large number of factors differentiate the construction industry from other industries, and from manufacturing in particular (Ball 1988). These are what Koskela (1992) called the inherent peculiarities of construction; they are summarized in Table 3.1. The unique and complex environment of the construction industry represents a challenge for production management. As Riley and Clare-Brown (2001) have argued, it would not be possible to transfer management practices from manufacturing to construction unless necessary effort had been taken to modify the management tools or the culture in the construction industry. The critical question in assessing the potential for the application of the lean approach to construction is the extent to which the one-off nature of the construction process can be changed (Winch 2010). This suggests that the appropriate model of manufacturing for the construction industry depends on the analysis of the construction subsector, which can generally be grouped into four broad categories (Winch 2010), namely (1) large infrastructure works, typically civil engineering, (2) prestige building projects, (3) "routine" building projects that provide the bulk of new buildings, including schools, offices, and so on, and (4) housing. The first three categories are on-site, project-oriented construction projects (the majority of the work is site-based), so that production in volume can hardly be achieved. Construction projects will remain a one-off production process where design and execution on-site are intimately linked.

Furthermore, according to Winch (2003), the last category—housing production—is the most similar to lean production in terms of its underlying business process, where high-volume production allows the approach to be economically viable. The housing production company embraces the whole construction process, from design to the factory production of a complete house. This was often viewed as indistinguishable from conventional site-built housing. This does not imply that lean construction principles can only be implemented in the housing sector. Winch (2010) has outlined that lean-inspired improvement activities such as TQM and JIT have their place in effectively managing construction projects, because they provide a valuable tool set for improving process capability. In addition, some structural modifications of the manufacturing management concepts (e.g. the concept of JIT) are indispensable for the concept to be applied in the construction industry (Low and Chan 1997).

3.4.3 Criticisms of Lean Construction

It has been argued that the lean construction community has been slow in surmounting the critical debate over what lean is and is not (Green 1999). Limited efforts have been made by researchers to enhance the credibility of lean construction by addressing or challenging its shortcomings (Jørgensen and Emmitt 2008). Green (1999, 2002) is one of the few academics to claim that researchers have ignored a crucial argument concerning lean construction. Basically, most of Green's (1999, 2002) concerns are related to the potential effects on the quality of working life that the lean method could bring about. This warning has been expressed in the literature on lean production. However, the lean construction literature has repeatedly failed to recognize or has appeared to ignore the consideration of the human resource management (HRM) implications of lean construction. Green (1999, 2002) highlighted a number of issues that the lean construction literature has so far failed to address:

Table 3.1 Differences between manufacturing and construction industry

Aspects	Construction industry	Manufacturing industry
Duration (life cycle)	Short	Long
Nature	One-of-a-kind nature	Repetitive
Work Station	Transient	Stable
Material components	Non-standardized	Standardized
Material supply	Schedule-driven	Order-driven
Safety provision	Less enforced	Highly enforced
Labour force	Seasonal, low job security	Not seasonal, higher employment security
Wages	Vary depending on skill, experience, and employers	More stable wage policies
Environment	Productivity influenced by the change in environment	Productivity less influenced by the change in environment
Assembly and production	final production is assembled in situ	Within the factory
Technology	Low level of automation, prefer not to use	Better and advanced
Quality	Related to product conformance; Rework is common	More closely to process control; Rework is generally avoided
Owner involvement	Highly involved	Less involved
Culture	Ill-defined, site personnel know nothing of company's management philosophy	Clearly defined so that staff are conscious of it
Regulatory intervention	Design solution and many work phases in a construction project are subject to checks and approvals by regulatory authorities	Less subject to checks and approvals

Source: Ball (1988); Koskela (1992); Low and Chan (1997); Oglesby, et al. (1989); Salem et al. (2006); Riley and Clare-Brown (2001); Winch (2003)

1. The institutional requirements for lean production. This requires an overall understanding of Japanese industrial structure as the business practice where lean production evolved.
2. The human cost of lean production, such as long working hours, loss of individual freedom, *karoshi* (sudden deaths), and severe stress.
3. Criticisms are not limited to production plants in Japan but also extended to Japanese overseas plants. Criticisms arising from overseas plants included increased management control instead of empowerment, intensification of work, etc.
4. The association of lean methods with totalitarian management regimes in the global context.

Moreover, Green (2002) warned that if the construction industry overwhelmingly focused on waste elimination and improving efficiency—without, however, explicitly considering the HRM implications—construction companies and professional firms would find it increasingly difficult to attract intelligent, creative young professionals to join the industry. Green's (2002) criticism has had meaningful impact on the lean construction community. A number of lean construction frameworks will be discussed later, all of which underscore the importance of HRM. Furthermore, this study uses Liker's (2004) Toyota Way model, which is a direct adaptation of Toyota's philosophy, to provide an alternative framework. The Toyota Way model concerns people management in lean organizations. More importantly, the discussion in Chap. 4 revealed that "people-related" principles were positioned higher than the "process" management aspects that could be described as the amalgamation of TQM with the JIT concept.

3.4.4 Lean Construction: Tools, Techniques, and Implementation Frameworks

3.4.4.1 Tools and Techniques

Lean construction does not simply adapt techniques from manufacturing into construction. Instead, as Paez et al. (2005) explain, lean construction tools were developed in three forms:

1. On the first level: lean construction *adapted* techniques from manufacturing for construction. *Kanban* cards are an example of this. This first level implies that lean construction has succeeded in overcoming the contextual difference and implemented the principles.
2. On the second level: lean construction has *expanded* the scope of lean manufacturing techniques. For example, lean construction has extended the scope of visual inspection (*andon*) of defective parts to the visualization of material and work-flow.
3. On the third level: lean construction has introduced new techniques for its unique context. For example, the Last Planner System can be viewed as a combination of production smoothing and the *kanban* system. In manufacturing, production smoothening prepares the product sequence that is driven by *kanban* cards. Similarly, in the construction industry, the Last Planner System prepares a reverse-phase schedule that pulls assignments.

Salem et al. (2006) discussed the transferability of lean manufacturing techniques to construction despite the differences in their working environment and process. An assessment tool was subsequently developed to evaluate the impact of the lean construction tools on the performance of the construction project. These lean construction tools include the Last Planner System, increased visualization, huddle meetings, first-run studies, 5-S, and fail safe for quality. Moreover, Picchi and Granja (2004) also considered that lean implementation is viable for the construction sector and summarized a few examples of lean tools used on job sites by employing lean-thinking principles. Unlike Salem et al. (2006) and Paez et al. (2005), who only focused on discussing the feasibility of lean production tools in the construction industry, Picchi and Granja (2004) instead studied the extent of construction firms dealing with lean tools and identified three common lean implementation scenarios as follows:

1. Scenario 1: This shows the most frequent application pattern to be adopted by construction firms so far. In most cases, the building professionals only took up one lean construction tool, and failed to connect it with other tools or other lean principles (Picchi and Granja 2004).
2. Scenario 2: This concerns situations in which there is systematic interpretation of the lean-thinking principles and the integrated application of lean tools on the job site.
3. Scenario 3: This represents a larger application of the lean approach to the job site and as company-wide transformation. Picchi and Granja (2004) noted that Scenario 3 cannot be achieved without involving several aspects of the company, such as product development, suppliers and customer relationships. It also indicates that lean implementation would be most effective when considering a construction firm as a whole and should be considered as a long-term goal. This is in line with Liker's (2004) Toyota Way model, which promoted the application of the fourteen principles as a whole and on a long-term basis.

3.4.4.2 Implementation Frameworks

Although lean construction is still in its infancy, there is already a set of practices that have been proposed, tested and implemented (Paez et al. 2005). The following section discusses the various frameworks of lean construction that have been adopted. These include lean construction as a socio-technical system (Paez et al. 2005) which equally concerns the human and technical aspects; lean initiatives in eight areas of the construction business (Johansen and Walter 2007); the lean construction wheel (Diekmann et al. 2004), and the three dominant models of lean construction as summarized by Green and May (2005) from the perspective of policy-makers in the industry.

Socio-Technical System of Lean Construction Socio-technical design is defined as the combination of a technical and human subsystem into the same work design. Lean construction and lean manufacturing are part of the same socio-technological design with the same goals, activities, and workforce capabilities but with different technical systems (Paez et al. 2005). Based on this, Paez et al. (2005) introduced a higher view of lean construction and lean manufacturing as a socio-technical system in Fig. 3.3. This model implies that the operational improvement will always rely on the joint effort of the technical and human elements that characterize the lean enterprise (Paez et al. 2005). It also concluded that the tools presented in the context of the lean manufacturing scenario can fit the construction industry to support the same principles (e.g. JIT, production smoothening and autonomation).

Scope: Design, supply and production			
Goals: Zero waste, flow and pull			
Human system	Problem-solving focus		
	Team		
	Creative thinking		
Technical subsystem		Construction Project-based	Manufacturing Process-based
	Just-in-time	Material kanban cards Last Planner	Kanban system Minimum batches
	Production smoothening	Concurrent engineering Daily huddle meeting	Production levelling Multi-functional layout Standard operations Preventive maintenance
	Autonomation	Quality management tools Visual inspection	Total Quality Management Autonomous control

Fig. 3.3 Lean construction as socio-technological design

Lean Construction in Eight Areas Given that an understanding of the complexity of the construction industry, as well as of the human aspects of lean construction has improved, Johansen and Walter (2007) proposed a conceptual model of lean construction, which can be employed to assess the level of awareness of lean construction. They conducted a questionnaire to study eight areas in an organization, namely design, procurement, planning/control, supply, installation, collaboration, behaviour, and management. The chosen areas were highlighted by Johansen and Walter (2007) as being fundamental in developing a lean culture. Later, the German construction industry was chosen as a pilot study, in which the results showed that a holistic understanding of construction activities has not been developed yet, because the majority employed only a few management concepts from the framework (Johansen and Walter 2007).

Lean Construction Wheel In a similar vein, efforts have been made by Diekmann et al. (2004), a team of researchers from the Construction Industry Institute (CII) in the USA, to help construction companies self-assess the extent to which they conform to lean behaviours. This framework looks similar to that of Tapping et al. (2002) in terms of structure and contains five key principles. The five principles are as follows:

1. Customer focus
2. Culture/people
3. Workplace standardization
4. Waste elimination
5. Continuous improvement/built-in quality

Yet there are many differences in terms of the sub-principles listed. This is because each sub-principle was evaluated for its applicability to the construction industry.

Lean Construction Models from the Perspective of Policy-Makers Green and May (2005) discussed three dominant models represent the practical adoption of lean in construction. The finding was derived from an investigation of the perception on lean amongst the industry's policy-makers. First of all, Green and May (2005) outlined that Model 1 is limited to the hardware of lean production and places little focus on human resource practices. It aims to sort the inefficiency within the industry by changing the term from "waste elimination" to "cut out unnecessary cost" (Green and May 2005). Secondly, Model 2 advocates "project partnering" and "strategic partnering" but is nevertheless concerned that the partnership would be become one sided. Moreover, it does not advocate the need for supporting human resource practices. Thirdly, Model 3 combines elements from the previous two models and is regarded as more sophisticated than the previous two (Green and May 2005). It has much stronger emphasis on the institutional context within which projects are delivered. Moreover, Model 3 places far greater emphasis on social and technological aspects such as technology and training at all levels than the two previous models.

In summary, this section discusses various models of the implementation of lean construction. The first model focuses on the human and technological aspects of lean construction and asserts that tools from the manufacturing scenario can work in the construction industry to realize the same goals. The second model examines opportunities for exercising the lean approach in various functions that a normal construction firm would deal with. This begins with the procurement strategy, extending from site planning and control to the supply chain, and also including organizational considerations. The third framework focuses on five key principles of lean construction and uses these principles as guidelines to assess the extent to which a company conforms to lean practice. The fourth framework summarizes the current practices within the industry from the perspective of policy-makers and highlights how waste elimination is the first step towards true lean transformation. The above-mentioned frameworks offer good practical references, which can be used to assess the awareness of construction practitioners, the level of leanness of a company, and also to give a holistic picture of lean implementation.

3.4.5 Lean Construction and Enhanced Performance

Many management initiatives do not go well at the early stage, due to the fact that the people involved fail to see the benefits associated with the changes (Song and Liang 2011). In the lean construction context, an increase has been seen in the number of empirical studies, originating from many countries, which have argued that the successful application of lean principles in construction can improve cost structure (Salem et al. 2006), productivity (Agbulos et al. 2006; Alex et al. 2008; Al-Sudairi 2007), delivery times (Diekmann et al. 2004), plan reliability (Ballard 2000; Cho and Ballard 2011; Liu et al. 2011), quality (Leonard 2006), relationship between working partners (Miller et al. 2002; Salem et al. 2006; Turner Construction Company 2012), and job satisfaction (Nahmens et al. 2012). These are good reasons for implementing lean construction. To be more specific, the evidence includes: studies using lean for managing drainage operations (Agbulos et al. 2006) and sewer installation (Alex et al. 2008) in the construction sector achieved, respectively, 4 and 35 % improvement in productivity in Canada. In a similar vein, Song and Liang (2011) reported potential productivity improvement for formwork installation using simulation techniques. Al-Sudairi (2007) reported 21 and 50 % increase in process efficiency for block-laying and plastering from 13 cases of low-rise residential buildings in Saudi Arabia. This suggests that lean principles are effective not only in complicated processes but also in simple processes (Al-Sudairi 2007). In the UK, Balfour Beatty (2011), a leading British contractor, reported its experience with sports stadium construction (the Emirates Stadium): it applied JIT delivery of the pipe reinforcement cages and saw a 20% improvement in productivity.

As for plan reliability, the application of the Last Planner System of production control to projects has been demonstrated to increase plan reliability (Ballard 2000). Following Ballard's (2000) work, Cho and Ballard (2011) surveyed a list of participants with experience of the project production system and found that there is a significant correlation between the implementation of the Last Planner System and project performance, as measured by cost and schedule reduction. In addition, research using 134 weeks of production data in the 10 working areas of a pipe installation project conducted by Liu et al. (2011) revealed that work-flow reliability and labour productivity are significantly correlated, and that the Last Planner System can reduce work-flow variation, which can help improve labour productivity.

Yu et al. (2013) reported a case study of a US modular building producer, which had been implementing lean for only 6 months, but which witnessed an improvement of 50 % in production throughout and of 10 % in labour efficiency, as well as 18 % decrease in labour costs without laying off a single worker. Leonard (2006) conducted research of kaizen activities for a homer builder in the USA and found that inspection time and cost were reduced by 50 % after kaizen improvement efforts were made. Also in the USA, Salem et al. (2006) noted that the benefits of lean construction implementation were tangible: a car park project in Ohio implemented lean construction techniques and was completed under budget and 3 weeks ahead of schedule. The subcontractors were also more satisfied with their relationships with the general contractor. This is consistent with the finding of Song and Liang's (2011) study that time saving was among the greatest benefits of using lean construction concepts, because these helped to generate teamwork among the subcontractors. Maturana et al. (2007) also investigated whether the benefits of lean construction can be extended to subcontractors. Maturana et al. (2007) developed an on-site subcontractor evaluation methods based on lean principles and partnering practices. This tool allows main contractors in Chile to help subcontractors improve their performance by providing them with periodic feedback on such tools. Furthermore, a case study conducted by Nahmens et al. (2012) in the US industrialized homebuilding sector revealed an increase of 11% in job satisfaction after lean implementation.

3.4.6 Lean Construction in Developing Countries

In developing countries, research is concerned with three major issues in the area of lean construction:

1. Feasibility studies and investigation into awareness of lean construction.
2. Exploring the barriers to the implementation of lean construction (Forbes et al. 2002; Alinaitwe 2009).
3. Implementing lean starts with identifying the waste in construction, which corresponds to the first stage of Green and May's (2005) framework of lean construction implementation.

Feasibility Study and Awareness of Lean Lean construction can be viewed as a strategic option when its implementation is placed in a new setting (Senaratne and Wijesiri 2008). Senaratne and Wijesiri (2008) used suitability and acceptability tests to conclude that lean construction is suitable and acceptable in the Sri Lankan context. They revealed that:

1. A number of controlled waste flows and their root causes were identified. This waste hindered the performance of the Sri Lankan construction industry. Lean construction is suitable in the Sri Lankan context as a strategic approach, because it can be used to eliminate this major waste.
2. The construction workforce in Sri Lanka accepts the core principles of lean construction and has a continuous improvement or *kaizen* mentality.

Moreover, Abduh and Roza (2006) revealed that the large Indonesian contractors have already implemented "macro" lean construction principles (such as the policy of continuous improvement and promoting transparency). Yet when it comes to the "micro" principles of lean construction (such as the reduction of cycle time and variability), the large Indonesian contractors still lack awareness and the ability to implement principles and techniques, due to their inadequate capability to plan work-flow well.

Barriers for Implementation of Lean Forbes et al (2002) conceded that the main obstacle to the implementation of lean in developing nations is that construction firms do not emphasize productivity and quality initiatives. Alinaitwe (2009) identified many barriers to lean production under different management concepts, including JIT, TQM, concurrent engineering, etc. in Uganda's construction industry. Alinaitwe (2009) detailed the barriers by grouping these into (1) barriers that strongly influence workers' productivity, and (2) barriers that are easier to overcome. Olatunji (2008) interviewed a total of 10 clients made up of five private and five public construction firms in Nigeria. In Olatunji's (2008) work, the barriers to implementation of lean were categorized under seven groups, namely: (1) skills and knowledge related, (2) management related, (3) government related, (4) attitude related, (5) resource related, (6) logistics related, and (7) others. Low and Gao (2011b) discussed the potential impediments to implementing the concept of JIT in the Chinese construction industry from a project life-cycle perspective. The impediments identified fall into the areas of design, procurement, construction, and inspection.

Elimination of Waste Ramaswamy and Kalidindi (2009) quantified different categories of waste in terms of cost in the Indian construction industry. They found that wastes due to non-value-adding activities carried out by labour or due to equipment were much higher compared to material wastes generated in the site. Polat and Ballard (2004) identified the material and time wastes in areas of design, procurement, material handling, operations, residual, and others, which were classified by Bossink and Brouwers (1996) as the main causes of waste in construction. Polat and Ballard (2004) highlighted that the material and time wastes are major problems in the Turkish construction industry and that the construction practitioners have failed to appreciate the adaptation of lean construction techniques. Contractors were therefore advised to do their part and to employ lean construction techniques through adopting the lean philosophy. It was further suggested to concentrate on eliminating the causes of waste, rather than only reacting to these problems.

The literature on the application of lean construction has received considerable attention in the developed countries, such as in the UK, USA, Singapore, and Nordic countries. One of the remarkable facilitators in promoting the lean concept is the government in these developed countries. In the developing countries, the implementation of lean construction is still in the very early stage. It began with investigation into the awareness of lean concepts among construction practitioners, and with waste elimination in the construction processes. It turned out that the awareness level of lean in the construction industry was low. For example, some contractors were already practicing either one or more lean concepts every day, but they were not consciously aware that this was in conformance with lean construction until the lean frameworks were introduced to them. Moreover, it is beneficial for the developing countries to explore the barriers that can hamper the adoption of lean in construction. This indicates that support from the government in the developing countries needs to be enhanced. Without clear objectives set by the government, the awareness of lean will be adversely affected. Given that China is still a developing country and the Chinese construction industry is operating in a very similar manner (i.e. traditional procurement, labour intensive, etc.) compared to other developing countries, the literature relating to such a similar context would bear implications for this current study.

3.5 Knowledge Gap Analysis

A number of scholars have suggested that the concept of lean should be built on the socio-technical system with a balanced view of process-related (hard) and people-related (soft) elements (Emiliani 2006; Liker 2004; Paez et al. 2005; Shah and Ward 2007; Low and Gao, 2011). Toyota Way is a good example which has to two main pillars namely "continuous improvement" and "respect for people" (Liker 2004). However, the principle of "respect for people" sometimes gives less attention than it should. Commonly, when we think of lean, we typically first consider process flow, pull *kanban* system, standardization, visual control tools, and opportunities for significant financial improvement. All of these fall under the "hard" part of lean, which greatly limits the amount of improvement that can be achieved (Bhasin and Burcher 2006; Emiliani 2006). Seeing this, Emiliani (2006) wrote that "*the people-related principle has long been unrecognized, ignored, or misunderstood by most senior managers outside Toyota and its affiliated suppliers*" (p. 4).

Similarly, in the domain of lean construction, the ignorance of the human resource management aspect of lean was constantly criticized. Examination of a number of lean construction frameworks supports this assertion, as made by Green (1999; 2002). As shown in Chap. 3, a majority of the above-mentioned frameworks provide answers to questions such as "what constitutes lean construction" (Diekmann et al. 2004; Salem et al. 2006). Some offer good practical references, which can be used to assess the awareness of lean construction among the construction practitioners (Santos 1999; Johansen and Walter 2007), or the status quo of their lean construction implementation (Diekmann et al. 2004; Salem et al. 2006). However, the examination also highlights the fact that the development of lean construction is uneven, because it is evident that "process-focus" thinking is still the dominant theme among all the lean construction framework examined. This is because with the shop floor-focused mindset, implementing lean tools can result in immediate improvements, but it would never come close to the benefits arising from implementation of the whole system (Liker 2004). In addition, only three of the frameworks selected (Diekmann et al. 2004; Paez et al. 2004; Green and May 2005) consider the soft side of the lean approach, while the remaining do not consider HRM implications at all. The knowledge gap could be fulfilled if the lean construction paradigm were to move forward to shift its present focus onto the issues of human part (the human resources), which is the soft aspect of lean.

3.6 Summary

This chapter reviewed three major paradigms of industrial production over the last 100 years and argued that lean production, as a dominant paradigm in recent years, combines the advantages of both craft production and mass production in concert with the principles of JIT and elimination of waste in order to minimize the total cost of producing a product (Crowly 1998; Womack et al. 1990). Lean production is an innovative way of manufacturing distilled from the two former paradigms. In order to evaluate the evolution of the lean production model, it is best to start by considering the evolution of leading firms such as Toyota, whose experience was the basis for the model itself. In addition, this chapter discussed the concept, elements and its theoretical support—five principles of lean thinking—contributed by Womack and Jones (1996) as well as various implementation frameworks of lean production. It can be argued that most implementation frameworks of lean production (e.g. Karlsson and Åhlström 1996; Koskela 1992; Ohno 1988; Shah and Ward 2007; Womack and Jones 1996) were process focused. Implementing a few lean tools that focused on the shop floor could result in some improvements, but it would never come close to the benefits that were possible from implementing the whole system (Liker 2004). Moreover, this chapter also discussed various issues in lean construction. It started with an overview of what lean construction is, followed by an outline of the differences between the construction and the manufacturing industry, and addressing the difficulties of transferring the lean principles to the construction industry. This chapter also reviewed the various implementation frameworks for lean construction and summarized how developing countries have adopted the lean principles. In addition, the role of government in promoting lean in the construction industry was mentioned.

4 The Toyota Way

4.1 Why Toyota Way Model

4.1.1 Why Toyota and Its Production System

Toyota is worthy of in-depth study because the company is good at manufacturing (Liker 2004; Sobek and Smalley 2008). Every automotive insider and many consumers are aware of and familiar with Toyota's stunning success in terms of its reliable products, high productivity and increasing profitability every year (Liker 2004). Toyota became the world's leading auto manufacturer with approximately 15 % of the global market share in 2005. Toyota's market value (US$177 billion in 2005) exceeds the combined value of General Motors (GM), Chrysler, and Ford (Morgan and Liker 2006). One reason for this success is the quality of Toyota products. According to the recent U.S. Vehicle Dependability Study published by J.D. Powers and Associates (2013),[1] Toyota motor models earn seven segment awards for higher quality and less problems experienced by car owners. Moreover, Toyota is the most productive company, according to the Oliver Wyman's (2008) Harbour report on the North American automobile industry. Toyota and Chrysler led the six largest multi plant auto-makers in total manufacturing productivity, averaging 30.37 labour hours per vehicle (GM averages 32.29 h per vehicle, while Ford averages 33.88 h per vehicle). It is worth noting that Toyota fabricates and assembles a greater percentage of its vehicle parts with its own employees, while the big Three (GM, Ford and Chrysler) purchase many modules and subassemblies from suppliers, thus saving labour. Although productivity gains are essential to a company's success, it is profitability that keeps the business growing. According to Oliver Wyman (2008), Toyota earned US$922 pre-tax profits per vehicle produced in North America in 2007. In contrast, Ford, GM and Chrysler lost US$1,467, US$729 and US$412, respectively.

[1] J.D. Powers and Associates is a global marketing information services firm, which is best known for its customer satisfaction research on new-car quality and long-term dependability.

DOI 10.1007/978-981-287-014-8_4, © Springer Science+Business Media Singapore 2014

Toyota's success has made the TPS the new paradigm in the manufacturing industry. The advantages of the TPS have been demonstrated tangibly enough in practice and widely cited in the literature. That is the fundamental reason why Toyota attracted so many audiences. Ideally, it seems logical to adopt the TPS as a model in a study dealing with improving operations in organizations, even when the firms are considered to be operating in a very different environment and have distinguished themselves differently from the manufacturing industry.

4.1.2 Why the Lean Paradigm Is Not Enough

For the production system that most firms have currently adopted, Liker (2004) advocated the use of the Toyota Way management principles as a whole system because: (1) the Toyota Production System is historically accepted as the mother platform where lean concepts originated; and (2) a very normal phenomenon was recognized by Liker (2004) that most U.S. companies have been learning the TPS and lean practices for decades without understanding what made them work together in a system. These organizations claimed to be advanced practitioners of lean methods, but when compared with Toyota actually showed otherwise. Although the lean programmes are effective in cutting down costs and have helped these companies enjoyed cost saving, Liker (2004) however highlighted that the problem persists because these companies have mistaken a particular set of lean tools for deep "lean thinking". This has been pointed out in Liker's (2004) Toyota Way model, where most companies are still stagnant at the "process" level. Although obvious progress have already been made and such companies have performed better than their non-lean counterparts, without adopting the other practices and philosophies, they will do little more than dabble in the same place, and their performance will continue to lag behind those companies that adopt a true culture of continuous improvement (Liker 2004). Therefore, it is very critical to introduce the Toyota Way principles as a whole. This has become another reason for studying the Toyota Way model.

4.1.3 Disadvantage of TPS

Recently, the limits of the lean model have been examined, both in Japan, the original context of the model, and in other industrial and cultural contexts (Cusumano 1994). Cusumano (1994) pointed out Toyota's practice of having suppliers make or deliver components just in time to assembly lines, requires suppliers to diligently increase the delivery frequency each day. This can result in traffic problems especially in congested urban areas as well environment problems (Katayama and Bennett 1996), wasting time when people are stranded in the traffic. Apart from the disadvantages of the process part of lean principles, the human resource aspects have also revealed side effects resulting from lean production practices (see Conti et al. 2006; Williams et al. 1992). The JIT philosophy involves the workers working much harder at continuous tasks, and Toyota has more recently been accused of making front-line workers work in conditions that could cause repetitive strain injuries (Crowther and Green 2004). In a similar vein, Williams et al. (1992) suggested that lean production is de-humanizing and exploitative. However, Conti et al. (2006) indicated that lean production is not inherently stressful and it depends heavily on management choices in designing and operating lean system. Hines et al. (2004) noted that the key criticisms of the main critics of lean thinking are the lack of contingency and ability to cope with variability, the lack of consideration of human aspects, and the narrow operational focus on the shop floor. Additionally, Hines et al. (2004) argued that the emerging shortcoming of lean was the outcome of organizations who have progressed on their learning curves, as well as the extension of lean thinking into new sectors with different settings and constraints.

4.2 From TPS to the Toyota Way Model

Gaining a true understanding of Toyota's philosophy perhaps requires going beyond TPS and lean to understand the Toyota Way (Liker 2004). In describing the Toyota Way, the Toyota Motor Corporation's (2003) internal training document with the same title [see Fig. 4.1 as adapted from Toyota Motor Corporation (2003)] is examined, which sums up five key principles of the Toyota employee conduct guidelines based on the dual pillars of "Continuous Improvement" and "Respect for People".

Liker (2004) incorporated and correlated these high-level guiding principles from Toyota's internal document with his Toyota Way model [see Fig. 4.2, as adapted from Liker (2004)]. He used a pyramidal model, which comprises a synopsis of the 14 principles to outline the Toyota Way principles and highlighted that this is the true source for Toyota's success. According to Gary Convis,[2] who was cited in Liker (2004, p.xi), "*the Toyota Way, along with the Toyota Production System, make up Toyota's DNA*". The principles are grouped in four broad categories and each category contains relevant sub-principles:

1. Long-term *P*hilosophy (Philosophy).
2. The right *P*rocess will produce the right results (Process).
3. Add value to the organization by developing your *P*eople and *P*artners (People/Partners).
4. Continuously solving root *P*roblems drives organizational learning (Problem-Solving).

The foundation of the pyramid is the management philosophy which bases its decisions on a long-term philosophy, even at the expense of short-term financial targets. The next level in the pyramid deals with the right processes such that

[2] Gary Convis is a managing officer of Toyota and President, Toyota Motor Manufacturing, Kentucky.

Continuous Improvement	Respect for People
Challenge We form a long-term vision, meeting challenges with courage and creativity to realize our dreams.	Respect We respect others, make every effort to understand each other, take responsibility, and do our best to build mutual trust.
Genchi Genbutsu We go to the source to find the facts to make correct decisions, build consensus, and achieve our goals.	Team work We stimulate personal and professional growth, share the opportunities of development, and maximize individual and team experience.
Kaizen We improve our business operations continuously, always driving for innovation and evolution.	

Fig. 4.1 The Toyota Way 2001

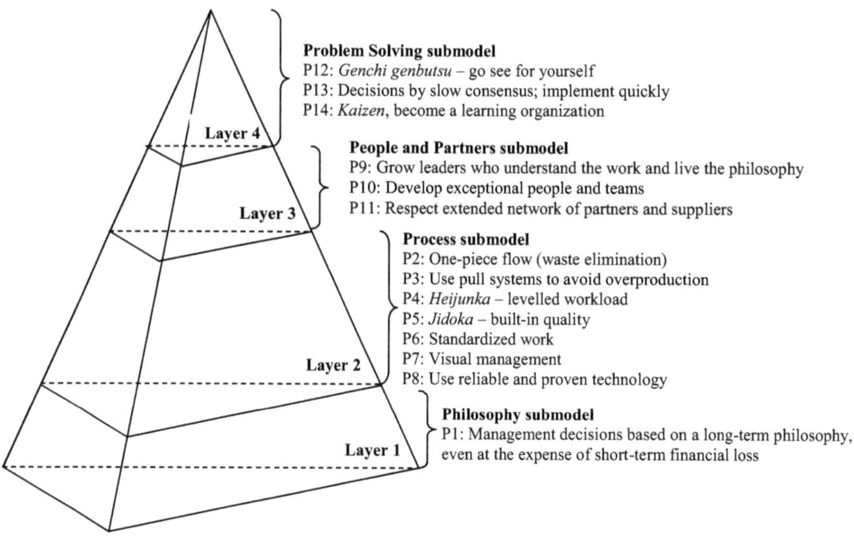

Fig. 4.2 "4P" model of the Toyota Way

production flow is levelled, pulled, standardized and visualized for everyone to identify problems. The next layer places respect on people and partners, while challenging and growing them. The last step of the pyramid in the Toyota Way model is the problem-solving philosophy by using various improvement tools such as *kaizen* and *genchi genbutsu*.

The Toyota Way actually supersedes the TPS and is, in fact, quite different in its emphasis (Liker and Hoseus 2008). In the TPS (see Fig. 3.1), the core pillars are just-in-time and *jidoka* (autonomation)—both technical concepts. People are at the centre of the TPS house but most lean applications implemented outside of Toyota focused specifically on the tools used to take waste out of the process. In contrast, the Toyota Way model focused on people and their way of solving problems, their continuous improvement and respect for others. Liker (2004) put the "process" part containing the principles of JIT and other lean tools in the middle of the pyramidal Toyota Way model, one layer down as sub-methodologies supporting "people" in creating more value to the final products in Toyota.

The 4P model of the Toyota Way provides a picture of the values that constitute the foundation of the Toyota Production System and how these principles are applied in practice. The principles together create a totality which has made Toyota an enormously successful and profitable company. These broad four categories of philosophy, process, people/partners, and problem-solving can be used to construct any organization. Therefore in the following sections, each of the 14 principles will be detailed and discussed at two levels: principles and tool/practice if any. It can be noticed that in Liker's (2004) book—*The Toyota Way*—the 14 principles were reviewed with stories of the challenges Toyota overcame to develop the Lexus and the Prius. Therefore, efforts should be made to distil the key principles and tools from these stories.

4.3 Toyota Way Philosophy Model

4.3.1 Principle 1: Long-Term Philosophy

"*Based your management decision on the long-term philosophy, even at the expense of short-term financial goals*" has been outlined as the first principle of the Toyota Way (Liker 2004, p.37). The first principle is the most philosophical foundation since it does not define hard action items but focuses a number of guiding principles of the company that Toyota firmly believed and stuck to it. It includes four sub-principles that have a very strong influence on the overall philosophy of the organization which permeate down to influence other principles and tools in other layers of the Toyota Way model. Each sub-principle is described and analysed below, namely: (1) sense of purpose, (2) long-term perspective, (3) self-reliance and responsibility, and (4) customer focus.

4.3.1.1 Sense of Purpose

One major characteristic of corporate purpose is that it embodies the ultimate priority (objective) of the organization (Basu 1999). It is the case that Toyota as a company, wants and needs to make a profit, but that is not the driving purpose of the company. Liker (2004, p.72) revealed that "*the Toyota people have a sense of purpose greater than earning a paycheck*". Toyota's purpose is to be a contributing member of society by taking care of its employees and local communities (Basu 1999; Liker 2004) and achieving long-term prosperity for all employees and partners (Liker and Hoseus 2008). There are changes in Toyota's purpose as revealed by Basu (1999), who reviewed the evolution in the guiding principles of

Toyota over the past decades and noted that what changed with time is the geographic scope of purpose, which from the earlier days focused only on the Japanese employees and the Japanese society to the global employees and the global society in more recent years. Liker (2004, p.72) summarized the first principle of the Toyota Way, by stating that: "*Toyota's strong sense of mission and commitment to its customers, employees, and society is the foundation for all the other principles*". Moreover, to develop the "constancy of purpose" is one of the strategies that Toyota had adopted and operated for a long-term purpose, which explained why Toyota enjoys a steady growth trend in the sales and profits for every year except during the economic recession caused by the Wall Street meltdown in 2008–2009.

4.3.1.2 Long-Term Perspective

The Toyota people commonly agreed that they are very cost-conscious (Liker 2004). However, cost deduction is not the overriding principle that drives Toyota. Most people would agree that focusing on short-term financial results at the expense of the long-term health of the organization is not the best approach. However, taking a long-term view is not easy in the presence of pressure to perform in the short-term. In addition, the company, according to Liker (2004), is like an organism nurturing itself, constantly protecting and growing its offsprings (this will be discussed in Principles 9 and 10) on a long-term basis, so that it can continue to grow and stay strong. A company that focuses as much as Toyota does on eliminating waste in manufacturing might find it pragmatic to lay off employees during the slow period. Toyota would not dismiss its employees because of a temporary downturn. Moreover, sustaining a long-term relationship with the suppliers is also essential to the Toyota Way and reflects its long-term philosophy of being committed.

4.3.1.3 Be Self-Reliance and Responsible

Liker (2004, p.78) outlined that self-reliance and "let's do it ourselves" is a unique spirit of Toyota which was implemented at a corporate institutional level. The concept of self-reliance plays an important role in developing Toyota's core competitiveness because although Toyota outsourced 70 % of the vehicles to suppliers, it never transferred all the core knowledge and responsibility in any key area to the suppliers (Liker 2004). Toyota distinguishes itself by endeavouring to be an expert and the best in the world at mastering certain core technologies. Moreover, at Toyota, the champion to self-reliance is responsibility for its own successes and failures. The Toyota Way 2001 document, cited in Liker (2004, p.80), states that: "*We strive to decide our own fate. We act with self-reliance, trusting in our own abilities. We accept responsibility for our conduct and for maintaining and improving the skills that enable us to produce added value*".

4.3.1.4 Customer Focus

The Toyota Way is about adding value to customers, employees, and society. According to Basu (1999), another purpose of Toyota is to enhance customer satisfaction. Understanding what the customer wants and needs is the first step for an organization to have a customer focus. Profits are generated by satisfying customer needs by providing valuable products and/or services. That is why the TPS starts with the customer, asking "*what value are we adding from the customer's perspective*" (Liker 2004, p.9). Moreover, to ensure that the organization always does the right thing for the customer, it is critical for Toyota to develop a culture that puts customer's interests above all. Convis (2001) outlined that the "customer-first" philosophy is one of the fundamental elements of the TPS that management must be fully committed to. In contrast to other organizations' envision of customers in terms of that person who purchases the final product, the TPS views each succeeding process, workstation or department as the customer (Convis 2001). Convis (2001) revealed that in a Toyota plant, the management endeavours to ensure that all team members and departments realize their dual roles, namely that they are not only the customers of the previous operation but also the suppliers to the next operation downstream.

4.3.2 Summary of Principle 1

Table 4.1 summarizes the operationalized measurables from the Toyota Way Philosophy model.

4.4 Toyota Way Process Model

Performing processes is what allows people to transform the organization's inputs into outcomes which customers are willing to pay for. Every activity performed that is able to directly or indirectly support this transformation can be seen as a process. Achieving an effective and efficient process is the most fundamental determinant of how successful the company is in satisfying the customer's needs today. Principles 2–8 of the Toyota Way are part of the second broad category of Process (refers to Fig. 4.2), which involve a number of TPS tools for improving the manufacturing process as well as to achieve a stable production flow (Liker 2004; Moore 2007). These tools and processes are important and powerful, but they are only the "tactical" or "operations" aspects of the Toyota Way. They can be far more effective when they are supported by a company-wide, long-term management philosophy (Principle 1) (Liker 2004). As Liker (2004, p.87) outlined that: "*Toyota leaders truly believe that if they create the right process the results will follow*". So what is the right process? What does a right process look like? To answer these questions, the rest of this section focuses on defining how to achieve this right process with detailed examination of each principle along with the relevant tools.

Table 4.1 Operationalized measurables from the Toyota Way Philosophy model

The Toyota Way Philosophy model	Operationalized measurables
Sense of purpose	A high purpose to generate value towards customers, employees and society
	Sustain a constant purpose
Long-term perspective	Long-term philosophy supersede short-term financial loss
	Develop a long-term vision and make a plan to achieve it
Self-reliance and responsible	Self-reliance attitude
	Be responsible
Customer focus	Understand what customer wants
	Customer first spirit
	Extend customer focus internally (e.g. employees, suppliers)

4.4.1 Principle 2: One-Piece Flow

One-piece flow is also called "continuous flow". It means products that move continuously through the processing steps with minimal waiting time in between them, and the shortest distance travelled, will be produced with the highest efficiency (Liker and Meier 2006). Although the JIT concept has not been explicitly mentioned in Principle 2, achieving the one-piece flow is widely accepted as one of the goals of JIT. Aggarwal (1985) viewed JIT as an approach for providing smoother production flows and making continual improvements in processes and products. Liker (2004) noted that the ideal JIT aims to eliminate the physical buffers (materials or time) between production processes. To achieve that, cutting out wasted effort and time that is not adding value (waste) is required. Therefore, experts (e.g. Liker 2004; Liker and Meier 2006) suggested that companies should begin their lean journey with waste reduction, discussed seven major types of non-value-adding activities in the manufacturing process.

Toyota leaders believe that if they create the right process, the results will follow. This principle focuses on optimizing a flow process to identify the rooted problems. When operations are linked together, there is more teamwork, rapid feedback on earlier quality problems, control over the process, and direct pressure for workers to solve the problems and think and grow. Ohno (1988) discovered that if he reduced the inventory, the problems surfaced, and people were forced to solve them or the system was forced to stop producing. In essence, the creation of flow forces the correction of problems, resulting in reduced waste.

4.4.1.1 Elements of One-Piece Flow

To accomplish a one-piece flow as a distinct production system, Miltenburg (2001) called for the following five unique elements to be present: (1) *Takt* time, (2) Flow manufacturing on U-shaped production lines, (3) Pull production control, (4) *Jidoka*, and (5) Standardized work. Each of these five elements is examined in details as follows:

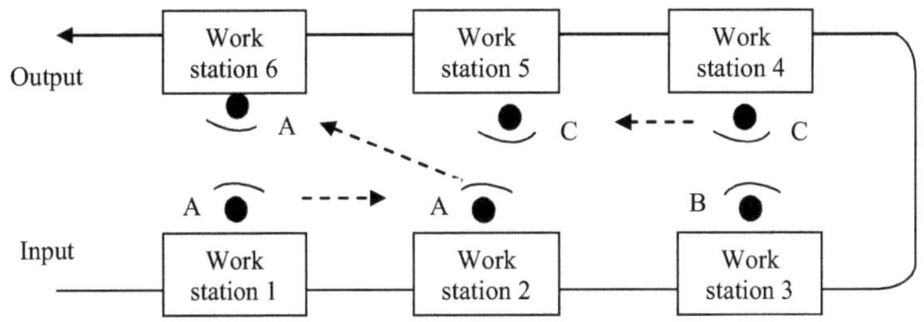

Fig. 4.3 U-shaped one-piece flow cell

1. *Takt* time

Takt is a German word for rhythm or metre, which is defined as the rate of customer demand for the group of product produced by one process (Liker 2004). *Takt* time is calculated by dividing the effective operating time of a process by the quantity of items customers require from the process in a certain time period. It does not automatically mean that every item should be produced at a rate of every *takt* time. Usually, there are two scenarios: if the production flow does not follow the *takt* time, when the production line is either going faster or slower. If they are going faster, they will overproduce; if they are going slower, they will create bottlenecks. *Takt* can be used to set the pace of production and advise the workers whenever they are getting ahead or are behind schedule.

2. U-shaped production lines

In a U-shaped production line, machines are arranged around the edge of a U-shape, allowing workers to walk the shortest distance from process to process, loading and unloading parts, and performing other manual operations [see Fig. 4.3 as adapted from Liker (2004)]. The advantages of the U-shape over a linear flow line are: firstly, it assists communication, since workers on a particular line are physically closer to each other; and secondly, the layout allows the workers access to a number of machines and to be able to operate several machines. It can be seen from the U-shape flow line example that a flexible multi-skilled workforce is required for efficient operation of the flow line.

3. Pull production control, 4. *Jidoka*, and 5. Standard work

These three principles are discussed individually as Principles 3, 5 and 6 of the Toyota Way. This infers that in order to achieve a flow manufacturing process, it requires an integration of all these principles working together.

4.4.2 Principle 3: Pull *Kanban* System

There are two primary ways of regulating work-flow in production systems: pull and push. Hopp and Spearman (2000) distinguished push from pull by the different mechanisms that trigger the movement of work in the system: a push system releases a job (e.g. materials or information) into a production process (e.g. factory, line, or workstation) precisely based on pre-assigned due dates. In contrast, a pull system releases a job into a system based on the state of the system (e.g. the amount of work in process and the quality of available) in addition to due dates. This avoids excess inventory resulting from bad guesswork. Liker (2004) highlighted that in the Toyota Way, the pull system means the ideal state of just-in-time manufacturing: giving the customer what he or she wants, when he or she wants it, and in the amount he or she wants. In construction, pull is ultimately driven by target completion dates, but specifically applies to the internal customer of each process (Ballard 2000).

4.4.2.1 Pull Concept

According to Liker (2004), in the mode of mass production or push system, each department will decide for all the largest volume items to follow their independent schedule in advance, without any coordination between the departments, before it changes over. That is the strategy which most mass production companies will take to minimize the equipment changeovers that are necessary for making different types of products with the same equipment. The disadvantage is well recognized by Ohno (1988) as such production systems will lead to overproduction and create large banks of inventory. Hence, Ohno (1988) decided to create small "store" of parts between operations to control the inventory. When the customer takes away specific items, they are replenished. If a customer does not use an item, it sits in the store but it is not replenished. This prevents overproduction and there are at least some direct connections between what customers want and what the company produces.

4.4.2.2 Use Visual Control: *Kanban*

Kanban is a tool to achieve JIT production (Monden 1998, p.16). Ohno (1988) invented simple signals—cards, empty bins, carts, called *kanban*, as a vehicle for signalling the assembly line to produce the specific number of parts. *Kanban* identifies the part number, container capacity, and certain other information. Because there are rich instructions listed in *kanban*, *kanban* is regarded as an effective and advanced visual control system focusing primarily on eliminating overproduction, increasing flexibility to respond to customer demand, and reducing costs by eliminating waste. Monden (1998, p.16) outlined that the TPS employs two kinds of *kanban*: one *kanban*, the production-ordering *kanban*, specifies the kind and quantity of product which the preceding process must produce; the other, called withdrawal *kanban*, specifies the kind and quantity of product which the subsequent process should withdraw from the preceding process. A *kanban* system therefore consists of a set of cards that travel between preceding and subsequent processes, communicating what parts are needed in the subsequent processes. In the processes controlled by *kanbans*, the operators produce products based on actual usage rather than forecasted usage. Therefore, the production process it controls must:

1. Only produce products to replace the products consumed by its customer(s).
2. Only produce products based on the signals sent by its customer(s).

The implementation of the pull system has resulted in striking improvements (Sepheri 1986) and has been identified as one of the characteristic elements of the just-in-time philosophy (Monden 1983; Low and Chan 1997). Following the discussion above, one of the first priorities of the pull production system is to achieve reliability of processes and smooth and synchronized flows involving a stable product mix (Monden 1983). Monden (1983) also listed the necessary conditions for *kanban* to work well: (1) Set-up time will need to be shortened in order to allow rapid response since there will be small or no butter stocks; (2) Proper machine layout can facilitate the flow; (3) Standardization of jobs; (4) Improvement activities; and (5) Autonomation (autonomous defects control).

4.4.3 Principle 4: Level Out the Workload (*Heijunka*)

Principle 4: Level out the workload (*heijunka*) can help to achieve the benefits of continuous flow. It focuses on strategies by levelling product volume, mixing and, most importantly, levelling out the demand on people, equipment, and suppliers.

4.4.3.1 The Principle of *Heijunka* and Its Benefits

Heijunka is a Japanese term used to describe a mixed production system, where various and changeable sequences of mixed models are produced in the same production line (Coleman and Vaghefi 1994; Liker 2004). It also referred to as production smoothing or levelling the production schedule. *Heijunka* does not build products according to the actual flow of customer orders, but takes the total volume of orders in a period and levels them out so that the same amount and mix are being made each day. In contrast to mass production, which would be to dedicate the line first to one model, then to another, the Toyota Way-styled *heijunka* allows various models can be produced in the same line on the same day, with quick changeovers (Hampson 1999). It reflects the Toyota's philosophy against speculative production and the idea that customers do not order in a stable and predictable way which therefore creates inventories (Coleman and Vaghefi 1994). According to Shingo (1988) and Coleman and Vaghefi (1994), the concept of *heijunka* incorporates the concepts of levelling and line balancing. Levelling is the term describing the effort to balance the work load to be performed to the capacity or capability of the process (machine and operators) to complete that work (Shingo 1988). The principle of line balancing attempts to equate workloads (production rate) at each process to each other (Shingo 1988). In summary, *heijunka* is a production planning method with two objectives (Coleman and Vaghefi 1994):

1. To reduce the inventories by setting up mixed-model production lines to produce small batches.
2. To equate workloads in each production process to each other and to capacity.

Liker (2004, pp.118–119) and Coleman and Vaghefi (1994) revealed the four benefits of levelling the schedule as follows:

1. *Flexibility to make what the customer wants when they want it*. This reduces the plant's inventory and its associated problems.
2. *Reduced risk of unsold goods*. If the plant makes only what the customer orders, it reduces the costs arising from owning and storing inventory.
3. *Balanced use of labour and machines*. The plant can create standardized work and level out production by taking into account that different machines have different level of manufacturing capabilities.
4. *Smoothed demand on upstream processes and the plant's suppliers*. If the plant uses a JIT system for upstream process and the suppliers deliver goods multiple times in a day, the suppliers will get a stable and level set of orders. This will allow them to reduce inventory.

In addition, Coleman and Vaghefi (1994) summarized four requisites for *heijunka* implementation:

1. Quick setups, the small-lot production sequences.
2. Cross-trained and flexible employees, with flexible machinery and equipment.
3. Effective quality assurance system.
4. Components/parts must also be supplied to the assembly process in very small lots, without delays. It necessitates the use of *kanban*, to attain the smooth shop-floor characteristic of JIT system.

4.4.3.2 Elimination of *Muda, Muri,* and *Mura*

According to Liker (2004), many companies failed to stabilize the manufacturing system and create evenness. Achieving *heijunka* is a key to TPS (Coleman and Vaghefi 1994) and fundamental to eliminating *mura*, which is fundamental to eliminating *muri* and *muda*.

1. *Muda* (non-value added): It includes the seven types of waste mentioned earlier. These are wasteful activities that lengthen lead time, cause extra movement, and create unnecessary inventories and so on.
2. *Muri* (overburdening people or equipment): *Muri* translates as "overburden—when workers or machines are pushed beyond their capacity" (Oliver and Wilkinson 1992). This may reduce the production life of both human beings and machines (Hampson 1999). "Overburdened jobs" resulted in extreme worker stress, repetitive strain injuries, and potential quality problems. Overburdening equipment causes breakdowns and defects.
3. *Mura* (unevenness): Unevenness results from an irregular production schedule or fluctuating production volumes due to internal problems, like downtime or missing parts or defects. *Muda* will automatically result in some varieties of *mura* (Hampson 1999; Liker 2004). This is because some workers and machines will be working below capacity for some of the time, while others may overproduce.

Eliminating *muda* is just one-third of the equation of making lean successful (Liker 2004), while eliminating overburden to people and equipment (*Muri*) and eliminating unevenness in the production schedule are equally important, yet not fully appreciated in companies attempting to implementing lean practices. Furthermore, Principle 4 allows a pre-determined level of finished goods inventory. It seems wasteful and contradicts lean Principle 2 in creating a continuous one-piece flow. A small inventory of finished goods is, however, often necessary to protect a supplier's level of production schedule from being jerked by sudden spikes in demand (Liker 2004).

4.4.4 Principle 5: Built-In Quality (*Jidoka*)

Quality is built into the product, according to Principle 5 of the Toyota Way model, by stopping the process when defects are encountered. This is done to prevent defects from being made and used by downstream operations. The Japanese term *Jidoka* is in agreement with Principle 5—stopping the process to build in quality. The term *jidoka*, translated as "autonomation" in English, is the second pillar of the TPS (Ohno 1988; Liker 2004). The origins of *jidoka* can be traced back to Sakichi Toyoda, founder of Toyota, whose approach was to achieve efficiency gains in the use of Toyota automatic looms. This brilliant method of enabling automatic looms would stop them when a thread broke (Liker 2004, p.129). Like many elements of the TPS, the concept of *jidoka* places further emphasis on control with the people actually doing the work allowing them to halt production to fix a problem as it arises. It is predominately a technique for detecting and correcting production defects and always incorporates the following devices: a mechanism to detect abnormalities or defects (Monden 1983). Monden (1998) highlighted that *jidoka* also has other equally important components and effects including cost reduction, adaptable production, and increased respect for humanity:

1. Cost reduction through decrease in the workforce: Autonomation utilizes worker's ability to handle more than one machine at a time because as equipment is designed to stop automatically when a defect occurs, there is no need for the employee to oversee machine operations.
2. Adaptable production: Since all machines halt automatically when they have produced the required number of parts and produce only good parts, autonomation eliminates excess inventory and thus makes possible JIT production and ready adaptability to changes in demand.
3. Increase respect for humanity: Autonomation calls immediate attention to defects or problems in the production process. It promotes improvement activities and thus increases respect for humanity.

4.4.4.1 The *Andon* System

The human aspect of *jidoka*, the decision to stop producing based on "things that cannot be express exactly", is collectively known as *andon*—literally translated from the Japanese means "traditional rice paper lantern" (Everett and Sohal 1991). In Toyota's shop floor, *andon* usually works with the rope call switch. When any team member judges an abnormal situation, whoever can take the rope call switch will light the *andon* to indicate the location of the problems. But the line will continue moving. The team leader proceeds immediately to the site to investigate the situation and provides assistance. If the problem has been fixed, the team leader pulls the rope again to switch off the *andon* informing all the production workers that the line has returned to normal. If the abnormality cannot be corrected within the prescribed section of the line, it will automatically stop and, the *andon* light further changes to red. In this way, the colour of the *andon* light informs all the workers of the condition of the problem in the production line with a glance. Under this principle, quality is guaranteed as it will be very uncommon for defects to move into the next station because the operatives are highly responsible for every product moving into their station, and they are empowered to examine the quality issue.

4.4.4.2 Built-In Quality Tools

Liker and Meier (2006) depicted the Toyota Way of stopping the line to fix problems by beginning with a focus on waste elimination. The core of this principle is to develop a system that emphasizes getting the quality right the first time. Toyota has developed an extensive support system to provide people with the tools and resources to identify problems and solve them. Monden (1998) detailed two approaches that can be used in general to stop and fix production when abnormalities occur: by relying on human judgment and by means of automatic devices.

1. By human judgment: All workers are empowered and given the responsibility to stop the line if all operations are not or cannot be performed in accordance with the standard operations routine. With line stoppage due to the above-mentioned two causes, the supervisor's responsibility is twofold: First, he should teach the workers to stop the line whenever the defects occur. Second, he must discover and correct the cause of the defects that have stopped the line at source. To reiterate, the key to preventing defects via human judgment is that each employee has the power to stop the line (Monden 1998).
2. By means of automatic devices:
   - Visual controls: implementing autonomation, various visual controls monitor the state of the line and the flow of production. For example, *andon* is one of the widely used visual controls in Toyota.
   - Foolproof systems: foolproof systems are used to eliminate defects that may occur due to an oversight on the worker's part, not due to a lack of time in the cycle or unwillingness to stop the line. While these detecting instruments sense abnormalities or deviations in the work-piece or the process, the restricting tool stops the line and the signalling device will sound a buzzer or lights a lamp to notify the workers.

Although it is impossible to reach zero defects in reality, this can be approached through the use of above-mentioned methods and tools. At Toyota they keep things simple and use very few statistical tools. Their quality specialists and team members have just four key tools (Liker 2004): (1) Go and see, (2) Analyse the situation, (3) Use one-piece flow and *andon* to surface problem, and (4) “5 Whys” techniques.

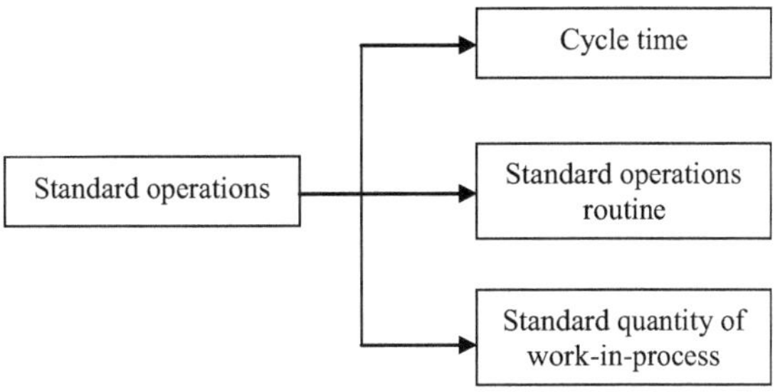

Fig. 4.4 Elements of standard operations

4.4.4.3 Autonomation and People

In mass production, the traditional attitude of management towards the operator is to tell him/her what to do instead of involving the operatives in decision-making. However, *andon*, or autonomation, when applied, changes the order of control in the workplace, as it is the operative who eventually controls the quality via the system, albeit through the demand for quality (Everett and Sohal 1991). In the construction industry, the likelihood that things go wrong and result in a defect are great. Hence it is necessary to have as many people as possible involved in detecting, analysing, and eliminating sources of defects. Workers on site, foremen, and project managers must all be part of this effort. Everett and Sohal (1991) summarized a number of key steps for the successful implementation of the *andon* system that are related to the people aspects.

1. Senior management are committed to the whole project is utterly vital from the outset. The ramifications must be clear and the organization should be in a stable mode of growth or a strong position to withstand the shock of possible product loss to the market during the early phase.
2. Trade union officials must be fully briefed, their cooperation gained and their concerns for their members addressed long before start-up.
3. Shop-floor supervisors and operatives must be well briefed on their roles.
4. Quality standards must be clearly agreed upon by all parties and displayed in the production line. All existing known quality problems should be solved prior to start-up.
5. All equipment in the area must be in, or brought up to, a satisfactory condition.

4.4.5 Principle 6: Standardized Tasks

Gibb and Isack (2001) identified standardization as the extensive use of processes or procedures, products or components, in which there is regularity, repetition and a record of successful practice. It involves the development of pre-set procedures and reference materials for performing a particular process or operation. Companies have methods they call "standards", but it is not what Toyota means by using the term "standardized work" to define the method used to perform work. In Toyota or lean terminology, standardized work is to reduce costs relating to production.

According to Monden (1998), standard operations in Toyota have three main goals, namely (1) to achieve high productivity through efficient work; (2) to achieve line balancing among all processes in terms of production timing; and (3) to limit the work-in-process to a minimum amount, helping to eliminate excessive in-process inventories. Figure 4.4, adapted from Monden (1998), summarizes the three key elements of standard operations in Toyota.

1. Cycle time: the cycle time or *takt* time represents the time span in which one unit of a product must be produced. It is deduced from the monthly market demand forecast and thus follows a push system. In this context, the cycle time is determined by the daily quantities of outputs required and the effective daily operating time (Monden 1998; Ohno 1988). Based on this information, management derives the minimum staffing level needed.
2. Standard operations routine: This can be defined as the order of actions that each worker must perform within a given cycle time (Monden 1998). According to Monden (1998), the setting of this routine serves two purposes. First, it provides the worker with the order or routine to pick up work tasks according to a routine sequence. Second, it is to give the sequence of operations that the multi-functioned worker must perform at various machines within a cycle time.
3. Standardized work-in-process: According to Monden (1998), the standard quantity of work-in-process consists of the work laid out and held between machines.

The traditional manufacturing model has an initial focus on achieving the lowest possible unit cost by utilizing time and motion studies to determine the most efficient work procedure, and a standard time is therefore allotted for the designated task (Liker and Meier 2006). The Toyota Way seeks the same objective as the traditional manufacturing model in terms of low cost but the primary focus is on reducing the waste within the system. In most organizations, there is substantial amount of waste that is caused by random activities and inconsistent methods. To eliminate waste, reduction or elimination of variation within processes is needed. The isolation of variation is a key to the establishment of standardized work methods and procedures (Liker and Meier 2006). By definition, variation implies the inability to standardize. Imai (1997) highlighted that if variability occurs because of the lack of standards, then one should develop new standards. If variability occurs even when people have adopted standards, it will be necessary to determine the causes and either revise/upgrade the existing standards, or verify that the existing standards are clear to the workforce. Standardized work has four main advantages according to Hall (1995):

1. It prevents overproduction.
2. It is an aid in achieving higher quality.
3. It lowers cost.
4. It provides a basis to judge the normal from the abnormal.

Once standard operations are set, it is the task of the supervisors and workers to continuously improve these standards. Liker (2004) explained that standardization is the basis for continuous improvement, because as Imai (1986) pointed out, it is impossible to improve any process until it is standardized. This means the standards do not remain the same forever, but future results are expected to improve from the standard (Liker and Meier 2006).

4.4.5.1 Organization Structure Supports Standardization

Standardization is primarily associated with Taylor's scientific management philosophy. In mass production, productivity gains can be achieved using Taylor's scientific management principles (e.g. time and motion studies). It, however, created very rigid bureaucracies in which workers were to blindly and simply follow the standardized procedure; whist managers play the role of doing the thinking. This results in a number of disadvantages such as red tape, hierarchical organizational structures, top-down control, resistance to change, poor communication, slow and cumbersome implementation and application, static and inefficient rules and procedures, and piles of written rules and procedures (Liker 2004). New United Motor Manufacturing, Inc. (NUMMI), Toyota's first joint venture in North America, has all the characteristics of bureaucracy and is a very "mechanistic" organization (Adler 1999; Liker 2004). Liker (2004) adopted Adler's (1999) study on Toyota's organizational practice and compared the coercive bureaucracy uses standards to control people and Toyota's enabling systems that help people control their own work (Table 4.2).

Based on the above observation, Liker (2004, p.145) noted that the key difference between Taylorism and the Toyota Way is that the Toyota Way preaches that the worker is the most valuable resource—not just a pair of hands taking orders, but an analyst and problem-solver. The Toyota Way also shows that, to maintain long-term competitiveness, a company must have viable and enabling standards so that it can continually improve upon repeatable processes (Liker 2004).

4.4.5.2 Empowered Employees

The critical task when implementing standardization is to explore the balance between providing employees with rigid procedures to follow and providing the freedom to innovate and be creative to meet challenging targets consistently for cost, quality and delivery. Liker (2004) offered the key to the balance, which lies in the following considerations:

1. The standards have to be specific enough to be useful guides, yet general enough to allow for some flexibility.
2. The people doing the work have to improve the standards. For a production person to be able to write a standard work sheet that other workers can understand, he or she must be convinced of its importance (Ohno 1988) and empowered to make his or her contribution to their own improvement.

Principle 6 can be summarized as:

1. Standardized work consists of three elements namely *takt* time, standard operations routine and standard quantity of work-in-process.
2. Rules and procedures are used as enabling tools—performance standards are used in parallel with information on best practices for achieving them.
3. Empowered employees to participate in the writing of standard procedures.

Table 4.2 Coercive versus enabling design of systems and standards

Coercive bureaucracies system	Enabling bureaucracies system
Top down control	Empowered employees
Minimum written rules and procedures	Rules and procedures as enabling tools
Hierarchy controls	Hierarchy supports organizational learning

Source: Liker (2004) and Adler (1999)

4.4.6 Principle 7: Visual Management

Principle 7 of the Toyota Way is to use visual control to improve flow. This principle can be shortened to "make it visual". The goal of visual management is to make waste, problems, and abnormal conditions readily apparent to employees and managers so that problems can be fixed. The principle behind visual management is based on the assumption that people are usually attracted what they see (Ho 1999). Liker (2004) outlined that the visual aspect means being able to examine the process, a piece of equipment, inventory, information or a worker performing a job and immediately seeing the standard being used to perform the task and if there is a deviation from the standard. It has gone beyond capturing deviations from a target or goal on charts and graphs and posting them publicly. The control aspect means an immediate approach will be adopted to fix any examined deviation to smooth the process. It is opposed to the old mindset of hiding problems to make things look good. Caravaggio (cited in Levinson and Rerick 2002, pp.134–135) summarized that a visual control system has five aspects:

1. Communication: written communications are easily accessible.
2. Visibility: communication with pictures and signs.
3. Consistency: each activity adopts the same conventions.
4. Detection: alarms and warnings will work when abnormalities occur.
5. Fail-safing: these activities prevent abnormalities and mistake.

To sum up, visual control tools play an important part of the communication process can tell us in a very straightforward way how work should be done and whether it is deviating from the standard just through a glance (Liker 2004).

4.4.6.1 The Practice of 5-S

"5-S" programme was developed by the Japanese that comprises a series of activities or guidelines regarding how to effectively organize a workplace or production process. According to Hirano (1995), the 5-S approach is a simple but powerful method for workplace improvement that yields impressive results. It has become so familiar in Japan that it is hard to find a factory or office that has not borrowed at least some of its ideas (Hirano 1995). Hirano (1995) summarized the five stages of a 5-S programme that consists of [see Fig. 4.5 as adapted from Liker (2004)]:

1. *Seiri (sort)*: during the *seiri* process, all materials and tools are sorted, and only needed to be kept for continued use. Everything else unneeded should be stored

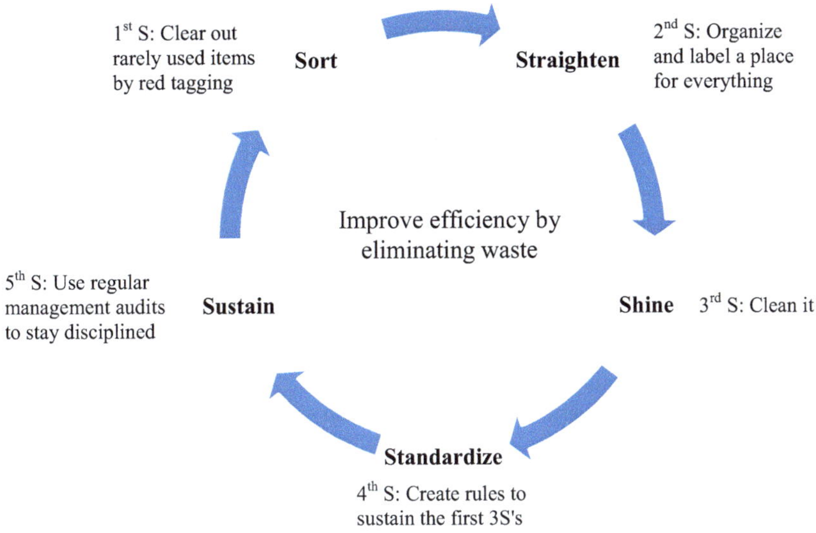

Fig. 4.5 The 5-S

or discarded. This process leads to fewer hazards and less clutter that might interfere with productive work.

2. *Seiton (set in order)*: *seiton* refers to organizing the way needed for things to be kept so that anyone can find and use them easily.
3. *Seiso (clean)*: *Seiso* stands for sweeping and cleanliness. It means to sweeping floors and keeping things in order. The key point is that maintaining cleanliness should be part of daily work—not an occasional activity that is initiated only when things get too messy (Haghirian 2010).
4. *Seiketsu (standardize)*: *Seiketsu* translates as "standards". It means making all the cleanliness, orderliness, and improvement processes a regular activity in the workplace.
5. *Shitsuke (sustain)*: *Shitsuke* means "sustaining discipline". It also means always following specified (and standardized) procedures to support long-term *kaizen* goals. The first four S's can be implemented smoothly if the employees are committed to maintain discipline.

In practice, 5-S in manufacturing ensures that all raw materials, work in progress, and finished products are located neatly on the well-labelled racks or spaces. Liker (2004) highlighted that the integration of the 5-S can create a continuous process for improving the work environment.

4.4.6.2 Relationship with Other Toyota Way Principles

Monden (1998) noted that the visual control system monitors the status of the line and the flow of production along with the implementation of autonomation. Toyota used an integrated set of visual controls or a visual control system to create a transparent and waste-free environment. In the discussion of the earlier Toyota Way principles, different forms of visual controls have been highlighted that played different roles with the same aim of ruling out the waste in the process. For example, the practice of *andon* was discussed in Principle 5. The *andon* cord, when pulled by an operator, lights up a display and shows a signal unique to the station. This communicates that a problem exists and indicates its location in the line so that it can be solved before it becomes necessary to stop production. *Kanban*, discussed in Principle 3, is another visual control tool which indicates to the downstream operatives to manufacture the exact number of products to rule out overproduction. In addition, the standard operations sheet as highlighted as a tool in Principle 6 is another visual control tool in practice. The use of visual control is regarded as the most important step in the process of developing standardization (Liker and Meier 2006).

Principle 7 can be summarized as:

1. Clean it up, Make it visual—use simple visual control systems (e.g. 5-S).
2. Integrate the visual control systems with other principles to the value-added work—use visual control to improve flow.

4.4.7 Principle 8: Use of Only Reliable, Thoroughly Tested Technology

Toyota has rigorous requirements on acquiring technology because it is not an easy task to search and choose appropriate tools and technology in the "techno-jungle". In Toyota, new technology cannot be introduced until it is proven through direct experiment along with the involvement of experts from cross-section departments (Liker 2004). This means that the technology would have been tested and evaluated to ensure it can add value to support people and the manufacturing process. Simply put, technology in Toyota is pulled by process, not pushed by process. Moreover, the technology is also designed to complement rather than substitute for production workers. For example, according to Pil and Fujimoto (2007), Toyota abandoned full automation efforts in the assembly line but focused instead on "in-line mechanical" automation, which consists of equipment and component jig-pallets synchronized with the auto bodies moving on the conventional continuous conveyers. This allows automation zones and manual assembly zones to coexist in the same assembly line. Morgan and Liker (2006) highlighted how Toyota implements five highly effective sub-principles as guidelines derived from Principle 8 for its production development system:

1. Technologies must be seamlessly integrated.
2. Technologies should support the process, not drive it: from Toyota's perspective, changing the process to conform to technology leads to instability, drives massive process variation, confuses people, and creates waste. What Toyota adopted is the opposite approach, not to acquire the next technological fad in a rush, but to focus on its potential that can enhance the process.
3. Technologies should enhance people, not replace them: since Toyota values their employees as the most value assets, it is best for Toyota to choose tools and

technology that make the best use of engineering time and talents. It is the impact of Principle 8 on people.

4. Specific solution oriented, not a silver bullet: Toyota people believe that technology is never a substitute for the head work, the potential of a required technology only lies in supporting and accelerating that hard work once a lean process is in place and highly skilled people are appropriately trained and organized.
5. Right size: it is a misconception to procure the biggest, fastest, and newest tools in the market. Toyota engineers still employ used notebooks for their engineering checklists, while their competitors developed an impressive online and fully integrated database. But the data was vacuous and rarely used. The point is that Toyota only uses simple tools that can greatly facilitate their employees' work and ensure that things are being done properly.

Toyota's attitude towards new technology is to adapt it appropriately. It is critical to Toyota, but Toyota looks at technology as a tool that, like any other tool, exists to support the people and the process. Some departments in Toyota still employ, and will continue to use, old in-house developed software under simpler circumstances. Because it has continuously evolved over the years and does exactly what is needed, there is no need to upgrade the system because according to the principle explained earlier, it will not add extra improvement to the process.

4.4.8 Summary of Principles 2–8

As the "process" title indicates, all the seven sub-principles along with their related tools aim to eliminate the waste and improve the flow. Liker (2004) highlighted the pitfall found in some firms even though the tools have been implemented; they have missed out the focus on flow, thus are still performing mass production with a couple of lean tools. Another observation is that each principle can be viewed as part of an integrated system to add value to the production flow. This part originates and evolves from the shop floor, and become the core of the Toyota Production System. Table 4.3 summarizes a number of operationalized measurables that are identified in this layer of the Toyota Way model.

4.5 Toyota Way People and Partners Model

The Toyota Way 2001 document indicates that "respect for people" and "continuous improvement" should go hand in hand. The basic tenet of the TPS is that people are the most important asset in Toyota (Convis 2001; Liker 2004); very few companies actually behave in a way that supports this basic tenet. If Liker's (2004) 4P model is examined carefully on the People and Partner layers, three key words, namely "*Respect*", "*Challenge*", *and* "*Grow them*" are highlighted (see Fig. 4.2). These are the core principles from the Toyota Way 2001. Here people can be extended to broadly mean leaders, teams, and the network of partners and

Table 4.3 Operationalized measurable from the Toyota Way Process model

Toyota Way Process model and its sub-principles	Operationalized measurables
P2: One-piece flow to bring problems to the surface	• Waste elimination • *Takt* time • Use flow oriented layout (U-shape) • Synchronize production activities so that one does not start until the previous activity has finished (pull) • Standardized work to stabilized flow
P3: Use pull system to avoid overproduction	• Pull from customer end—including both internal and external customers • Use visual control—*kanban* system
P4: Level out the workload (*Heijunka*)	• Eliminate overburden to people and equipment (*muri*) • Eliminate unevenness in the production schedule (*mura*) • Level out the workload of all manufacturing and service process
P5: Build a culture of stopping to fix problems	• Deliver perfect first time quality • Reveal and solve problems at the source as they occur • Keep quality control simple • Create culture—involve and empower employees to continuously improve
P6: Standardized tasks are the foundation for continuous improvement and employee empowerment	• Standardized operating procedure (SOP) • Continuously improve the standardization • Empowered employees to participate in the writing of standard procures
P7: Use visual control so no problems are hidden	• Practice of 5-S • Integrate the visual control systems to the value-added work
P8: Use only reliable technology that serves people and process	• Thoroughly test new technology • Technology must support people • Technology must improve flow • Technology must support the company values

suppliers. The following three principles reveal how Toyota selects, develops, and motivates people to become committed to the goal of building high-quality products in Toyota.

4.5.1 Principle 9: Leaders and Leadership

Throughout Toyota's history, key leaders have been found within the company, at the right time, to shape the next step in Toyota's development. Toyota does not go shopping for "successful" CEOs and presidents because their leaders must understand the Toyota culture and philosophy well. These include the former president

Fujio Cho, who grew up in Toyota and was a student of Taiichi Ohno, where he and Ohno created the theoretical basis for the Toyota Production System, to the current president Akio Toyoda, who has also worked for Toyota for approximately four decades. These leaders lived and thoroughly understood the Toyota culture day by day (Liker 2004). For this reason, Toyota cannot readily recruit leaders; they must take people who have some natural leadership abilities and develop them to think and act in the Toyota Way every day—a process which easily can take decades or more to home (Liker and Hoseus 2008).

4.5.1.1 Thoughtful Leaders and Servant Leadership

Toyota's internal document, *the Toyota Way 2001* cited in Liker and Hoseus (2008, p.319), defined thoughtful leaders as: "*having the ability to energize and invigorate others, willingly giving realistic challenges and development opportunities and fostering a sense of accomplishment in subordinates. Thoughtful leaders monitor individual and team performance, holding people accountable for their actions and taking responsibility for their activities*". Unlike the traditional manager's image as a monitor and controller under a command structure, Toyota leaders focus on confirming that all the works are followed by a set of defined rules of *takt* time, operations-standard work, 5-S, etc. rather than catching people make mistakes and blame them. Management should have a shop-floor focus because they are taught that all value-added activities start on the shop floor and their job is to support the team members (Convis 2001). Moreover, thoughtful leaders do not assume that the right rewards and punishments will produce impact on the behaviours of their employees. Rather, thoughtful leaders develop a culture in which they can effectively delegate to and trust their team members to produce excellent results (Liker and Hoseus 2008). Servant leadership is the concept formalized by Mikio Kitano, TMMK's[3] second president (Liker and Hoseus 2008), who prioritized the team members at the top and put himself (and other leaders) at the bottom in an upside-down pyramidal model [see Fig. 4.6 as adapted from Liker and Hoseus (2008)].

In this model, the group leader is the first level of "management", who leads a small group of approximately 5–7 people. Both team leader and group leader has three basic responsibilities: (1) support the operations, (2) promotion of the system, and (3) leading change (Liker and Meier 2006). The key concept of servant leadership recognizes that the value-adding work is the process of building cars where team members can directly add value. Leaders only add value by supporting those who are most actively adding value to the process, and therefore, leaders are posted to the bottom of the pyramid. This is unlike in a traditional top-down organization, where the capacity and imagination is limited to a few leaders at the top.

[3] TMMK refers to Toyota Motor Manufacturing, Kentucky in the United States of America.

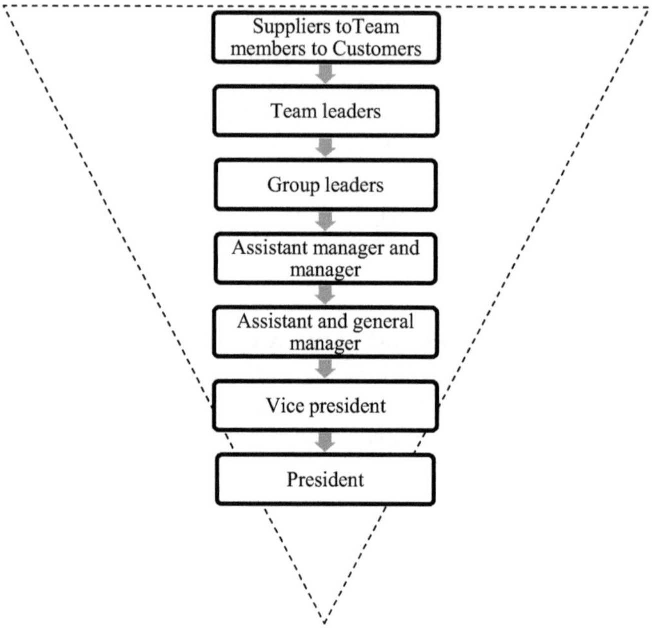

Fig. 4.6 Servant leadership in Toyota plants

4.5.1.2 Toyota Leadership Model

Table 4.4 summarizes some of the key differences of the Toyota Way leadership which is contrary to the practices of western managers. It reveals the distinctive leadership in Toyota from other companies. The Toyota leaders must have a combination of in-depth understanding of the work and the ability to develop, mentor and lead people. The expectation of leadership in Toyota is to effectively develop people so that performance is constantly improved. This is accomplished by instilling the Toyota culture in all employees, by continuously developing and growing capable people, and by focusing the efforts on strengthening the Toyota Production System (Liker and Meier 2006).

In summary, the current leadership tenet of the Toyota Way can be summarized as:

1. Support the culture: Toyota invests years to develop leaders who carry the DNA of the company in their thoughts, words and actions. In turn, the leaders should make efforts to support the culture in order to create the environment for a learning organization.
2. Support the people doing the work: The absolute core of the Toyota philosophy is that the culture must support the people doing the work (Liker 2004, p.176). This feature of Toyota leadership is sometimes described as "servant leadership". The higher leaders go, the less direct power they have.
3. Toyota wants leaders who live the core values, including the spirit of challenge, *kaizen*, *genchi genbutsu*, respect, and teamwork.

Table 4.4 Traditional Western leadership compared to Toyota leadership

Traditional Western leadership	Toyota leadership
Quick results	Patient
Proud	Humble
Climb ladder rapidly	Learn deeply and horizontally and gradually work the way up the ladder
Results at all costs	The right process will lead to right results
Accomplish objectives through people	Develop people
Overcome barriers	Take time to deeply understand problem and root cause before acting
Manage by numbers and graphs	Deeply understand the process

Source: Liker and Hoseus (2008)

4.5.2 Principle 10: People Management

A famous saying by Toyota leaders can be borrowed to describe the theme of Principle 10, which states that "*Toyota does not just build cars. They build people*". Basically, the rationale for Principle 10 lies in Toyota's philosophy that people are truly the greatest asset. This is the fundamental principle that has guided the company to decide on a series of human resource policies to cultivate its employees and to grow with them. Liker and Hoseus (2008) highlighted that the Toyota Production System would not function well without high performance teams on the shop floor. The main mechanism for transmitting the Toyota culture is the basic work team. Principle 10 in short addresses three issues: first is to develop excellent individual work, second is to promote effective teamwork within Toyota, and the last is Toyota's unique organizational structure.

4.5.2.1 Develop Individual Work and Training

Toyota spends an immense amount of time and effort in screening prospective employees and carefully develops them. Different forms of training and development programmes have been established at Toyota aiming to nurture the workers to perform well and to deliver better-quality products. A number of unique Toyota training programmes are outlined in the next sections following the process of training a Toyota's production worker as summarized in Fig. 4.7 [as adapted from Liker and Hoseus (2008)].

Normally it is an ongoing long-term process. Prior to the team members being assigned to the team, they receive training in what Toyota calls "fundamental skills". Workers are then assigned to a team with a team leader and group leader who introduce them to the first job they need to learn. This job has been broken down into tiny work elements that are taught piece by piece using the Toyota job instruction training method. In addition, the individual member continues to be supported full time until he or she is comfortable with doing the job (Liker and Hoseus 2008).

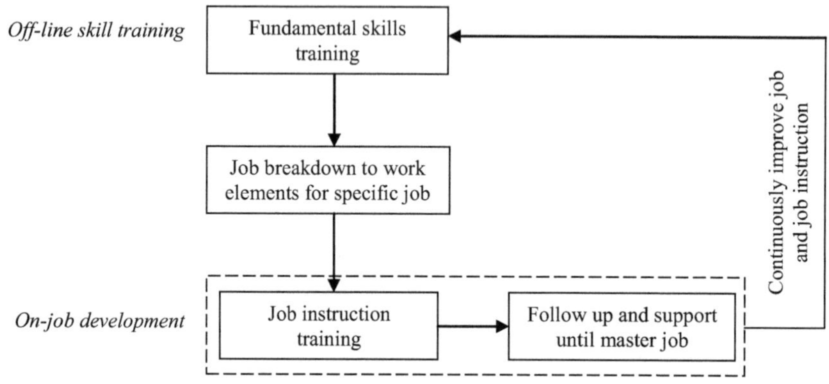

Fig. 4.7 Step-by-step progression to stable job performance

1. Fundamental skills training
   Fundamental skills are necessary for people to be able to successfully perform their work. This is the basis for the further development of additional capabilities, such as improving communication skills and leadership abilities, planning, and developing new methods or procedures.
2. Job instruction training
   Job instruction training is the key to developing the employee's exceptional skills, which is also known as the four-step method. The four steps are to prepare workers, present operations, try out performance and follow up. It is based on two main processes: the training material and the training method. The training method is developed within a stable environment, where the work has been broken down to the tiniest details under the principle of standardization. Liker and Hoseus (2008) advised that because most companies cannot have the high degree of stability or level of standardization as Toyota does, it would be more appropriate for them to be selective about processes or areas and develop some level of stability in that area for training purposes. The learning in these pilot areas can then be extended to other areas and eventually across the plant.
3. Building team associates for the long term
   Some work done on a day-to-day basis can decrease the level of "excitement" after some time. This is especially true in a repetitive working environment that does not require a high skills level. The Toyota Way promotes the growth and development of all employees by maintaining their morale and commitment, and by offering various kinds of optional and required courses. Moreover, all the Toyota employees are encouraged to participate in activities and programs including suggestion programmes, quality circles, leadership development programmes, and *kaizen* teams.
   - *Toyota suggestion programmes*: Suggestion schemes represent an individual mechanism for capturing worker knowledge and thereby improve the quality of the product and manufacturing process. It differs from most traditional suggestion programmes based on the premise that people inherently want to

improve their work environment, and that the contributions of the employees provide long-term continuous improvement (Liker and Meier 2006). Moreover, the benefits of the Japanese system are often measured by looking at the number of suggestions per employee and year (Karlsson and Åhlström 1996). This is an important element of the lean production model.
- *Quality circles*: Quality circles are an essential part of *kaizen* in Toyota. These have been an ongoing management tool for productivity and quality improvement for decades and are excellent to promote teamwork and develop the capacity of individuals to make continuous improvement. The circle is responsible for setting goals and meeting schedules, and the group leader acts in an advisory role. Most circles deal with issues in the work area, where meetings are conducted.
- *Developing team associates for leadership roles*: Selecting and developing team members for leadership roles is a critical matter in Toyota (Liker and Meier 2006). The leaders are responsible for teaching and coaching others in the Toyota Way. They must convey the message to the next generation and also be responsible for sustaining the daily operation and for continuous improvement.

4.5.2.2 Multi-functional Teams

Karlsson and Åhlström (1996) highlighted that the most salient feature of the work organization in lean production is the extensive use of multi-functional teams. According to Olivella et al. (2008), the possession of multiple-skills implies flexibility, provides team members with an overall vision of the work to be done, and facilitates learning and continuous improvement. The aim of forming multi-functional teams is to have employees who are able to perform more than one task in the team. These teams, in the manufacturing context, are often organized along a cell-based part of the product flow and each of them is given the responsibility of performing all the tasks along this part of the product flow. Monden (1998) noted that Toyota uses a job rotation system to cultivate multi-functional workers, according to which each worker rotates through and performs every job in the workshop. One of the benefits of utilizing multi-functional teams is that the number of job classifications decreases, as the employees perform many different tasks during a single day. Dependence on a single person automatically decreases. Achieving this multi-functionality, however, requires efforts to be made in staff training (Karlsson and Åhlström 1996), as well as overcoming resistance to increase the number of tasks they perform (Sanchez and Perez 2001).

4.5.2.3 Teamwork

Toyota sees teamwork as the foundation of the company and that all systems are there to support the teams doing value-added work (Liker 2004, p.185). The role and use of team working is an important element of lean production (Womack et al. 1990). Table 4.5 compares the effects of flow in team function in the mass and lean environment.

Table 4.5 Effects of flow in team function in mass and lean environment

	Mass environment	One-piece flow environment (Toyota)
Work nature	Mass production process enables the workers to work individually with focus only on their individual tasks	Flow involves tight coordination between each step in the process and this coordination facilitates building teamwork (Liker 2004)
Operatives' responsibility	Only the white-collar or skilled technician is responsible for problem-solving, quality assurance, equipment maintenance, and productivity	The front-line workers are developed, empowered and motivated to discover and solve the problems in their daily work

4.5.2.4 Types of Work Groups in Toyota

Toyota builds its culture and organization around the basic unit of the work group (Liker and Hoseus 2008). There are two roles for teams, according to Liker and Hoseus (2008): one is to support individuals as they do their work and the other is to solve problems to improve how the work is done. These are referred to as work groups and problem-solving groups, respectively. *Work groups* are usually comprised of a small group of 5–7 people on the shop floor in Toyota. These workers report to a team leader, while several of these small groups report to a group leader. They are mainly responsible for the daily work to be done, and in the process, they find opportunities to continuously improve the way the work is done. They can be seen on the organizational chart as part of the formal reporting structure. *Problem-solving groups* are often temporary and usually do not appear on the organizational chart. These teams work as task force, quality circles, and temporary cross-functional teams to solve particular problems. They often deal with problems that occur across different departments of the organization. For Toyota, all of these types of teams are essential. In them, workers feel a sense of belonging, and the small groups help them feel connected (Liker and Hoseus 2008).

4.5.2.5 Organization Structure

Toyota strives to establish a relatively flat organizational structure while still maintaining the right group size so that people can effectively work together in solving problems. Toyota discourages making the organization completely flat, through a large span of control; instead Toyota's system heavily relies on teams led by highly skilled leaders. Hence, in the Toyota Production System, shop-floor work groups are the focal point for problem-solving and consist of team members, team leaders and group leaders [see Fig. 4.8, as adapted from Liker and Hoseus (2008)].

In Toyota language, workers are called "team member" who perform manual jobs to specified standards and are responsible for problem-solving and continuous improvement (Liker 2004; Liker and Hoseus 2008). Team members are managed by "team leaders" who themselves report to "group leaders". Team leaders has a number of important roles including responding to *andon* pulls, auditing the standardized work, ensuring that safety and ergonomic procedures are followed, and facilitating the process of solving problems. The team leader is an hourly

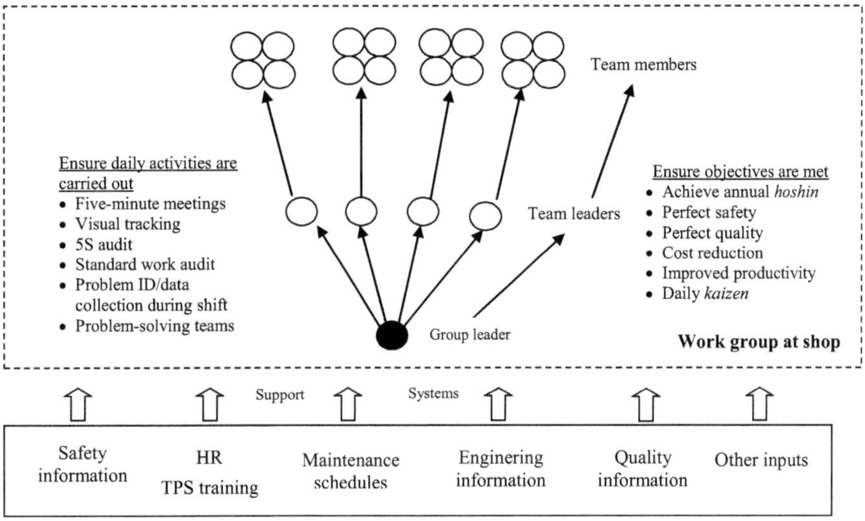

Fig. 4.8 A typical work group in Toyota

employee who has learned all the jobs in the team and has a set of off-line responsibilities, but whose first duty is to respond to *andon* calls. The group leader is the first-line supervisor, who has broader planning responsibilities as well as overall production responsibility for several work teams. Liker and Hoseus (2008) summarized three key differences between Toyota's organizational structure and those in the conventional organizations:

1. The team leader role as practised in Toyota is virtually nonexistent in most companies, in which there may be a "utility" person who can substitute for others when he or she is absent, or on leave, but there is no functioning team leader role to respond to *andon* and support daily production and problem-solving.
2. Team leaders at other companies support departmental mandates and enforce policies, rather than providing the daily support seen in Toyota.
3. First-line supervisors at other companies act as fire fighters, disciplinarians, and an arm of management, rather than providing leadership, coaching, teaching, and planning support for team members.

4.5.3 Principle 11: Partner Relationships

The concept of supplier networks implies that a company has strategic and long-term supplier relationships. It is widely known that Toyota has invested heavily in supplier partnerships over many decades. The primary reason for Toyota to sacrifice the short-term cost reduction for longer term supplier partnership is driven by quality (Liker and Meier 2006). Toyota has the ability to show other organizations how to efficiently operate to achieve good quality products. As a result, suppliers can win initial contracts with Toyota; however, to be rewarded more contracts, suppliers have to adopt Toyota philosophies and efficiencies (Liker and Choi 2004). Like the associates who work inside Toyota, suppliers eventually became part of the extended learning enterprise who grew and learned the TPS (Liker 2004).

The strategy that Toyota adopted for its partners and suppliers is in line with the way it treats its employees within Toyota. Principle 11 states that "*respect your extended network of partners and suppliers by challenging and helping them improve*" (Liker 2004). Liker and Choi (2004) revealed that Toyota and Honda shared a common supplier relationship strategy by following the six distinct steps:

1. Understanding how the suppliers work
2. Turn supplier rivalry into opportunity
3. Supervise the suppliers
4. Develop suppliers' technical capacities
5. Share information intensively but selectively
6. Conduct joint improvement activities

It is worth noting that Toyota or Honda were recognized by most vendors as the best and toughest customers was not because only they exercised one or two elements but because they used all six elements together as a system (Liker and Choi 2004). Inspired by the general guidelines highlighted by Liker and Choi (2004) as well as Toyota Way principles, a number of key characteristic of Toyota's partnering principles can be summarized as given below:

1. *Respect*: As pointed out in the Toyota Way, fairness, high expectations and challenge characterized how Toyota treats its suppliers, which is the definition of respect from Toyota's perspective (Liker 2004). This is because the business goal is always to maximize profits, but not at the expense of suppliers. As Ohno, cited in Liker and Meier (2006): "*the achievement of business performance by the parent company through bullying suppliers is totally alien to the spirit of the Toyota Production System*". As Toyota challenges its own people to improve, the challenges are extended to its suppliers which include a series of aggressive targets to meet.
2. *Reduced supplier base*: Lean production requires close coordination with suppliers to achieve the desired levels of quality. Toyota has done well at this, which under a general rule, have sole or two suppliers for every component; it also shows Toyota's trust in them. Most importantly, this partnership is established on a long-term basis except for the most egregious behaviour of the suppliers. Moreover, Toyota has developed a tiered structure, in which Toyota works most closely with the top tier suppliers and a few critical lower-tier suppliers, which supply major subassemblies or modules and major raw materials to Toyota's manufacturing factory, respectively. In addition, Toyota expects its top tier suppliers to in turn manage the lower-tier suppliers.
3. *Direct involvement*: Good examples exist in Toyota that have applied the direct-involvement approach to their suppliers by working extensively to teach them the way at building and delivering high-quality components just-in-time, often by sending their own employees into supplier plants for weeks or months to

reorganize the process flow, modify equipment, and establish problem-solving groups (MacDuffie and Helper 1997; Dyer and Nobeoka 2000; Liker 2004).

4. *Communication*: All kinds of communication approaches are mentioned in the supplier-partnering hierarchy, including setting specific times, places, and agendas for meetings; using rigid formats for sharing information; insisting on accurate data collection and sharing information in a structured fashion.
5. *Long term relationship*: Toyota's relationship with its parts suppliers can be explained by a long-term implicit contract. Once Toyota opens trade with a certain supplier, it is accepted that this business relationship will last over a long period. This saves transaction and information costs, and suppliers can specifically invest in equipment for the production of Toyota parts. Tsuji (2003) reasoned that monitoring and incentive schemes are the two factors required to sustain this long-term relationship.

On the other hand, like other Japanese auto-makers, Toyota is an extensive outsourcer, which outsourced 70 % of the components of the vehicles to its partners and suppliers (Liker 2004). However, Toyota is very careful when deciding what to outsource and what to do in-house. In addition, Toyota has a clear view of its core competency and endeavours to become an expert and the best in certain technology that is core to the vehicle. Liker (2004) noted that Toyota is humble to learn from the suppliers, but never transfers all the core knowledge and responsibility in any key area to the suppliers.

4.5.4 Summary of Principles 9–11

Principles 9–11 were less discussed in the earlier TPS principles, which mainly focused on the shop floor practice. Liker's (2004) elaboration on Toyota's leaders, employees, teams, and its partners and networks vividly outlined how Toyota manages people—the most important asset in Toyota. All the efforts towards people and partners have indirectly contributed to add value in the production flow. Based on the discussions above, Table 4.6 breaks down the key features of Principle 10 of the Toyota Way model as operationalized measurables.

4.6 Toyota Way Problem-Solving Model

The problem-solving approach used in Toyota today comes from multiple sources (Sobek and Smalley 2008). Firstly, this was simply the culture of the company tracing back to its founders, Sakichi Toyota and his son, Kiichiro Toyoda, who established the early culture of Toyota and its inventive spirit. Liker (2004) highlighted that the practice of *genchi genbutsu* is deeply rooted in the country's culture. Nisbett and his associates, cited in Liker (2004, p.235), concluded that "*Westerners prefer abstract universal principles. East Asians seek rules appropriate to a situation, and the East Asians see the same situation in more details than the westerners*". This explained why Toyota's people are keen to go and see.

Table 4.6 Operationalized measurable from the Toyota Way People and Partner model

Toyota Way People and Partners model	Operationalized measurables
P9: Leaders and leadership	• *Genchi genbutsu* practice • In-depth knowledge of their work • Support the people doing the work (ability to develop, mentor and lead people) • Support the culture
P10: People management	• Carefully screening prospective • Various training opportunities • Multi-functional skills • Teamwork • Activities are organized to improve team cohesion • Level of hierarchies in the organizational structure • Forms of work group
P11: Partner relationships	• Degree of challenging the work partners • Number of supplier base • The degree of direct involvement of people • Communication • Based on a long-term relationship

This also implies that it would be more difficult for the westerners to emulate. The second input was the personality of Taiichi Ohno and his insistence on going to the shop floor in order to investigate the root cause of a problem. Thirdly, Toyota was deeply influenced by a high-level methodology initially developed by Walter Shewhart of Bell Laboratories in the 1930s, and later adopted by Edwards Deming who became its biggest proponent (Sobek and Smalley 2008). The methodology is the PDCA cycle, also called the Deming Cycle. The PDCA cycle begins with the Plan step, in which the problem-solver thoroughly studies a problem or opportunity to understand it from as many viewpoints as possible, and then analyses it to find the root causes.

The last three principles of the Toyota Way work hand in hand in Toyota. The problem-solving methodology is a skill that runs deep and strong at all levels of the organization within Toyota and across all functions, from manufacturing to purchasing to sales and the rest of Toyota. Without a practical and continuous problem-solving process that is used on a daily basis, there will be a gap in any company's lean transformation (Liker and Hoseus 2008). The problem-solving method starts with Toyota's people seeing problems as opportunities, solving the problems aggressively and systematically to find a better way to do things, and then rigorously verifying that the better way is indeed better. This method encompasses a critical and logical thinking process. It requires thorough evaluation and reflection (*genchi genbutsu*) (Principle 12), careful consideration of various options to reach a consensus (Principle 13), and a high level of continuous improvement (Principle 14). This section examines the problem-solving methodology in Toyota as well as its tools and techniques.

4.6.1 Principle 12: *Genchi Genbutsu*

Principle 12 can be summarized as comprising the following three elements:

1. Solving problems and improve processes by going to the source, which in Japanese is referred to as *genchi genbutsu*.
2. Think and speak based on personally verified data rather than theorizing on the basis of what others or the computer screen tell you.
3. Top management is encouraged to go and see things for themselves. They will accomplish more when they have the skills to analyse and thoroughly understand the situation.

The first requirement of problem-solving is to determine the merit of solving the problems. The first step of the Toyota Way in dealing with the problem is *genchi genbutsu*. It is viewed as one of the founder's philosophies in Toyota. Osono et al. (2008) infer that the root cause of problems is revealed by on-site investigation and inquiry. Hence, when there is a problem on the production floor, the practice of *genchi genbutsu* requires that the Toyota leaders must "go and see" the shop floor first-hand, and really understand the actual situation at the shop floor level. It also means talking with the people involved to find out exactly where the problem occurred, when, and under what circumstances. According to the Toyota Way, a superficial impression of the current situation in any division of Toyota will lead to ineffective decision-making and leadership. Moreover, this fundamental philosophy has been mentioned earlier in Principle 9, in which The Toyota Way requires that the employees and managers must "deeply" understand the processes of flow, standardized work, etc. More importantly, the Toyota Way emphasizes staff's abilities to critically evaluate and analyse the work in the shop floor (Liker 2004, p.224). The advantages of *genchi genbutsu* include (1) enhanced communications, (2) increased level of trust, and (3) it reflects the commitment from management on quality control.

Making decisions based on facts, instead of intuition is one of the foundations for effectively reducing variability in a production system. Many problems can be solved readily using a combination of production management principles and experience without any assistance from data collection. However, non-trivial situations, such as the solutions for complex engineering difficulties, or the introduction of a new technology require substantial preparation and consideration. In such cases, data from all possible angles is a fundamental requisite to obtain effective solutions (Imai 1997).

4.6.2 Principle 13: Consensus Decision-Making

Making decisions based on facts may seem such an obvious statement that is not even worth discussing. However many managers in most companies make decisions based on very limited information (e.g. reports generated and summarized by subordinates). For Toyota, how the decision being made is just as important as the quality of the decision (Liker 2004). The management knows that people do make mistakes. Hence, instead of focusing on the correction of any particular decisions, the management also values the process through which the decision was made even though the decision-making process yielded a bad result. As Liker (2004, p.138) commented: "*management will forgive a decision that does not work out as expected, if the process used was the right one*". There are five critical elements that form the decision-making process in Toyota, which are summarized by Liker (2004):

1. Finding out what is really going on, including *genchi genbutsu.*
2. Understanding underlying causes that explain surface appearance—asking "Why" five times.
3. Broadly considering alternative solutions and developing a detailed rationale for the preferred solution.
4. Building consensus within the team, including Toyota employees and outside partners.
5. Using every efficient communication vehicle to do the above 1–4, preferably on one side of one sheet of paper (A3 paper).

It can also be concluded that the last three principles of the Toyota Way model, forming the problem-solving philosophy, are structured based on these five critical processes in sequence (Liker 2004). Apart from the principle of *genchi genbutsu* discussed in Principle 12, the rest of the four elements will be examined in detail below.

4.6.2.1 5 Whys

"5 whys" is a tool used to keep asking "why" until the root cause(s) are determined. Ohno (1988) pointed out the Toyota Production System has been built on the practice and evolution of this scientific approach. The problem-solver would start the question with "why is this problem occurring?" Upon answering it, he or she would have identified a cause to the observed effect. The problem-solver asks the same questions again, aiming to turn the cause into an effect, to identify a deeper cause. The problem-solver continues this inquiry until the root cause can be probed. When completed, the problem-solver has a clear and coherent cause-effect chain that demonstrates an in-depth understanding of the problem in context, noting how the root cause is linked to the observed phenomenon (Sobek and Smalley 2008). Take countermeasures at the deepest level of the cause that is feasible and at the level that will prevent reoccurrence of the problem. In cases that the root cause cannot be analysed from the 5-why techniques, Sobek and Smalley (2008) commented that structured tests or experiments can otherwise be used to eliminate possible causes and in problems related to organizational processes.

4.6.2.2 Alternative Solutions

Compared not only with the U.S. automobile manufacturers but also with other Japanese car manufacturers, senior engineers and managers in Toyota were trained to think in sets of alternative solutions as well as to think in a set-based concurrent engineering scenario (Liker 2004). In the course of devising countermeasures for a given problem, problem-solvers are strongly urged to consider multiple alternative

Fig. 4.9 Alternative Toyota decision-making methods

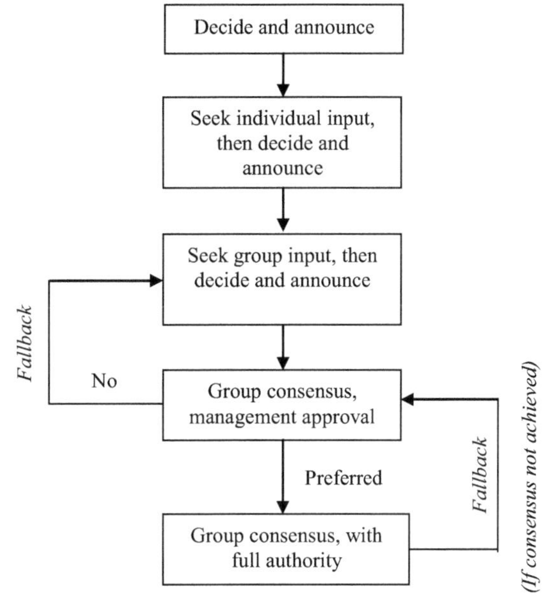

countermeasures even at the risk of delaying decisions. Not only does this approach foster creativity in problem-solving among the employees, but it also offers the other participants more tangible inputs into the final agreement.

4.6.2.3 Consensus

The Japanese management style requires a consensus to be reached before any decisions are made (Ouchi 1981). As noted by Keys and Miller (1984), the consensus decision process appears to be an application of the American concept of participative management. Consensus is generated by taking inputs from many people and evaluating many alternatives. Toyota Way's Principle 13 also includes the important process of *nemawashi: making decision slowly by consensus, thoroughly considering all options, implement rapidly* (Liker 2004, p.241). *Nemawashi* literally means binding the roots of a plant before pulling it out and refers to the practice of broad consultation before taking action (Vogel 1975). By the time the formal proposal is ready for a high-level approval, the decision is probably finalized. Figure 4.9, adapted from Liker (2004), describes the decision-making methods used in Toyota in different circumstances. Once a plan is made, everyone knows what to do and how to complete their task. This has been practiced by Japanese construction organizations where less planning and monitoring meeting within the project team are needed compared with their U.K. and U.S. counterparties (Xiao and Proverbs 2002). Toyota also welcomes the suppliers and other stakeholders to contribute their inputs to enrich the alternative solutions. This principle can be applied in the construction industry where efforts are required from various stakeholders that include the designers, contractors, sub-contractors, and other practitioners to reach a consensus on certain problems that occurred. In this way, information is shared in a transparent way.

4.6.2.4 Efficient Communication Vehicles

The A3 report, particularly referred to in this problem-solving principle, is a key part of the process of efficiently getting consensus on complex decisions. The A3 report is so named because it fits on one side of an A3-sized sheet of paper, which is roughly equivalent to an 11 × 17 in. sheet. A3 is a powerful lean tool that is based on, and supported by the Deming Cycle (Liker 2004; Sobek and Smalley 2008). It serves as guidelines for addressing the root causes of problems that arise in and around the workplace. Shook (2009) concluded that the ultimate goal of the A3 report is not to simply solve the problem, but to make the process of problem-solving transparent and teachable that fosters learning for problem-solvers.

4.6.3 Principle 14: Reflection and Continuous Improvement

The Toyota Way's Principle 14, sitting at the peak of the 4P model, is an ongoing process in Toyota, which aims to create a learning organization through relentless reflection (*Hansei*) and continuous improvement (*Kaizen*). Liker (2004, p.251) highlighted that "*to become a true learning organization, the very learning capacity of the organization should be developing and growing over time*". To learn means having the capacity to build on your past and move forward incrementally (Liker 2004). Toyota's learning organization has four key elements according to Liker (2004): (1) identify the root causes and develop countermeasures, (2) use *hansei*, (3) utilize policy deployment (*hoshin kanri*), and (4) *kaizen*.

4.6.3.1 Identify Root Causes and Develop Countermeasures

Toyota identifies root causes primarily using a very simple method called the "5 whys". This simply entails asking the question "why" as many times as possible to determine the root cause of a problem. This problem-solving tool has been elaborated in Principles 12 and 13.

4.6.3.2 Use *Hansei*

Hansei is a unique Japanese culture that roughly means "reflection on mistakes/weaknesses and devising ways to improve". Liker and Hoseus (2008) highlighted that there are three key components of *hansei*:

1. The individual must recognize that there is a problem—a gap between expectations and achievement—and be opened to negative feedback.
2. The individual must voluntarily take personal responsibility and feel deep regret.
3. The individual must commit to a specific course of action to improve.

Hansei is one of the most difficult things Toyota has ever had to teach, but it is an integral ingredient in Toyota's organizational learning (Liker 2004). *Hansei* is not only a philosophical belief system in Toyota but also a practical tool for improvement. *Hansei* (reflection) can be used at key milestones and after completing a project to openly identify all the shortcomings of the project and to follow up by developing countermeasures to avoid the same mistakes again. Within Toyota culture, *hansei* is considered essential for *kaizen* (Liker and Hoseus 2008) and it is the "check" stage of PDCA.

4.6.3.3 Utilize Policy Deployment (*Hoshin Kanri*)

According to Lee and Dale (1998), *hoshin kanri*, sometimes called "policy deployment", was developed in Japan in the early 1960s to communicate a company's policy, goals and objectives throughout its hierarchy. Lee and Dale (1998) noted that Toyota adopts this process and sets aggressive objectives at the executive level and cascades the objectives down to the work group level (Liker 2004). Each work group level in turn would develop measurable objectives for the year, designed to support the executive-level goals. The PDCA cycle is extensively applied to the planning and execution of a few critical strategic organization objectives (Lee and Dale 1998; Liker 2004). Furthermore, it is usual for the plant manager to audit the process being made by each team member towards the *hoshin kanri* objectives on a quarterly basis. Liker (2004) highlighted that the "check" and "act" part of PDCA are critical to turn the planned goals into effective action.

4.6.3.4 *Kaizen* (Continuous Improvement)

Most of what is discussed today about continuous improvement comes from interpretation of the Japanese practice called "*kaizen*". *Kaizen* means continuous improvement involving everyone—top management, managers, and workers (Imai 1986). Brunet and New (2003) summarized three key characteristics of *kaizen*: (1) *kaizen* is continuous—which is used to describe this unique nature of the practice and also its place in a never-ending journey towards quality and efficiency; (2) incremental in nature—in contrast to organizational or technological innovation; and (3) participative feature—which entails the involvement and intelligence of the workforce. In the *kaizen* philosophy, improvement in all areas of business serves to enhance the quality of the firm. Thus Evans and Lindsay (2008) outlined that any activities directed towards improvement falls under the *kaizen* umbrella. According to Evans and Lindsay (2008), a successful *kaizen* program usually consists of three basic elements: (1) operating practices, (2) total involvement, and (3) training. First, the activities discussed under the principles of the Toyota Way (Process part) such as the implementation of the just-in-time production system, standardized work, and visual management to reveal waste and inefficiency as well as poor quality will lead to improvement. Second, in *kaizen*, every employee strives for improvement and top management sees improvement as an inherent component of the company strategy and provides support to improvement activities (Evans and Lindsay 2008; Imai 1997). Finally, training is required both in the philosophy and in the tools and techniques for the employees. It includes suggestion system, and self-development programs that teach practical problem-solving techniques.

4.6.3.5 *Kaizen* and Problem-Solving

Kaizen is a holistic approach to problem-solving and is people-centred rather than system-centred (Huda and Preston 1992). The starting point for improvement is to recognize the need. This comes from recognition of a problem (Imai 1986; Huda and Preston 1992). If no problem is being recognized, there is no recognition of the need for improvement. Therefore, *kaizen* emphasizes on problem-awareness and provides clues for identifying problems (Imai 1986). Once problems have been identified, these problems must be solved consequently. Thus Imai (1986) highlighted that *kaizen* is also a problem-solving process which requires the use of various problem-solving tools. When improvement reaches new heights with every problems solved, such improvement must be standardized in order to consolidate the new level of awareness. It explained why Liker (2004) and Imai (1986) commented that: "*Kaizen* cannot be achieved without standardization". This is where the concept of involving people becomes crucial, because instead of just following instructions, the worker is able to explore and think and assume responsibility for improvement. Besides the standardized process, there are another two key prerequisites for implementing *kaizen*, namely management support and culture change (Evans and Lindsay 2008):

1. Management support: It does not necessarily call for great investments to implement *kaizen*; it does, however, call for a great deal of continuous effort and commitment.
2. Culture change: The difference between *kaizen* and innovation has been well elaborated by Imai (1986). In short, the West tends to think of each innovation in building practice as a major step change, while the Toyota people, or more broadly speaking the Japanese, constantly look at what they are doing, look for problems, and compare themselves with what others are doing, and try to better their performance.
3. Standardized process: The other key prerequisite for implementing *kaizen* is to stabilize and standardize the process beforehand. In another word, once a stable process has been established, continuous improvement tools can be employed to determine the root cause of inefficiencies and to apply effective countermeasures.

4.6.4 Summary of Principles 12–14

Problems, if not solved, may adversely affect organizational production and processes, leading to defects, high costs, safety issues, customer dissatisfaction, decreasing competitiveness, and so on. Goetsch and Davis (2009) outlined two models for solving and preventing problems, namely the PDCA cycle and the Toyota model. The Toyota Way for problem-solving has been elaborated earlier with a practical problem-solving process. Moreover, the PDCA cycle is a series of activities pursued for improvement, and it is also understood as a process through which new standards are set to be challenged, revised, and replaced by newer and better standards (Imai 1986).

Table 4.7 Operationalized measurable from the Toyota Way Problem-Solving model

Toyota Way Problem-Solving model	Operationalized measurables
P12: Go and see for yourself to thoroughly understand the situation (*Genchi Genbutsu*)	• Processes the skills to analyse and thoroughly understand the situation • Think and speak based on personally verified data • Solve problems and improve processes by going to the source
P13: Make decisions slowly by consensus, thoroughly considering all options, and implement rapidly	• Practice of 5 whys • Alternative solutions • Practice of consensus • Effective communication
P14: Becoming a learning organization through relentless reflection (*Hansei*) and continuous improvement (*Kaizen*)	• Identify the root causes and develop countermeasures • Reflection • Policy deployment • *Kaizen* activities (continuous improvement) • Learning organization culture

Problem-solving and decision-making are fundamental to total quality. On the one hand, good decisions and solutions will decrease the number of problems that occur. On the other hand, the workplace will never be completely problem-free (Goetsch and Davis 2009). Even the best-managed organizations have problems. Toyota's recent recall incident indicates that even though Toyota was committed to quality control and problem-solving, problems can still emerge. The last three principles of the Toyota Way form the core of Toyota's problem-solving approach. These are concerned with problems that can impact the organization or its customers in some way—usually negative. It starts with the Toyota leaders' *genchi genbutsu* initiative, which places great commitment on problem-solving and objective data for decision-making. Next, the Toyota Way on how consensus can be achieved is considered. Lastly, Toyota as a learning organization emphasizes on continuous improvement (*kaizen*) and reflection (*hansei*). Table 4.7 lists a number of operationalized measurables from the "Problem-Solving" layer of the Toyota Way model.

4.7 Development of the Toyota Way Model for Construction

4.7.1 Introduction

In this section, as a starting point, the Toyota Way model is proposed to solve some of the inherent limitations of the frameworks that are currently available in the lean construction domain. The Toyota Way model was briefly mentioned earlier: to reiterate, it comprises of 14 principles in four layers. Each layer can be viewed as an individual model. The first task with respect to the Toyota Way practices is to operationalize each principle into actionable attributes within the construction industry. Under the Toyota Way model, the underlying principles should have positive implications for the construction industry. In theory, the model equally values the "process" and "people" aspects, as well as other. This would be an appropriate choice, since most lean construction frameworks have a strong technical focus, with limited attention on the human dimensions.

4.7.2 Implementation of Process Model in Construction

The lean construction literature reports a number of construction projects that have already interpreted and exercised some principles from the Toyota Production System in the construction industry. However, as Picchi and Granja (2004) observed, in most cases the building professionals only put one lean construction tool into practice and missed the need to interact with other lean tools. In this section, efforts are made to discuss the implementation of the eight core Toyota Way principles under the "Process" category in the construction industry. Moreover, to overcome the difficulties due to the differences between the two sectors, Ballard's (2000) Last Planner System (LPS) is employed to interpret a few Toyota Way principles from an alternative perspective. The LPS has perhaps achieved a greater degree of industrial penetration (Green and May 2005), and shares some common grounds with a number of the Toyota Way philosophy in the "Process" layer of Liker's (2004) Toyota Way model.

4.7.2.1 Linking Toyota Way to the Last Planner System

As discussed earlier, Last Planner System serves as one of the theoretical foundations of lean construction (Koskela et al. 2002), and in some circumstances the LPS turns out to be synonymous with lean construction (Green and May 2005). The LPS is now regarded as the most powerful and well-known planning and control system from all the lean construction techniques and tools (Kenley and Seppänen 2010). According to Ballard (2000), the LPS builds on the principle of systematic reactive work planning executed on the lowest possible level in the hierarchy of planners—the last planner. The underlying philosophy is to ensure that all the prerequisites needed for performing distinct construction work are in place before it is assigned to a work group (Ballard 2000; Ala-Risku and Kärkkäinen 2006). It uses the overall project plan as the general framework, but suggests that the daily activities of the production should be managed by a more flexible approach that is cognizant of the actual progress of the project. There are four main categories for any executable project task, namely SHOULD, CAN, WILL, and DID [see Fig. 4.10 as adapted from Ballard (2000)]:

1. SHOULD: tasks that need to be performed in the near future according to the overall project plan.
2. CAN: tasks that have all their prerequisites ready: e.g. previous project steps are completed, necessary materials are at hand, and workforce is available.
3. WILL: the tasks that are commenced before the next planning round.

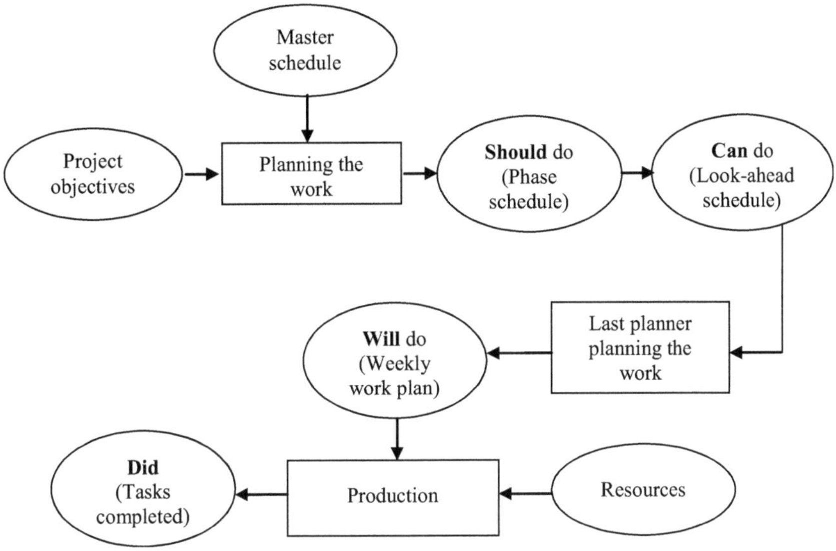

Fig. 4.10 The Last Planner System

4. DID: the tasks that are completed.

The overall objective of the LPS is to increase plan reliability, and thus to serve as a framework for addressing waste deriving from uncertainty and plan deviance. The Last Planner System employs a four-level hierarchy of schedules and planning tools: master plan, phase (pull) plan, look-ahead plan, and weekly work plan (Ballard 2000; Kenley and Seppänen 2010, p.110).

Master and Phase Plan

The master schedule is the overall project schedule, which is developed from the design criteria and supports the client's project objectives. It consists of milestones and items with long lead times. Milestone dates are determined by using the "pull" process from successor milestones (Pappas 1999). The plan is then developed by those responsible for building the phase together with subcontractors, starting backward from the planned phase completion date (Kenley and Seppänen 2010, p.110). The process reveals what must be done to release work for production.

Look-Ahead Plan

The Look-Ahead Plan represents an intermediate level of planning. It is a schedule of potential assignments, typically for the next 6–8 weeks (Ballard 2000). The number of weeks over which a look-ahead process extends is determined by project characteristics, the reliability of the planning system, and the lead times for acquiring information, materials, labour, and equipment (Ballard 2000). The work is planned on assignment level, which means something that can be communicated to workers (Kenley and Seppänen 2010, p.112). Management continues to break down the activities into more details and screen the resulting smaller activities throughout the look-ahead window, until the activities are essentially assignment-level tasks (Pappas 1999).

Weekly Work Plan

The weekly work plan is an assignment-level schedule. Detailed schedules are derived from the look-ahead plans on a weekly basis. The weekly work plan is formed based on the mechanism of Last Planner System, which aims to transform what SHOULD be done into what CAN be done, thus forming an inventory of ready work. In the meanwhile, examination of the prerequisites can take place when this level of detailed schedule can be achieved (Kenley and Seppänen 2010, p.112). A typical weekly work planning procedure proposed by Ballard and Howell (1998a) should follow the principles including definition, soundness, sequence, size and learning.

Percent Plan Complete

This is another key feature of the LPS, which tracks what is known as Percent Plan Complete (PPC). It is calculated by dividing the number of completed assignments (what "did" get done) by the total number of assignments each week (what was projected "will" get done) and reasons are identified and acted on for failures to complete assignments. A high PPC means that the LPS allows for reliable forecasting of work, and that tasks made ready are being completed on schedule.

The Last Planner System nicely works out some application problems in terms of the inherent differences between the manufacturing and construction sector. When it is argued by construction practitioners that the Toyota Way or lean principles cannot be adopted in the construction industry, the LPS offers a number of similar grounds to facilitate this. Principles 2–5 of the Toyota Way can in particular find their application template in construction with the last planner in the area of planning and control (see Table 4.8).

4.7.2.2 Implementation of Toyota Way Principles (Principles 2–8)

Apart from the Last Planner as an important template with which Toyota Way principles can be implemented in the construction industry with a planning and control focus, attempts had been made to seek the alternative means of implementing of the Toyota Way principles in the construction context. As a production philosophy, it seems that the seven principles of the Toyota Way model have much in common with the JIT philosophy, especially in areas of material management, site layout, and so on.

One-Piece Flow (P2)

As discussed earlier, one-piece flow can be achieved through collective efforts such as *takt* time, U-shape machine layout, pull systems, and so on. Continuous flow and *takt* time are most easily applied in repetitive manufacturing and service operations (Liker 2004). In construction, creating a continuous process flow on-site is a huge challenge due to its fragmented nature, low standardization patterns of activities, the one-of-a-kind features of construction products, and so on (Koskela 2000).

Table 4.8 Linking a number of Toyota Way principles to the Last Planner System

Toyota Way principles	The Last Planner System	Explanation
Principle 2: One-piece flow—creating a flow stability	Weekly planning	• Controlling planning reliability as the approach to improve work-flow reliability, and finally to achieve the flow stability (Ballard and Howell 1998a). • The weekly planning process is deliberately intended to shield production from poor planning. • The LPS prevents work which cannot be completed from being scheduled, thus shielding the crews from waste generated by interruption.
Principle 3: Pull system	Look-ahead plan	• LPS is a type of pull system, for example a look-ahead plan ensuring assignments are ready is explicitly an application of pull techniques (Ballard 2000). • Pull can be understood as ultimately a derivative from target completion dates, but specifically applies to the internal customer of each process (Ballard 2000).
Principle 4: Level-out workload (*Heijunka*)	Weekly planning and look-ahead plan	• Developing a weekly plan is very much similar to application of "*heijunka*" in construction. • The look-ahead plan is supposed to maintain a backlog of workable assignments for each production unit, which requires estimating the amount of load and capacity of the production unit. The weekly plan specified who is to do what during each week as regards planning and control. • Work is selected in the right sequence—that is, so as to best move the project towards its objective. Sequencing decisions can also be made by last planners (foremen) based on their intimate knowledge of working conditions and constructability issues. • The right amount of work is selected: that amount of work that uses the labour and equipment capacity as directed by the schedule. By reviewing and signing off on quality plans beforehand, management validates quality plans and can then focus on controlling execution of the plans.
Principle 5: Built-in quality	Weekly planning	• The construction analogy to stopping production rather than passing on a defective product is to make only quality assignments (Ballard and Howell 1998a). • Weekly work plans are effective when they meet specific quality requirements for definition, soundness, sequence, size, and learning. Quality assignments shield production from work-flow uncertainty (Ballard and Howell 1998a).

However, repetitive operations can also be reflected partially in construction; hence, this principle can be applied accordingly, with necessary treatment. For example, in a house construction project, Yu et al. (2009) defined *takt* time as the rate at which a home builder must build the house to satisfy customer demand. More specifically, the *takt* time is determined by the average volume of sales in the previous months and the available workdays in a certain month. Based on the *takt* time on hand, the number of *kanban* for each of the tasks in the value stream can be determined; hence the production paces of working stations are synchronized. This research, however, does not intend to investigate whether *takt* time is or can be used in the Chinese construction industry. Instead, there are three elements, namely waste elimination (Sowards 2007; Polat and Ballard 2004), labour flow (Thomas et al. 2003) and material flow (Akintoye 1995), that can reflect this one-piece flow principle in the context of construction industry.

Pull *Kanban* System (P3)

Potential applications of the pull (*kanban*) system have already made their appearance in the construction industry. More specifically, the *kanban* or pull system can be used in procurement of materials at the right time and in the right quantities based on the actual demands on-site (Low and Chan 1996; Low and Mok 1999; Khalfan et al. 2008). Arbulu et al. (2003) developed the *kanban* strategy to manage the replenishment of certain types of made-to-stock products from preferred suppliers to site. The *kanban* system enables construction teams to get products from the marketplace on a daily basis, according to site needs. This *kanban* strategy cannot be realized until some components are available, including (1) marketplaces or main site store, (2) collection vehicles or "milk runs", (3) satellite stores, and (4) an inventory management system (Arbulu et al. 2003). It criticized the traditional approach that generates orders in big batches on a weekly or biweekly basis. However, the big-batch-mentality is one of the most important challenges in the implementation of the *kanban* strategy (Arbulu et al. 2003). Apart from the on-site practice, the pull mechanism is also suggested as one of the principles within the JIT framework that would assist building professionals in the construction supply chain to enjoy the similar benefits. In this regard, Low and Choong (2001) suggested that each empty truck returning to the plant was a *kanban*. In a case study reported by Khalfan et al. (2008) in the UK construction industry, an e-procurement system was employed as *kanban* signals by the client, where they placed an order for the required product on-site with specific dates and specifications, as well as delivery time. Due to the differences between the manufacturing and construction sector, the *kanban* principles, most of the time, serve as a supplier *kanban* to help contractors better control their materials and inventory requirements. Furthermore, Jang and Kim (2007) acknowledged that *kanban* is able to fulfil the functions of visibility, production control, and progress monitoring. Moreover Jang and Kim (2007) suggested that the *kanban* can also be used as a safety control tool because safety information can be added on each *kanban*.

Heijunka (P4)

Heijunka, or levelling out the workload, is perhaps the hardest to implement in the construction industry. Compared to manufacturing, the key difference is that the elements of construction require different amount of time. It is worth mentioning that there are a number of points of commonality between the last planner system and the principle of *heijunka*. Both aim to achieve a stable and reliable work-flow. The last planner system is well documented in the literature and sometimes it has been used to represent lean construction. Apart from adopting Ballard's (2000) four levels of plans, including master plan, phase plan, looking-ahead plan, and weekly plan, one of the important issues here is that the foreman needs to be empowered to make his own commitment on what day-to-day or week-to-week tasks he can actually deliver in a given time. By doing this, foremen can have a sense of ownership of the project programme.

Built-In Quality (P5)

The principle of built-in quality can be interpreted as "*do it right the first time*", which is the overarching goal of Total Quality Management (TQM). However, the quality culture in the construction industry prefers to use the inspection period to fix occurring problems rather than to apply the Toyota Way's built-in quality approach to eliminate the defects in the first place. In this case, Principle 5 of the Toyota Way could be understood as the adoption of prevention approaches as well as whether employees' attitudes towards quality parallel to "stop-and-fix". To achieve "*do it right the first time*" requires a mindset change, as employees should be encouraged to expose problems as well as trained to upgrade their capacity to identify problems. In addition to these, the ISO9000 series has played a role in setting a quality assurance standard that also drives the construction firms' efforts towards quality management. Love and Li (2000), however, argue that quality assurance does not provide enough benefits to justify its implementation in terms of cost. To reap the benefits and improve the overall cost competitiveness of construction, one needs to implement a company-wide TQM system, which requires a culture built around ISO 9000 and continuous improvement. Moreover, similar to the *andon* practice which needs empowerment from the top management, employee involvement and empowerment from management were identified as key measurables to reflect TQM performance in the construction industry (Low and Teo 2004). Nesan and Holt (1999, p.220) outlined a three-phase empowerment for implementation in construction organizations and highlighted some of the crucial organizational factors such as organizational structure, process-control mechanisms, and employees' attitudes towards empowerment, as well as the dynamic nature of the built environment which is characterized by constant changes with respect to the workforce, client requirements, project goals, technology, and economy.

Standardized Work (P6)

Standardization implies that all work should be highly specified in terms of timing, content, sequence, and outcome. Creating standardized work requires identifying the repeatable elements of a process, assessing the best way to perform those elements, developing a reliable method to ensure the performance of those elements, and then performing the reliable method according to a required time. Construction practitioners may express scepticism that standard work is not possible in construction because each construction project is unique and is strongly affected by external factors. The implementation of standardization (Principle 6) in the construction industry can be treated as:

1. the promotion of standardized components such as using off-site techniques in the prefabrication sector.
2. standardized tasks in the construction process in which repeatable elements of a process in identified.

The rationale for prefabrication is that economies of large-scale production can be reaped by standardizing the components, along with other benefits including improved quality with better quality controls, reduction in wastage, less labour-intensive operation, and faster production of building components (Low and Chan 1996; Low and Choong 2001). Hence, industry experts advised the construction industry should use more offsite techniques and standardization in order to increase quality and reduce cost and time. Cooke and Williams (2009) emphasize that lean thinking can be successfully applied to various forms of assembly technology to smooth the construction process. However, standardization has changed over the years, with efforts being made to meet clients' needs and produce customized individual buildings, while still using standard components and employing standard processes to ensure success (Gibb and Isack 2001). Low and Choong (2001) advocated that, in order to meet the needs for greater variety, prefabrication should move towards lean production rather keep the old-fashion way of mass production, which refers to the production of smaller but economically viable volumes of standardized components, specially tailored for a project. This exactly mirrors Ohno's (1988) strategy for Toyota to produce fewer quantities of cars but with a greater range of products.

In terms of the standardized tasks, Santos et al. (2002b) examined two indicators with a focus on the "bricklaying" process, namely (1) the number of bricklayers who are aware of the written standard and (2) the number of revisions of the standards per year. It has been found that written standards were ineffective due to the lack of teamwork and problem-solving activities (Santos et al. 2002b). The construction firms have developed standards, but often failed to implement and maintain their standards in practice. Moreover, the written standards have little value if the workers do not have sufficient knowledge of their contents or motivation to apply them in practice.

Visual Control (P7)

Few studies report on visual management in construction, due to a number of factors such as the physical environment involved, construction technology, and contractual relations that result in difficulties visualizing the flow of work in progress on-site (Sacks et al. 2009; Tezel et al. 2010). As far as the construction practitioner is concerned, visual control can be achieved in many ways. In a technical report conducted by Salford University scholars (Tezel et al. 2010),

18 visual management practices were collectively reported in a construction context. These practices covered a range of areas and employed tools such as kanban, mistake proofing, good site practice such as “5-S”, prototyping and sampling, and others. Basically, the visual management practices can be classified into different layers of visual workplace framework (see Galsworth 2005), namely visual order, visual standards, visual measures, visual controls, and visual guarantees. For example, one company in Peru installed lights in the control trailer that corresponded to each floor of the building under construction. The crew working on that floor would set the light switch according to their requirements: green means that all is OK, yellow means that they are going to be out of materials in 30 min, and red means that work has stopped. Moreover, Sacks et al. (2009) demonstrated how building models can generate 3D visualization of a construction process, helping the construction practitioners to better manage their work. Although these software-oriented visual control tools have been found useful in improving flow processes on construction sites, a longer time might be needed for them to be employed in developing countries. At Toyota, as Liker (2004) revealed, the creative use of any means is the best available approach to create true visual control. The 5-S concepts can be readily applied in construction sites, as these do not need sophisticated software requirements. However, 5-S concepts are not well recognized by practitioners in the construction industry presently (Low and Ang 2003). Low and Ang (2003) reported that numerous improvements to site layout—including decreases in the demand for storage space, minimal movement and handling on-site, better site access—can be achieved by implementing the JIT and 5-S concepts concurrently.

Adoption of Reliable Technology (P8)

Most likely, every organization wants to be on the cutting edge of technology. This is also true in the construction industry, which is still very labour intensive. Building professionals have attempted various new technologies, in the hope of improving performance, in an industry which is known for its slow rate of adopting new technology. Yet most of the time, it ends up creating unrealistic or unsustainable expectations. Liker (2004) suggests a view of adoption of new technology in the Toyota Way thinking that includes principles such as:

1. new technology must be thoroughly tested and proven to be reliable.
2. new technology must support continuous flow in the operation (process).
3. new technology must help employees perform better (people).

The important implication of the above for the construction industry is obvious. It also provides valid justification for new technology adoption when cost is considered.

In summary, Table 4.9 translates seven Toyota Way principles that belonging to the “Process” layer of the Toyota Way model to the construction context. It implies that the implementation of Toyota Way in construction largely draws on lean construction practices, which are closely related to planning and control activities on the construction site. According to Ballard and Howell (1994), lean construction distinguishes itself from traditional construction management in two aspects:

Table 4.9 Operationalized measurables of the Toyota Way Process model in the construction industry

Sub-principles	Operationalized measurables in Toyota Way	Operationalized measurables in construction industry
P2: One-piece flow to bring problems to the surface	• Waste elimination • *Takt* time • Use flow oriented layout (U-shape) • Synchronize production activities so that one does not start until the previous activity has finished (pull) • Standardized work to stabilized flow	• Waste elimination • Workforce flow • Material flow • Work-flow (weekly plan)
P3: Use pull system to avoid overproduction	• Pull from customer end—including both internal and external customers • Use visual control—*kanban* system	• The level of material inventory • Look-ahead plan
P4: Level out the workload (*Heijunka*)	• Eliminate overburden to people and equipment (*muri*) • Eliminate unevenness in the production schedule (*mura*) • Level out the workload of all manufacturing and service process	• Weekly plan
P5: Build a culture of stopping to fix problems	• Deliver perfect first time quality • Reveal and solve problems at the source as they occur • Keep quality control simple • Create culture—involve and empower employees to continuously improve	• Deliver perfect first time quality • Reveal and solve problems at the source as they occur • Quality circle • People involvement and responsibility for quality
P6: Standardized tasks are the foundation for continuous improvement and employee empowerment	• Standardized operating procedure (SOP) • Continuously improve the standardization • Empowered employees to participate in the writing of standard procures	• Level of off-site technique usage • Level of standardized work • SOP • Level of employee empowerment to improvement the standards
P7: Use visual control so no problems are hidden	• Practice of 5-S • Integrate the visual control systems to the value-added work	• Practice of 5-S • Practice of visual control
P8: Use only reliable technology that serves people and process	• Thoroughly test new technology • Technology must support people • Technology must improve flow • Technology must support the company values	• Thoroughly test new technology • Technology must support people • Technology must improve flow • Technology must support the company values

(1) waste management and (2) flow management. Similarly, these seven principles work hand-in-hand to eliminate waste throughout the construction process. These include the introduction of the Just-in-time (JIT) process, 5-S, visual control, and others, in order to improve the construction process. On the other hand, the implementation of Toyota Way principles in the construction context can initiate a production control system, such as the adoption of the Last Planner System, to have a strategic procurement plan, where reliable production planning is required. Both reflect the operations and tactical aspects of lean construction and are also in agreement with the Toyota Way principles underlying the "Process" category.

4.7.3 Implementation of the People and Partner Model in Construction

The three principles dealing with "People and Partners" imply that companies should develop leaders who live the philosophy of the company and for whom mutual respect is present between the suppliers and the "main company", as well as between management and workers. This is applicable to the construction practitioners, namely the leaders, individuals, suppliers, and partners.

4.7.3.1 Leaders and Leadership

Similarly, Toyota leaders must have a combination of in-depth understanding of the work as well as the ability to develop, mentor, and lead. In the construction context, project managers are required to develop two types of skills, namely specific skills and general skills (Edum-Fotwe and McCaffer 2000). The former relate directly and only to construction projects, and reflect their specific skills; the latter are essential for the project manager to function effectively with his or her specialist knowledge, and include leading, communicating, negotiating, and problem-solving skills (Edum-Fotwe and McCaffer 2000). Other than the essential skills required of Toyota leaders in Toyota Way Principle 9, Toyota's unique organizational structure is highlighted by Liker (2004) and reflects Toyota's philosophy of adding value to its employees. In construction, Orr (2005) compared the traditional stakeholder's organization with its lean version. Traditionally, the stakeholder's organization may be drawn as a pyramid as illustrated in Fig. 4.11 [adapted from Orr (2005)] (the left pyramid), with the director at the top, and the tradesperson at the base.

The principle is that, in a traditional organization, directions are given from the top in a command-and-control manner. Inspired from the team leader concept in the lean (TPS) philosophy, another pyramid (this time inverted) was proposed by Orr (2005), as illustrated in Fig. 4.11, where the tradesperson is at the top and is supported by the rest of the stakeholders. This mirrors the servant-leadership concept in the construction context. If the upside-down pyramidal organization structure can be successfully implemented in the construction industry, it will enable the leaders to use all efforts, as well as all necessary resources, to help the front-line workers. In construction, leadership is of prime importance and is required to smooth and activate effective teamwork at the site level. For example,

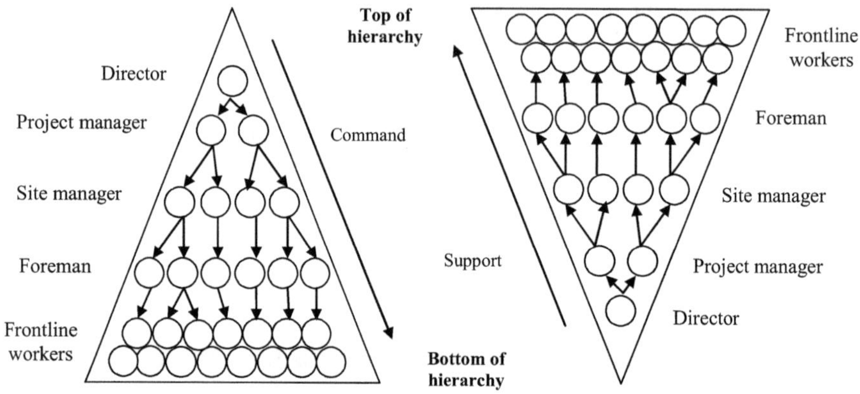

Fig. 4.11 Organizational structures in the traditional and lean perspective

site managers are no longer simply inspectors to check the work after its completion. Instead, they must devote time to solving problems by going and seeing the source first-hand, and being responsible for achieving and maintaining productivity excellence.

4.7.3.2 People Management

Toyota strives to develop excellent individuals, and this involves four major activities: careful selection, a variety of training, effective teamwork, and an upside-down organization. This reflects that the Principle 10 of the Toyota Way—people management—is a soft model of HRM that treats human resources as valued assets. According to Green (2002), the dominant culture of construction industry currently emphasizes the hard model of HRM, which sees humans as a resource to be "provided and deployed" as needed to achieve organizational objectives. Pursuing Principle 10 of Toyota Way in construction can be interpreted as encouraging the soft model of HRM. The literature has covered various elements of the soft model, including teamwork, training, empowerment, and so on, in the context of construction (Nesan and Holt 1999; Loosemore et al. 2003). According to Nesan and Holt (1999, p.51), the practice of teamwork through an organization is an essential component of lean production implementation, as it builds trust, improves communication, and develops interdependence. All members of the team should strive for continuous improvement and customer satisfaction through teamwork. Apart from extensive training, multi-skilled employees are not uncommon in companies that practise lean. However, the construction industry is characterized as craft-based, with extensive labour-only subcontracting prohibiting investment in training. Multi-skilled workers can be seen in Europe, such as in the construction industries of Germany and the Netherlands, as reported by Clarke and Wall (2000).

4.7.3.3 Partner Relationships

Partnering in the construction industry is not a new concept. In response to stimulating a radical change in the construction industry in terms of value for money, profitability, and reliability, partnering has become a common pretender requirement for government-funded capital building projects (Beach et al. 2005). Moreover, as there is increasing recognition of Just-in-time and Total Quality Management among building professionals, construction organizations have acknowledged the benefits of having a few good suppliers with whom they have worked closely on a long-term basis. Furthermore, manufacturing firms outsource their work through contractual arrangements with other parties. Similarly, most construction work undertaken by subcontractors can be categorized as being outsourced. In order not only to generate sustainable profit margins but also to add value throughout the construction process, harmonization between contractors and subcontractors was argued to be a prerequisite for encouraging lean construction (Miller et al. 2002). In this regard, the Toyota way of treating its partners includes mutual cooperation, trust, and the sharing of benefits on a long-term basis, all of which can foster harmony between construction firms and subcontractors.

4.7.4 Implementation of the Problem-Solving Model in Construction

Problem-solving characterizes much of the management activities in construction (Li and Love 1998). However, the contractors and the construction industry in general do not perform this in the way intended by Toyota. This is because construction problems are ill structured in terms of various uncertainties (variables) and variations, the lack of understanding of the interrelationships between these variations, and the multiple solutions to construction problems (Li and Love 1998). Most often, the contractors solve problems, but are not concerned about learning from the situation or finding the root cause of the problem. Experiential knowledge is what they rely upon heavily, even though this is not well codified in books and is weakly organized in memory (Li and Love 1998). Moreover, construction problem-solving often lacks a clear procedural structure (Li and Love 1998). In this regard, the Toyota Way principles of problem-solving have the potential to improving the problem-solving practices employed in the construction industry. Furthermore, becoming a learning organization is the ultimate goal of implementing the Toyota Way model (Liker 2004). However, the learning process in the continuous improvement philosophy (*kaizen*) is not adopted in a broad manner within the construction industry. According to Love et al. (2003), few construction organizations have a system for systematically acquiring, capturing, converting, and connecting the lessons they have learned, a few demonstrate any interest in doing so. Love et al. (2004) provided a conceptual framework of nurturing a learning organization in construction. This framework employs TQM as an enabler and embraces the concept of organizational learning.

4.8 Summary

This chapter provides an in-depth review of the Toyota Way model. Four building blocks of the Toyota Way model as well as correspondent principles were carefully examined. A number of measurable parameters therefore were derived consequently. It is important to note that the Toyota Way model is more than a set of methods for eliminating waste. In contrast, the Toyota Way can be viewed as a socio-technical system that recognizes the importance of people and the lean manufacturing tools. Moreover, this chapter took the Toyota Way model as a framework for discussing its application to the construction industry, following the 4Ps—philosophy, process, people/partner, and problem-solving philosophy. In the discussion of the "process" part of the Toyota Way model, an endeavour has been made to link a number of Toyota Way principles to the Last Planner System as a necessary change for better implementation.

Theoretical and Conceptual Framework 5

5.1 Overview

This chapter aims to explain the theoretical support from the mainstream theories towards these 14 principles. In order to better understand the Toyota Way model, each individual layer (i.e. Toyota philosophy, process, people/partner and problem-solving) will seek the theoretical support from the mainstream theories. The theory of the business, production theory, different schools of management thinkings, and theory of quality management seem to be relevant and can be used as the theoretical sources.

1. Theory of the business: Peter Drucker's (1994) theory of the business can be used to link the Toyota Way Philosophy model.
2. Production model: Koskela's (2000) interpretation of production has been elaborated earlier (see Chap. 2). In this part, it starts with discussing the Flow model of the TFV model with the process layer of the Toyota Way. Efforts are then made to connect the TFV model to the remaining layers of the Toyota Way model.
3. Human resource management: These will seek the theoretical support for the Toyota Way People and Partners model. Management theories in the production management domain have been reviewed in Chap. 2. It provided a general understanding of each school of management thought which are based on somewhat different assumptions about human beings and the organizations for which they work. However, this section discuses the leadership theory and motivation theories that the Toyota Way model uses to develop their leaders and employees, respectively.
4. Quality management paradigm: The problem-solving principles of the Toyota Way can be understood as Toyota's quality management philosophy. This is because Toyota is dedicated to continuously solve root problems that drive organizational learning. Hence, the theory of quality management and learning organization will be reviewed.

S. Gao and S.P. Low, *Lean Construction Management*,
DOI 10.1007/978-981-287-014-8_5,

5.2 The Toyota Way Philosophy Model: Linking the Theory of the Business

Peter Drucker (1994) contended that every organization operates to a "theory" of its business built upon an assumptive framework which guides and nurtures the organization's activities and that it can be a powerful catalyst for business sustainability and growth. A theory of the business has three parts, which are outlined below (Drucker 1994, pp.99–100):

1. There are assumptions about the environment of the organization: society and its structure, the market, the customer, and technology.
2. There are assumptions about the specific mission of the organization.
3. There are assumptions about the core competencies needed to accomplish the organization's mission.

The assumptions about environment define what an organization is paid for. The assumptions about mission define what an organization considers to be meaningful results; in other words, they point to how it envisions itself making a difference in the economy and in the society at large. Finally, the assumptions about core competencies define where an organization must excel in order to maintain leadership. The first principle of the Toyota Way under the theory of the business in terms of the above three sets of assumptions is examined below:

1. The Toyota Production System was invented in response to the severe external environment (e.g. oil crisis, recession and slow economic growth) and establishes "cost reduction" as its goal with advocating the absolute elimination of waste (Ohno 1988). These reflect that Toyota had made good assumptions about the external environment of the organization.
2. According to Liker (2004), the true mission of Toyota consists of three parts that are summarized as: First, contribute to the economic growth of the country in which it is located (external stakeholders); second, contribute to the stability and well-being of team members (internal stakeholders); and third, contribute to the overall growth of Toyota. This mission drives Toyota to make excellent products and to challenge its workers to contribute to Toyota and to create history (Liker 2004).
3. Toyota holds a strong sense of self-reliance, rather than to rely on outside business partners (Liker 2004). Toyota's strategy on partnering with the suppliers reflects the last assumption. For example, in the circumstance that Toyota outsourced 70 % of its work to the suppliers, Toyota still works hard to develop its core competencies by mastering the new technology if it is core to the vehicle production and to learn with the suppliers as well.

5.3 The Toyota Way Process Model: Linking Production Model

As mentioned in Chap. 3, lean production borrowed the idea of stable and standardized processes and time study from the work of Taylor on scientific management. Faced with an inadequate theory of production, the TFV model of production was proposed by Koskela (2000) as a main contribution to developing a new production philosophy. In this section, efforts need to be made to connect the TFV model to Toyota Way principles. Among all the 14 principles of the Toyota Way, the process-related principles were regarded as most relevant to production on the shop floor. Furthermore, the TFV model can also be extended to bridge the people and problem-solving principles of the Toyota Way model.

5.3.1 Flow Model and the Toyota Way Process Model

The comparison of Koskela's (2000) TFV model of production and Liker's (2004) Toyota Way Process model is needed, because both offer fundamental principles in the domain of production management. Moreover, since the advent of Lean thinking, there is no further theoretical improvement on lean production until Liker (2004) proposed the Toyota Way model. Both flow concepts of the TFV model and Toyota Way Process-related principles have been articulated earlier in Chaps. 2 and 4, respectively. Table 5.1 presents the links between the two.

5.3.2 TFV Model and the Toyota Way Model

The Flow concept of production finds a number of common grounds with the Process layer of the Toyota Way model. In addition, similarities between the TFV model of production and the remaining three "P" categories of the Toyota Way model, namely Philosophy, People and Partners, and Problem-solving are also sought in this section.

5.3.2.1 Philosophy

As Liker (2004) described, in Toyota, there are several ways Toyota pursues the idea of social contribution. "Customer focus" is the core of the Toyota Way Principle 1. "Customer first" is a basic belief and a lean principle that Toyota holds to serve the society. In lean thinking, delivering customer-defined quality becomes the core purpose of the organization (Morgan and Liker 2006). Toyota succeeds at this by focusing process, people and tools on clear objectives. Therefore, the "Customer first" philosophy that pervades in Toyota captures the principle of the value generation model, namely to ensure that all customers' requirements, both explicit and latent, have been captured.

5.3.2.2 People and Partners

As mentioned earlier in the Flow model, Koskela (2000) highlighted that training a multi-skilled workforce can help simplify a process and increase flexibility. The Toyota Way endeavours to develop the employees into a multi-tasking workforce and to encourage them to work in a team to raise their work productivity and efficiency.

Table 5.1 Linking the Flow model of TFV to The Toyota Way Process model

	Flow model of TFV paradigm	Principles of The Toyota Way
Objectives	Reducing the share of non-value-adding activities (waste)	Waste elimination is the chief aim of the Toyota Way Process model
Reduce the lead time	Eliminating non-value-adding time is a basic improvement rationale to compress the lead time. To achieve that, a number of heuristic implementation approaches were listed (Koskela 1992, 2000), including: (1) Reduction of batch size (2) Reduction of work-in-progress (3) Minimization of distance (4) Isolation of value-adding activities from supporting activities (5) Changing the order of the process (6) Synchronization and smoothness of flows (7) Solution of control problems and constraints to a speedy flow (8) Reduction of variability	Principle 2 of the Toyota Way—create continuous process flow to bring problems to the surface—demonstrates that the Toyota production process is mainly created based on the flow concepts. It highlighted that continuously seeking to remove the non-value-adding wastes is fundamental in this principles. Principle 2 employs the similar concepts and techniques (see the following five critical elements) which are in line with the flow concept of the TFV paradigm to achieve the same goal (1) *Takt* time (2) Flow manufacturing on U-shaped production lines (3) Pull production control (4) *Jidoka* (5) Standard work
Reduce variability	The practical approach to decreasing variability is made up of the well-known procedures of the statistical control theory. The basic heuristic approaches to reduce variability are: (1) Standardization of activities (2) Installing *poka-yoka* devices into the process (3) Measuring, detecting and eliminating the root cause of a problem (4) More expensive transformation technology will provide for less variability (Koskela 2000)	Principles 5, 6, 7, 8 and 12 of the Toyota Way can be connected to the sub-principle (reduce variability) of the Flow model (1) Standardized task (P6) is the foundation for continuous improvement (2) In building a culture of stopping to fix problems (P5), *poka-yoka* is one of the lean tools that help the employees to detect the defects and halt the process (3) Measuring, detecting and eliminating the root cause forms a major discussion in the P12 (4) Principle 8 of the Toyota Way indicates that Toyota only uses reliable technology that serves people and adds value to the process
Simplify	Koskela (2000) proposed two means that can promote the concept of simplification: (1) By eliminating non-value-adding activities from the production process, and by reconfiguring value-adding parts or steps (2) Organizational change can also bring about simplification. Multi-skilled or autonomous teams can eliminate non-value-adding activities	Principles 2, 6 and 10 of the Toyota Way can be connected to the sub-principle (simplify) of the Flow model (1) One-piece flow (P2) is in line with the concept of simplification (see elements of P2 above) (2) Standardization (P6) is viewed as a practical approach to simplification (3) Providing training to the employees to become multi-skilled front-line workers who can handle more work tasks and increase their productivity is a key component of P10

(continued)

Table 5.1 (continued)

	Flow model of TFV paradigm	Principles of The Toyota Way
Increase flexibility	Koskela (1992) submitted that practical approaches to increase flexibility, including: (1) Minimize lot size to closely match demand (2) Reducing the difficulty of setup and changeovers (3) Customizing as late in the process as possible (4) Training multi-skilled workforce	Principles 2 and 10 of the Toyota Way can be connected to the sub-principle (increase flexibility) of the Flow model (1) Minimizing the lot size and reducing the difficulty of setup and changeovers are the necessary requirements of achieving one-piece flow. Moreover these are also the perquisites of the pull system (2) Principle 10 explained multi-skilled front-line workers who can handle more work tasks to increase their productivity
Increase transparency	Koskela (1992) offered practical approaches for enhanced transparency that include the following: (1) Maintain a clean/order workplace (5-S) (2) Making the process observable (3) Rendering invisible attributes of the process visible through measurement (4) Incorporating process information in work areas, tools, container, materials and information system (5) Utilizing visual controls to enable any person to immediately recognize standards and deviations from them Reduce the interdependence between work stations	Visual control is able to significantly enhance the transparency of the production system, which has been explained by Liker (2004) in three Toyota Way principles: *Kanban* (P3) works as a major instrument for communicating orders from downstream to upstream workstation; *andon* system (P5) and visual management tools (P7) (1) The first attribute of "increasing the transparency"—maintain a clean/orderly workplace is in line with the 5-S practice in Principle 7 (2) "Rendering invisible attributes of the process visible through measurement" can be achieved by utilizing other visual tools in Principle 7 as well (3) "Incorporating information into the processes" plays an important role on the workforce perception of effectiveness. The adoption of visual tools is able to help employees work smart. Incorporating process information in work areas was also discussed in Principle 7 such as the use of A3 papers "Reduction of interdependencies" may be achieved through improvements and innovations in design, production methods or simply by carrying out changes in the schedule

5.3.2.3 Problem-Solving

Koskela (2000, p.78) highlighted that during the value generation process, transformations and flows are controlled for the sake of the customers. This is because attributes of transformations and flows impact directly on the resultant value. In this regard, the major focus of the quality movement was the customer value. The Toyota Way sees quality problems as an opportunity that is subject to the PDCA cycle. It is also driven by continuous improvement and the problem-solving discipline to minimize waste and maximize value during the process.

5.4 The Toyota Way People and Partner Model: Linking HRM

Following the sequence of the "People/Partner" category as discussed in the Toyota Way model, this section accordingly seeks human resource management (e.g. Leadership theory and Motivation theory) and the conceptual framework concerning supplier management as the theoretical foundation to link this part of the Toyota Way model.

5.4.1 Linking Leadership Theory to Toyota Way

Various terms have been used to describe leadership as well as different views of how one can become a leader or what characteristics a leader possesses. Toyota's unique leadership, also sometimes known as servant leadership, has been discussed earlier in Liker's (2004) Toyota Way model (see Chap. 4). It seems that Liker and Hoseus (2008) borrowed the term "servant leadership", which was first coined by Greenleaf (1977), to describe the Toyota leaders, by having a combination of in-depth understanding of the work and the ability to develop, mentor, and lead people, are respected for their technical knowledge as well (Liker 2004, p.182). In contrast to leaders who utilize people as machines in the Taylorism period, servant leaders empower followers to "*grow healthier, wiser, freer, more autonomous, and more likely themselves to become servants*" (Greenleaf 1977, pp.13–14). Servant leadership theory prioritizes its focus on others rather than upon self and on understanding of the role of the leader as a servant (Greenleaf 1977). Russell and Stone (2002) established a practical model for servant leadership, which included nine functional attributes namely vision, honesty, integrity, trust, service, modelling, pioneering, appreciation of others, and empowerment. An interpretation of the servant leadership attributes that are in agreement with Toyota's leadership practice is presented in Table 5.2.

5.4.2 Linking Motivation Theory to Toyota Way

Motivation theory for the workplace was developed in the middle of the twentieth century. The main mechanism behind the theory was that in order to achieve organizational goals, managers endeavour to search a way of taking workers' "hearts and minds" along with them (Crowther and Green 2004, p.38). Many theories including human relations, "soft" human resource management approaches and neo-human relations ideas can be grouped under motivation theory. These

Table 5.2 Matching the functional attributes of servant leadership to Toyota's leadership practice

Functional attributes of servant leadership (Russell and Stone 2002)	Toyota's leadership practice (Liker 2004; Liker and Hoseus 2008)
Vision (communication)	Toyota leaders' vision is to focus on a long-term purpose for Toyota as a value-added contributor to society (Liker 2004)
Honesty and integrity (credibility)	Humbleness is a good trait of Toyota leaders (Liker and Hoseus 2008)
	Respect for the individuals is one of the five supporting values of the Toyota way (Liker 2004)
Trust (competence)	Toyota leaders take responsibility to do their best to build mutual trust (Liker and Hoseus 2008) which is what makes it possible for individual employees to admit problems and take responsibility for solving them
	Toyota employees trust their leaders also because the leaders possess solid technical knowledge that can help them solve problems
Empowerment (teaching and delegation)	Empowerment occurs when employees use the company's lean tools to improve the company (Liker 2004, p.39). However, Empowerment is not placed until individuals and teams really understand the Toyota Way and TPS (Liker 2004)
Service (stewardship)	Toyota leaders constantly work on shop floor and provide timely assistance in answering the *andon* call. Basically, the most significant responsibility of Toyota leaders is to add more value to the employees
Modelling (visibility)	Toyota leaders exhibited their dedication in quality by going to the *gemba*—the actual place where the real added-value work is done to modelling themselves to the employees
Pioneering (influence and persuasion)	Toyota leaders are willing to meet challenge with courage and creativity (Liker and Hoseus 2008). Leaders always challenge the employees to achieve improvement in a production area (e.g. double productivity; reduce the changeover time)
Appreciation of others (listening and encouragement)	Various motivation theories are practiced at Toyota

Note: Attributes in brackets are accompanying attributes of servant leadership model

theories share the common goals of increased productivity, profitability or cost-effectiveness for the organization.

5.4.2.1 Maslow's (1954) Hierarchy of Needs

The motivation theory has its basis in Maslow's (1954) ideas about factors that motivate people in general. Maslow's (1954) hierarchy of needs provides a model of human needs that range from the basic need to provide personal sustenance for oneself up to the fully functioning, self-actualized individual who achieves his fullest potential.

5.4.2.2 Herzberg's (1959) Hygiene Factors and Motivation Factors

Frederick Herzberg (1959) identified, based on humans as individuals at work, two groups of factors:

1. Hygiene factors: must be in place, otherwise people become dissatisfied; and
2. Motivation factors: help to increase satisfaction and affects motivation.

Herzberg (1959) explained that some aspects of work have the potential to satisfy higher order needs in employees while others do not. Herzberg's (1959) motivation-hygiene theory provided some insights into what actually motivate people at work. In general, practitioners believed that by improving employee "on the job" social satisfaction, they can also improve productivity. One of the techniques that eventually evolved from this movement in the USA was "job enlargement". It offers the opportunity for workers to rotate between tasks as a means of combating boredom and to develop a sense of personal competence and responsibility. Followed by the advent of "job enlargement", job enrichment was developed with the goal to make work more meaningful. This includes engaging workers in planning and control activities. Toyota has performed well in providing for hygiene factors through job security with a safe and attractive work environment (Liker 2004). Moreover, the TPS endeavours to make the tasks more motivating especially for those working along the assembly line. Examples include: (1) job rotation which gives the work group ownership over a subsystem of the vehicle; (2) feedback on how workers are doing at their jobs; (3) *andon* system allows the worker to be proactive in solving problems; and (4) good autonomy over the tasks.

5.4.2.3 McGregor's (1960) XY Theory

Theory Y propounded by McGregor (1960) created the basic foundations for the TPS, especially with regard to considering the most valuable assets of a corporation, its employees, whose expertise and potential should be developed continuously for the mutual benefits of the company and the individuals themselves. Hall (1995) noted Theory Y as a precondition in the design of learning organizations such as Toyota. In 1960, Douglas McGregor introduced his XY theory of management based on opposing assumptions about people at work. McGregor (1960) assigned the following characteristics to the two domains:

1. Theory X: Management assumes workers dislike work and use the efficiency-rating system as a weapon against laziness. The workers are motivated purely by financial considerations.
2. Theory Y: Takes a more positive view of human nature. Management assumes that workers enjoy productive work if permitted to participate in decision-making. Management pays workers a fixed monthly or daily wage and guarantees lifetime employment. Cooperative company unions are formed and quality control circle activities and other company-wide improvement programs are promoted.

The Theory X and Theory Y characteristics probably coexist in every person, but one or the other may tend to appear dominant in a given environment. It is the task of management to create the conditions which individuals may satisfy their motivational needs, and in which they achieve their own goals through meeting the goals of the organization (Mullins 2006). Shingo (1988) highlighted that historically in Japan, it has been understood that since the country is poor in natural resources, it cannot prosper unless its people work hard. This mentality encouraged a nation of Type Y individuals whose inclinations are strengthened by the lifetime employment system and by a welfare system based on a fixed daily or monthly payment.

5.4.2.4 Skinner's (1948) Behaviour Modification Theory

Behaviour modification is the most generalized approach of using rewards and punishments to motivate (Liker 2004). Any society may be designed in such a way that desirable behaviours are reinforced through the effects linked with behaviours (Skinner 1948). This application originated from Skinner's (1948) experiments on animals that showed the effects of rewards and punishments on animal learning. The critical principle is that the positive or negative reinforcement comes as quickly as possible after the action (Skinner 1948). In Toyota's plants, workers are encouraged to pull the *andon* cord to indicate to the supervisors when the problems occur. Leaders constantly work on the floor and provide timely assistance that raise employees' morale to a large extent and that builds up employees' confidence in detecting the problems.

5.4.2.5 Locke's (1968) Goal-Setting Theory

The goal theory or the theory of goal-setting is primarily based on the work of Locke (1968). The basic premise of the goal theory is that people's goals or intentions play an important part in determining behaviour (Mullins 2006). Goal setting consists of "purposefully directed action" in the process of developing and setting specific work goals or targets for employees to accomplish (Wren and Bedeian 2009). Locke and Latham (2002) explained that goals affect performance through four mechanisms:

1. Goals serve a directive function: they direct attention and effort towards goal-relevant activities and away from goal irrelevant activities.
2. Goals have an energizing function. High goals lead to greater effort than low goals.
3. Goals affect persistence: There is often a trade-off in work between time and intensity of effort. Tight deadlines lead to a more rapid work pace than loose deadlines.
4. Goals affect action indirectly by leading to the arousal, discovery, and/or use of task-relevant knowledge and strategies.

Put simply, people are motivated by challenging but attainable goals and measurement of progress towards those goals. Toyota sets goals that meet these standards through policy development (*hoshin kanri*). Policy development in Toyota means objective goals were developed in a measurable and concrete form from the top of the company that stretch down to the work level.

Furthermore Liker (2004) linked the five most prominent motivation theories discussed above to the Toyota's approach of developing people and team, and commented that each was used to great effect in Toyota. According to the way

Table 5.3 Various motivation theories and the Toyota Way

	Concepts	Toyota's approach
Maslow' (1954) hierarchy of needs	Satisfy lower level needs and move employees up the hierarchy towards self actualization	Job security, good pay, and safe working conditions satisfy lower level needs
		Culture of continuous improvement supports growth towards self actualization
Herzberg's (1959) motivation-hygiene theory	Eliminate "dis-satisfiers" (hygiene factors) and design work to create positive satisfiers (motivators)	5-S, ergonomic programs, visual management, and human resource policies address hygiene factors
		Continuous improvement, job rotation, and built-in feedback support motivators
McGregor's (1960) theory-X and theory-Y	Theory X assumes that employees are inherently lazy, indifferent, and uninterested in excelling on the job. Theory Y assumes that employees are self-motivated, willing to work hard, and rewarded by challenging work	Each employee has responsibility and is trusted for quality control and safety
		Various *kaizen* programmes are challenging work that appeals the employees
Skinner's (1948) behaviour modification theory	Reinforce behaviour on the spot when the behaviour naturally occurs	Continuous flow and *andon* create short-lead times for rapid feedback
		Leaders constantly work on the floor and provide reinforcement
Locke's (1968) goal setting theory	Set specific measurable, achievable challenging goals and measure progress	Sets goals that meet these criteria through *hoshin kanri* (policy deployment)
		Continuous measurements relative to targets

Source: Liker (2004)

people are motivated, Liker (2004) categorized these theories into internal motivations and external motivations, which are summarized in Table 5.3.

5.4.3 Theory Related to Supplier Relationships

Supplier management has been recognized as crucial to the firm's competitiveness. According to Kim and Michell (1999), two divergent views concerning supplier management have been identified. These are the contractual (arms-length) view and the relational view.

5.4.3.1 The Contractual View

This view is widely practiced in the West, in which minimal dependence was placed on suppliers, with the object of maximizing bargaining power and avoiding commitment (Kim and Michell 1999).

5.4.3.2 The Relational View

This view, on the other hand, plays a key aspect in relationship marketing, which has resulted in the success of Japanese firms. In this view, a buyer and a supplier establish and maintain close relationships on an ongoing basis. Based on Hunt and Morgan's (1994) trust-commitment theory, Kim and Michell (1999) pointed out that this view encourages the long-term relationships between suppliers and buyers to develop as partnerships. Hunt and Morgan (1994) posited that the relationship commitment and trust develop when firms attend to relationship by: (1) providing resources, opportunities, and benefits that are superior to the offerings of alternative partners; (2) maintaining high standards of corporate values and allying oneself with exchange partners having similar values; (3) communicating valuable information that includes expectations, market intelligence, and evaluation of the partner's performance; and (4) avoid taking advantage of exchange partners. Such actions will enable firms and their networks to enjoy sustainable competitive advantages over their rivals and their networks in the global marketplace (Kim and Michell 1999).

Moreover, Chen and Paulraj (2004) identified various supply chain initiatives and factors to develop key supply chain management constructs. A few critical elements of the buyer–supplier relationship were incorporated in their framework, including (1) supplier base reduction, (2) long-term relationships, (3) communication, (4) cross-functional team, and (5) supplier involvement. These elements can be theoretically supported by the relational view of relationship marketing. It is noteworthy that the key aspects identified earlier in Chap. 4 (the Toyota Way Principle 11) mirrors those constructs in Chen and Paulraj's (2004) framework.

5.5 The Toyota Way Problem-Solving Model: What Theories Can Be Linked

Since the last three principles of the Toyota Way focused on two key words, namely continuous improvement (*Kaizen*) and learning organization, efforts will be made to explore the theoretical underpinning these two areas. Zangwill and Kantor (1998) explained two origins of continuous improvement: the Toyota Way and statistical reasoning. The continuous improvement practice in Toyota has been discussed earlier in Chap. 4. The second trend underpinning continuous improvement lies with the quality movement and statistical reasoning, which were conceived in the 1920s by Shewhart (1931). Its contemporary contribution came from W. Edwards Deming's 1950 lectures to Japanese executives, during which Deming (1986) highlighted the importance of data collection and Shewhart's PDCA cycle (Zangwill and Kantor 1998). Hence, it appears that theories from the quality management domain can be used to support the last three principles of the Toyota Way model.

5.5.1 Linking Quality Management to the Toyota Way

Many authors have written about *kaizen* which was often presented as one of the underlying principles of lean production or Total Quality Management (TQM). According to Imai (1986), *kaizen* practices were heavily influenced in the beginning by the American statisticians W. Edwards Deming and Joseph M. Juran. Their teaching about rigorous statistical control, right after World War II, matched very well with the Japanese unique holistic, democratic and collaborative behaviour and the urgent need for improving quality in their products. Deming (1986) also introduced the "Deming circle", one of the crucial Quality Control (QC) tools for assuring continuous improvement, to Japan. Juran was invited to Japan for a seminar on the topic of quality control management, the concept of which was dealt with for the first time from the overall management perspective (Imai 1986). In 1962, a QC circle was started. This was formed by a small group of volunteers who performed quality control activities within the group. According to Levy (1990), the QC concept has caught the attention of the Japanese construction industry since 1970s.

Liker (2004) highlighted that the Deming Cycle or PDCA cycle is a cornerstone for continuous improvement. The PDCA cycle is used in Japanese companies to initiate, track and review improvements. Imai (1986) outlined that this began with a study of the current situation, during which data was gathered to be used in formulating a plan for improvement. Once this plan has been finalized, it was implemented. After that, the implementation was checked to see whether it has brought about the anticipated improvement. When the experiment has been successful, a final action such as methodological standardization was taken to ensure that the new methods introduced would be practised continuously for sustained improvement (Imai 1986). Sobek and Smalley (2008) also noted that the PDCA cycle is the heart of the problem-solving approach in Toyota. A close examination of the PDCA cycle and the last three principles of the Toyota Way model are tabulated in Table 5.4. In the Plan stage, the *genchi genbutsu* concept (P12) will enable the leaders to come to the workplace to work with the front-line worker to identify the root problems. Problem-solving techniques such as 5 whys (P13) may be employed. Alternative solutions will be discussed (P13) until a satisfactory consensus has been reached. Table 5.4 summarizes how these three principles work hand in hand with the PDCA methodology.

5.5.2 Learning Organization and the Toyota Way

Liker (2004) termed the last principle of the Toyota Way as "becoming a learning organization", and mentioned that by continuous improvement (*Kaizen*) and reflection (*Hansei*), the goal of becoming a learning organization can be achieved. Peter Senge (1990) popularized learning organizations as places where people continually expand their capacities to create the results they truly desire, where new and expansive patterns of thinking are nurtured, where collective aspiration is set free,

Table 5.4 Linking the PDCA cycle and problem-solving in Toyota

PDCA methodology	Toyota Way Principles 12–14
Plan: before any corrective action is taken on the problem at hand, a number of activities should be undertaken in the planning stage: (1) Define the problem (2) Gather relevant information (3) Identify the root cause of the problems (4) Develop and consider the possible solutions (5) Select the best alternative for implementation	The Plan methodology of PDCA corresponds to Principles 12 and 13 of the Toyota Way (1) The Toyota Way (P12) enables the leaders to go and see the actual situation for them to understand the problem source and identify the problem (2) Make the decision on the verified data (3) 5-whys methodology (P13) as a useful tool is employed to probe the rooted causes (4) Alternative solutions will be discussed (P13) until a consensus is agreed
Do: implement the solution chosen as the best	Solve the problem
Check: monitor the implemented solution and gather data of the effects of implementation and comparing these with the target or prediction	Principle 14 indicates that *hansei* is part of the check part of the PDCA cycle
Act: Establish the new process, solution, or system as the standard if the results are satisfactory, or taking remedial action if they are not	Principle 14: *Kaizen* activities

and where people are continually learning how to learn together. Based on McGill and Slocum's (1993) four types of organizational models (namely knowing organization, understanding organization, thinking organization and learning organization), Hines et al. (2004) outlined four stages of lean thinking evolvement, which are closely related to the stages of development of organizational learning as shown in Table 5.5.

Learning organization seeks to maximize the learning opportunities of employees, suppliers, customers and even competitors (McGill and Slocum 1993). Garvin (1993) suggested that to become a learning organization, companics need to be skilled in the following five activities:

1. *Systematic problem-solving*: Relates to the philosophy and methods of the quality movement, relying on scientific methods rather than guesswork; uses actual data rather than assumptions and simple statistical tools.
2. *Experimentation with new approaches*: Systematic searching for and testing new knowledge; but unlike problem-solving, experimentation is usually motivated by opportunities and new perspectives rather than current difficulties. It takes two main forms: ongoing programmes and one-of-a-kind demonstration projects.
3. *Learning from their experiences and past history*: A review of successes and failures and reflecting and self-analysis.
4. *Learning from experiences and best practices of others*: Benchmarking; looking outside the immediate environment; openness to the outside world; environmental scanning.

Table 5.5 Four stages of lean thinking evolvement

	Cell and line (1980–1990)	Shop-floor (1990–mid 1990)	Value stream (mid 1990–2000)	Value system (2000+)
Prescription	Tool-based approach	Practice approach	Lean principles	Customer value, policy deployment, technology
Organizational learning	Knowing organization	Understanding organization	Thinking organization	Learning organization

Source: Hines et al. (2004)

5. *Transferring knowledge quickly and efficiently throughout the organization*: Knowledge transferred quickly and efficiently throughout the organization; mechanisms in place to facilitate the process; written and oral reports; site visits; tours; rotation programmes; and education and training programmes.

These five activities are pertinent to the last three principles of the Toyota Way and extend to other principles as well. The first activity is a summary of the last three principles. More specifically, Principle 13—the Toyota Way of making decisions based on the verified data is more relevant. The second activity is in line with the *kaizen* activities that are carried out throughout Toyota, especially the small ongoing experiments that are usually common on the shop floor. The third activity reflects the basic meaning of *hansei* ("reflection" in Japanese), which focuses on learning from past experience. The fourth activity stretches to other principles of the Toyota Way, for example, the Toyota staff have the attitude of learning from others when they are working with suppliers. According to Liker (2004, p.210), learning by doing process is a typical Toyota Way style that is adopted by Toyota to learn with its suppliers. The last activity can be linked to Principle 10 of the Toyota Way, which suggests that training programmes are powerful tools for transferring knowledge and nurturing people.

5.6 Conceptual Framework

Guided by the research aim and objectives, the proposed conceptual framework adopts Liker's (2004) Toyota Way model as a base for development. However, before the 4"P" model can be established as a model for application to the large Chinese construction firms, different schools of management thoughts and their corresponding applications in the manufacturing industry have been reviewed. Following that, the TPS, and later the popular term lean production, and finally a systematic view of the Toyota Way were reviewed, respectively. In the course of reviewing the 4"P" model of the Toyota Way (14 principles in total), the distinguishing features in each principle have been identified. Efforts have also been made to seek the theoretical support behind the "4P" model. This started with linking Peter Drucker's (1994) theory of the business to the first principle of the Toyota Way model. Toyota's chief goal to reduce cost by eliminating all the non-value-adding activities (waste), Toyota's mission and Toyota's self-reliance culture have mapped well with the three assumptions that underpinned Drucker's (1994) theory of the business. In a similar way, the production model integrated with three different views of production, namely transformation view, flow view and value generation value, proposed by Koskela (2000) were linked to the Toyota Way model (see Table 5.1). This discussed the Flow model of the TFV paradigm with the Process category of the Toyota Way as both focused on waste elimination on the shop floor. Moreover, heuristic principles derived from the Flow model were well mapped with the "tactical" and "operational" aspects of the Toyota Way. Efforts were also made to connect the TFV model to the rest of the layers in the Toyota Way model. It is worth mentioning that the TFV model was espoused by Koskela (2000) as a "theory" of production that supports a variety of tools and techniques (e.g. the Last Planner System) to be applied to the management of construction projects. In addition, management theories in the production management domain have been reviewed earlier in Chap. 2, among which the servant leadership theory and various motivation theories under the Human Relations school in particular can be linked to the Toyota Way People and Partner model to reflect how Toyota develops and grows with its people and partners. Because the partners are also inclusive in this layer of the Toyota Way model, a few critical elements of the buyer–supplier relationship were adopted as the theoretical support to explain how Toyota treats, challenges and grows with its partners. Lastly, the problem-solving principles of the Toyota Way can be understood to be Toyota's quality management philosophy, because Toyota is dedicated to continuously solve root problems that drive organizational learning. From a practical point of view, the last three principles are constructed based on the Deming cycle or PDCA to achieve continuous improvement. From a theoretical point of view, they were linked with Anderson et al.'s (1994) paradigm of quality management that underlies the Deming management method. Finally, as Liker (2004) highlighted that the ultimate goal of becoming a learning organization cannot be achieved without multiple efforts including (1) a stable and standardized process that provide opportunity to learn continuously from the improvement made when waste and inefficiencies are visible, (2) a stable personnel, slow promotion, and very careful succession systems to protect the organizational knowledge base, and (3) having capacity to identify the root problems and develop countermeasures. Therefore, the learning organization concept was also examined. Not surprisingly, Toyota's problem-solving practices as well as the training programme provided to the employees and *kaizen* activities conducted in the shop floor were all pertinent to Garvin's (1993) underlying five activities that a learning organization must have.

It is worthwhile to note that the discussions above regarding linking the theories to the Toyota Way model are within the scope of production management. This is because concepts derived from two functions of production management, namely production and management can be directly connected to all four layers of the Toyota Way (see Fig. 5.1).

1. Philosophy: most companies have developed a statement of corporate philosophy, or mission, to which operating objectives are closely tied (Chase and Aquilano 1992). In Toyota Way model, the overall philosophy of the

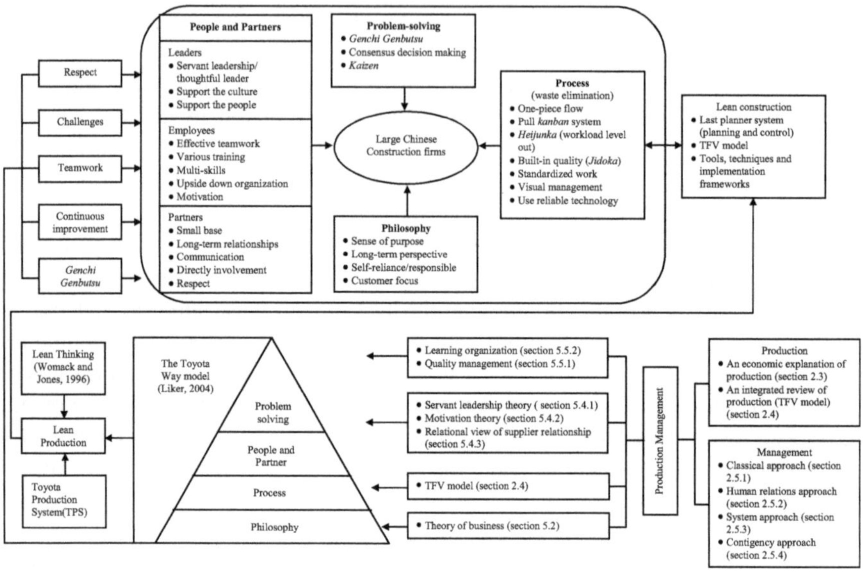

Fig. 5.1 Conceptual framework of this research

organization has a strong influence and permeates down to influence the principles and methods of the other three building blocks. In simple term, without organizational philosophy, it is less likely to achieve the production objectives.

2. Process: process is one of the five P's of production management discussed in Chap. 2. The process layer of Toyota Way model covers a number of key issues in production management literature. These include planning and control, inventory management, factory layout, quality management, and so on. Moreover, Shingo (1988) highlighted that process should be given top priority in improving production and pointed out that the Toyota Production System represented a pioneering attempt at a new production philosophy with a focus on process improvement. This further formed a basis of flow concept of Koskela's (2000) TFV model which aimed to eliminate the non-value-adding activities throughout the process. Lastly, it seems that the scientific management theory—time and motion studies in particular—had great impact on Toyota in increasing efficiency and decreasing waste.
3. People and Partners: production system is managed to achieve the production objectives, which are directly and indirectly affected by people—the employees, the leaders and the partners. This is because all people are different. Their abilities, personalities, interest, ambitions, training and experiences vary widely (Gaither and Frazier 1999). More importantly, how to make the employees to be more productive, how to grow leaders to be role models of the company's philosophy, as well as what approaches are appropriate for developing partnership that are becoming major concerns in the field of production management

pertaining to the people factor. These lead to discussion of teamwork, motivation, leadership and partnership in production management textbooks. Furthermore, the evolution of management theory highlights that since the Hawthorne studies, human relations school of management theory advocated that people were the key to productivity and technology. Management theorists of this school were firm believers that only motivated and skilled individuals who were committed to organizational objectives were the key to success. This was reinforced by the socio-technical system approach that human resource and technical aspects of the organization should be treated equally. The research in lean community also suggests that lean enterprise should be designed based on socio-technical system (Paez et al. 2004; Shah and Ward 2007).

4. Problem-solving: one of the most powerful aspects of Toyota Way is its focus on quality at first time. The mechanism behind which is that Toyota employees are trained to use PDCA framework to seek the root causes of problems, rather than be satisfied with quick solutions. This continuous improvement practices eventually lead to the organizational learning. Besides, Toyota Way extensively uses a number of useful problem-solving tools such as *genchi genbutsu*, 5-whys, quality circles, etc. (Liker 2004) as well as basic tools for quality control (Ishikawa 1990) that are widely encouraged in manufacturing production. They have indeed become an integral part of production management literature.

All the discussions undertaken above were within the context of the manufacturing industry. The literature review on the lean construction philosophy as well as the implementation of the Toyota Way principles in the context of construction in Chap. 4 explored the potential application of the Toyota Way framework to construction organizations. Because of the inherent differences between manufacturing and construction, efforts have been made to link a number of the Toyota Way Process-related principles to the Last Planner System as a necessary modification for better implementation. The remaining principles have remained unchanged. As illustrated in Fig. 5.1, the Toyota Way framework can be potentially implemented within Chinese construction firms and help to improve their organization and project performance.

5.7 Summary

This chapter discusses the theoretical basis of the Toyota Way model. A number of theories and paradigms concerning different areas, such as business, production, human resource management, and quality management are discussed. More importantly, the selected theories and paradigms are also demonstrated to have links to, and to support, different layers of the Toyota Way model. As is known, the Toyota Way model itself is practically oriented, and it is worth endeavouring to link different theories to the Toyota Way in order to allow a balanced understanding of this framework from a theoretical perspective. This chapter concludes with a conceptual framework that incorporates several concepts from earlier chapters, such as lean production and lean construction.

China's Construction Industry 6

6.1 Introduction

China has demonstrated an astonishing economic growth capability to the world. Since the introduction of the open door policy, China's Gross Domestic Product (GDP) growth has amounted to over 8 % annually in recent years (National Bureau of Statistics of China 2011). This economic success cannot be achieved without contributions from infrastructural and urban development, or without rapid growth in the construction industry. According to National Bureau of Statistics of China (NBSC 2011), the construction industry in China has accounted for approximately 6 % of GDP since 2006. Employment in the construction industry also accounted for about 7 % of the total permanent employment in the urban areas of China. A characteristic of China's construction industry is the large share of labour-only subcontracting, often on the basis of individual self-employed labourers from rural areas of China. As NBSC (2011) noted, by the end of 2010, there were 71,863 registered firms employing 41.6 million people in the construction industry, excluding the labour subcontractors. This created a total output of 9603.1 billion RMB (approximately US$1538.6 billion). Statistically, these figures clearly indicate the leading position of the construction industry and its enormous achievements. However, its performance over the past several years still suggests that the industry faces serious difficulties.

The following sections firstly examine China's construction industry, as well as its external environment. Next, this chapter investigates the performance of the Chinese construction industry and the large construction firms through the measurements of quality, health and safety, productivity, profitability, technology development, and its project management practices. A number of challenges that might potentially hamper the implementation of lean or Toyota Way in China's construction industry are also examined.

S. Gao and S.P. Low, *Lean Construction Management*,
DOI 10.1007/978-981-287-014-8_6,

6.2 China's Construction Market, Structure and Its Leading Construction Firms

6.2.1 Domestic Market

Benefiting from the rapid growth in GDP, China's construction market has developed dramatically in the past years. The construction market in China is expected to overtake the USA as the world's largest construction market by 2018, according to a forecast by Global Construction Perspectives and Oxford Economics (Xinhua 2009a). As estimated in the same report, the Chinese construction market is projected to be worth almost US$2.4 trillion in just 10 years, which represents 19.1 % of the global construction output (Xinhua 2009a). With such potential in the construction market, the China Construction Association (2009) highlighted that three outstanding features will occur during the expansion of domestic demands:

1. A larger scale of construction projects will be needed, given that China has accelerated its urbanization process towards building a moderately prosperous society. According to MOHURD (2009a, pp. 85–93), the Chinese construction industry will continually grow, with a focus on six major areas: large-scale infrastructure construction, residential buildings, "New Rural" projects, coastal development, industrial projects and post-disaster reconstruction projects.
2. A high degree of difficulty will be met with in construction tasks, as projects are becoming complex (Egbu 2006), which enables contractors bring in innovative systems and approaches to construction.
3. Shorter cycles will occur in construction periods, given that the clients both in public and private sectors require faster project delivery.

6.2.2 International Construction Markets

The global construction market in recent years has witnessed rapid expansion of Chinese international construction companies (CICCs). It is worth mentioning that, even during the economic downturn period (2008–2009), adverse effects were not seen in the CICCS. This is because, according to Hao (2010), the economic downturn has had less impact on many developing countries (African and Asia in particular), where CICCs have been undertaking numerous huge infrastructure construction projects.

According to Engineering News-Record (ENR 2012a), by the end of 2011, the CICCs gained total contracting revenue of US$64.76 billion from their overseas construction projects, and was closely followed by the American firms, which took up approximately 13.4 % of the total overseas market with an amount of more than US$60 billion. The latest statistics from the ENR saw 41 Chinese companies (ranked by total construction contracting revenue) in the ENR's Top 225 Global Contractors list, and half of the top ten international contractors are CICCs (ENR 2012b). The ENR 225 list is based on revenues. The companies are ranked by total revenues generated outside of each company's home country for each fiscal year.

The possible weakness is that using solely revenue as a measure—as in the ENR list—does not take into account other performance factors, such as project quality, profitability, and so on. China's Ministry of Commerce (2009), cited in China International Construction Association (CICA 2010), revealed that the contracting value contributed by the major governmental enterprises (known as State-Owned Enterprises) accounted for nearly half of the total contract value, led by China Railway Group Limited, China Petrochemical Corporation and China State Construction Engineering Corporation. Not only were these encouraged by the government to venture into the overseas construction market, but they were also attracted by the attractive project margins. Hao (2010) highlighted that the gross margin in most overseas construction projects is usually two percentile points higher than in domestic projects. The increase in overseas income and the attractive profit margins drew more attention from the top management of the CICCs. Hence, a number of large Chinese construction companies became determined to expand their overseas market with a 3- to 5-year strategic focus (Hao 2010).

6.2.3 Structure of Chinese Construction Industry

Before the study goes further into the analysis, this section shed light on the structure of the construction industry in China, as well as define the scope of the term "large Chinese construction firms". According to the NBSC (2011), there were 71,863 registered construction firms operating in the Chinese construction industry in 2010. Firms in the Chinese construction industry were organized into four categories in terms of their ownership: state-owned enterprises (SOEs), collective-owned enterprises (COEs), enterprises with shares and private enterprises (Chen 1998; Low and Jiang 2003; Wang et al. 2006; Zeng et al. 2005). During the recent years of transition—when reforms in Chinese enterprise were vigorously sought—the numbers of both SOEs and COEs decreased (Wang et al. 2006), but they have continued to maintain their leading positions in terms of undertaking the most construction work (Lu and Fox 2001). In 2010, among the 71,863 registered construction firms in China, there were 4,810 SOEs, accounting for 6.7 % of the total number of firms, but employing 5.76 million employees (13.8 %), earned 1814.8 billion RMB, or 19.9 % of the total output of construction value (9103 billion), and contributed 27.5 % of the total taxes to the central government (NBSC 2011). Moreover, research undertaken by Liu et al. (2013) revealed that the operating revenues of the four largest Chinese construction firms (all SOEs) increased from 10 % of total construction revenues in 2003 to 15 % in 2010. This seems to imply that the state-owned construction firms still hold their leading positions in the Chinese construction market though engaging a smaller numbers of firms and employees, but obtaining a good proportion of market share (NBSC 2011).

Alternatively, all the general contractors in China are classified as premier (highest level), first, second, or third grade, or below third grade, in terms of their financial abilities, management abilities and technological ability (NBSC 2011).

Table 6.1 Indicators of Chinese contractors with different grades in 2008

Category of grades	No.	%	Gross output value of construction (100 million RMB)	%	Total value of the contracting (100 million RMB)	%
Premier	260	0.07 %	8,701	19.6 %	18,531.7	24.8 %
First	2,846	8.3 %	1,7438.8	39.2 %	31,369.9	42 %
Second	9,961	29.2 %	1,0761.7	24.2 %	15,176.7	20.4 %
Third and below	21,004	61.6 %	7,517.6	17 %	9,554.8	12.8
Total	34,071	100 %	4,4419.3	100 %	7,4633.2	100 %

Source: NBSC (2008)

Contractors in different grades can only be allowed to bid for the relevant projects. The latest statistics available on the key indicators of the construction firms by different grades are from the NBSC (2008) and are illustrated in Table 6.1. The Chinese construction industry is characterized by the large number of small firms (with second, third, or below in terms of their grades). In 2008, less than 10 % of firms were premier (0.07 %) or first (8.3 %) grade firms and these categories are viewed as relatively larger-scale Chinese construction firms. However, these larger construction firms take up a fairly large portion of the construction market share in China, having 58.8 % of the output value and 66.8 % of the total contracted value, respectively.

Overall, this is consistent with the central government's policy to accelerate its efforts to revitalize a number of construction companies as future leaders in the industry with a competitive edge in their construction business (MOHURD 2008; Zeng et al. 2005). These leading Chinese construction firms have emerged with large-scale operations, solid construction capabilities and strong initiatives. In addition, they have been recognized in various rankings, including the Annual Engineering News-Record (ENR)/Construction Times China ranking. Among these large construction firms, the research focus of this study is on general contractors with the capability to plan, design, and research and develop, because these capabilities are similar to a typical manufacturing company in which lean concepts can be applied. This is supported by Bhasin's (2012) observation that the larger organizations view lean as an ideology and perform better. Secondly, as it may not be possible to implement lean concepts in all Chinese construction firms, it is necessary to narrow the targeted construction firms to those with sound financial capacities, equipped with advanced technology, and/or with relatively advanced management practices, because lean implementation requires a long-term philosophy, strong financial backup, and so on. For this reason, large Chinese construction firms appear to be suitable candidates for the study of lean implementation in the Chinese construction industry.

6.2.3.1 Subcontracting

With the rapid development of the construction sector in China, and the huge demand for the construction labour force, subcontracting is, however, becoming

Table 6.2 Changes in subcontractor indicators in the Chinese construction industry from 2003 to 2011

	No. of firms	No. of persons employed (Million)	Total revenue (Billion RMB)	Total profits (Billion RMB)
2011	6,443	2.45	129.74	3
2010	6,835	2.57	94.1	2.49
2009	6,756	2.15	74.28	2.05
2008	6,837	1.89	62.09	2.03
2007	4,357	1.47	37.89	0.89
2006	3,748	1.12	25.37	0.63
2005	3,101	0.86	18.27	0.6
2004	n.a.	n.a.	n.a.	n.a.
2003	2,021	0.57	11.05	0.39

Note: data for 2004 is not available
Source: NBSC (2011)

unavoidable. In fact, the Chinese general contractors have been reducing their reliance on a permanent workforce since 1984, when a reform programme entitled "separation of management from field operations" was launched (Lu and Fox 2001). For example, Table 6.2 shows the continuing changes in labour-only subcontractor indicators in China. The number of labour-only subcontractors increased approximately three times from 2,021 in 2003, up to 6,443 in 2011. The number of people employed in 2003 underwent a nearly five-fold increase, up to 2.45 million in 2011, which brought about 129.74 billion RMB revenue—an almost ten-fold increase over 2003.

These figures indicate that the subcontracting sector has been booming in recent years. However, the subcontracting system is problematic, according to Lan (1999) and Lan and Jackson (2002). Firstly, there is no clear relationship between contractors and subcontractors (Lan and Jackson 2002). Hence it is unclear what percentage of construction jobs is or should be undertaken by subcontractors. In most cases, the general contractor outsources almost all of the work to low-tier subcontractors, and also transfers risks and labour recruitment to them (Pun and Lu 2010). On projects, general contractors only take charge of project management and arrange equipment for the subcontractors (Pun and Lu 2010). Secondly, many contractors seek profits by illegally leasing their licenses or by subcontracting their jobs to unqualified firms (Lan and Jackson 2002). Arising from this, there was a terrifying mishap in Hang Zhou, China in 2008, where a subway tunnel collapsed as a result of construction work being illegally subcontracted out four times (Xinhua 2008). Thirdly, unlike the harmonious relationship that can be seen between a general contractor and subcontractors in a project using lean concepts, the subcontracting sector in China is sometimes still subject to violence, arguments and fights, due to the tense relationship between subcontractors and frontline workers, most commonly related to delays in payment (Pun and Lu 2010).

6.3 Status of the Chinese Construction Industry

6.3.1 Quality

Broadly, there are two views of quality in Chinese construction projects. On the one hand, a number of symbolic projects have helped China to win many accolades, and indeed these exemplary projects (such as the National Stadium, the Three Gorges Dam, and the Shanghai World Financial Centre) reflect the highest level of Chinese construction quality. Having observed these enormous achievements, Yung and Yip (2009) underlined that Chinese construction quality is expected to improve on a continuous basis, but at a decreasing rate as the economy develops. Moreover, Yung and Yip (2009) have highlighted that improved construction quality in China cannot be achieved without (1) the gradual implementation of mandatory construction supervision systems; (2) improved labour productivity; (3) the availability of resources, including machinery and labour and (4) the use of more plant or machinery.

However, these exemplary construction projects do not represent the average level of the Chinese construction industry in terms of quality. On the other end of spectrum, there are the criticisms and complaints relating to poor construction quality, which appear to continue unabated elsewhere in the country. Poor construction quality was recognized as one of the critical problems in China in the 1990s (Chen 1998; Lam and Cheng 2004) and it is still a major problem (China Daily 2010a). There has been an alarming increase in fatal accidents caused by bad construction quality across the country. For example, a large number of schools and hospitals collapsed during the Sichuan earthquake in 2008, resulting in thousands of students being killed or seriously injured. In June 2009, a 13-storey building in the Lotus Riverside residential complex in Shanghai toppled, killing one worker. One recent tragedy includes the collapse of a bridge in northern China, at Harbin, which killed three people and injured five (Xinhua 2012). This was at least the 18th collapse since 2007 (South China Morning Post 2012). The national news agency, Xinhua (2012), investigated the incident further and discovered that the real cause of the collapse was that the bridge was planned to take 3 years to complete, but was actually completed in 18 months. The blind pursuit of quick delivery at the expense of quality is to blame in this case. Earlier, the collapse of Hangzhou subway in 2008 was due to the malpractice in subcontracting practices. This irregular contracting has become a major cause of poor-quality housing and infrastructure projects in China (Xinhua 2012). Some projects have been outsourced repeatedly between many contractors and subcontractors, making it difficult to manage and supervise construction quality. Apart from the failing of the legal environment affecting Chinese construction industry, the use of inferior construction materials for projects is also common. For example, inspections by state officials have found raw, unprocessed sea sand in at least 15 buildings under construction in Shenzhen, including a building which, when finished, was set to become China's tallest.

6.3.1.1 How Quality Management Works in China

Although quality management is increasingly practised in China, its implementation appears to be uneven (Li et al. 2003). In China, construction quality is generally achieved under the supervision of (1) supervision firms (known as *jianli*), (2) relevant government authorities and (3) the construction firms' own project management teams.

1. Supervision firms: The roles and responsibilities of Chinese supervision engineers are in line with those of US design professional engineers (Wang et al. 2009). They act as the quality control team on site. However, Wang et al. (2009) have pointed out that the supervision professional's unclear scope of quality liability and safety liability in current laws, along with low level of competence seen in the practice of quality supervision, have become major causes of supervision liability risks, and ultimately would affect construction quality supervision.
2. Government authorities and the building quality check programme: According to the Ministry of Housing and Urban-Rural Development (MOHURD 2010a), a number of quality awareness activities (e.g. "Safety Year activity") have been organized, to continue to stress the importance of quality awareness. Nevertheless the expected improvement is slow and disappointing.
3. Construction firms: Attempts have been made by the Chinese government since the 1990s to implement the TQM framework, and to introduce ISO 9000 certification (Zeng et al. 2003). Tang et al. (2005) and Zeng et al. (2003) have pointed out that TQM has been accepted and applied in the construction industry in China, and there is a popular trend to obtain the ISO 9000 certification.

6.3.1.2 Role of Government

In more recent years, the Ministry of Housing and Urban-Rural Development (MOHURD 2008) realized that there is a need to fine-tune strategy for construction firms from simple the "pursuit of output expansion to enhancing quality improvement as part of their cooperation culture". Subsequently, the government carried out a series of measures to improve the quality of construction works, such as setting up quality supervision stations, establishing professional project management systems and promoting the implementation of ISO 9000. The construction industry is under constant scrutiny for its quality of work (Chini and Valdez 2003). According to MOHURD (2009b), the government ordered a nationwide quality inspection of housing construction immediately after a residential building under construction in Shanghai collapsed. This quality inspection programme involved 90 cities across China, including 180 ongoing projects. The results of the national quality check suggested that quality was under control in most projects (MOHURD 2009b).

6.3.2 Productivity

Low productivity is always an issue in the construction industry. The construction industry is still very much a crafts-based industry. Unlike some developed countries where labour shortage was largely a concern of their governments, China seldom worries about such issues in the construction sector. However among the millions of construction labour force, the outstanding issue that started to emerge is that skilled labourer and qualified project management teams are in demand. The basic definition of productivity as a ratio of output and input, where the calculation involves dividing the measure of output (i.e. gross value added) by labour input (number of workers) (Crawford and Vogl 2006). A number of countries and organizations (see Abdel-Wahab et al. 2008; Crawford and Vogl 2006) have preferred to measure productivity in this way, due to its two advantages, namely that (1) governments have an interest in both input and output, given that this measure shows the net value (output) added from construction activities to the economy, and (2) it is easy to calculate and estimation is less subject to data limitations. Using this indicator, construction productivity in China has improved from only US$500 per person in 1980, rising impressively to US$6,471 per person by 2010 (National Bureau of Statistics of China (NBSC) 2011). This revealed that the Chinese construction firms have made progress during the last 10 years. Figure 6.1 compares value added per person per year between the construction industry and the manufacturing industry in China since 2000.

The gap shows that the productivity of the construction industry was one third that of the manufacturing industry in 2000 and was about one fifth of the industry in 2007. Xue et al. (2008) were concerned that there remained large gaps in the productivity levels between different regions and recommended that the Chinese government should adopt effective policies and measures to improve productivity. Furthermore, the construction industry in China is much less productive than in other countries. Xu et al.'s (2005) study shows that China's construction productivity in year 2000 lagged behind three developed countries that include the USA, Japan and the United Kingdom, with the output per person being only one twentieth of the general level in the three developed countries. However, the research undertaken by Shen et al. (2011) seems to be not consistent with the international productivity comparison using China against developed countries. By comparing the labour productivity of US and Chinese counterpart at the activity level (i.e. quantity installed per labour hour using published national average productivity data), Shen et al. (2011) found out that in terms of labour-intensive activities, small labour productivity gaps are found between China and the USA. Moreover, Singapore's Minister of National Development, Khaw Boon Wan, has reflected on the productivity differences between Singapore, Hong Kong and China, and noted that the Chinese construction productivity is not very high, but has the advantage that its ample resources can be quickly mobilized to achieve fast construction, often 24 h round the clock (Khaw 2012). Khaw (2012) mentioned that Singapore's BCA visited Changsha to glean some learning points, since CNN featured a construction project there that proceeded with record-breaking speed in Changsha, in which a

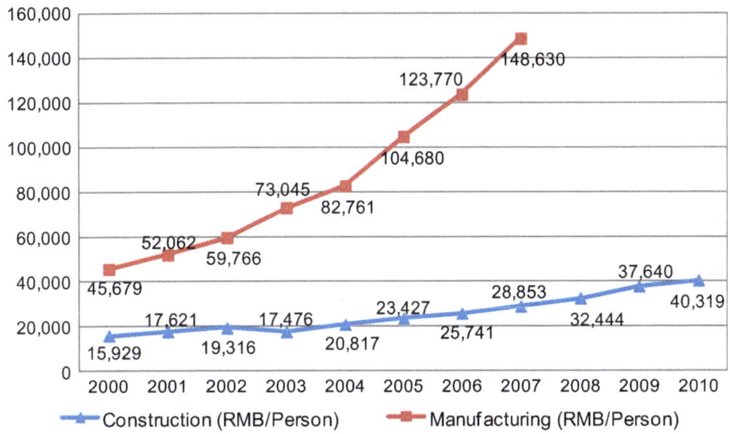

Fig. 6.1 Productivity in the construction and manufacturing sectors, 2000–2010. *Note*: data for productivity of manufacturing industry from 2008 onwards is not available. *Source*: adapted from NBSC (2011)

new hotel of 30 storeys was built in 15 days. This was a unique case, but implies that if the right technology (prefabrication) is used, plans, teamwork, and precise execution is in place, high levels of productivity can be achieved.

6.3.3 Profitability

Under the previous planned economy, construction output in China was planned at the national level (Huang et al. 2012). Construction firms were given the freedom to determine their profit margins by tendering and other methods until the economy transited to a market-oriented economy (Wang et al. 2006). The competitive environment has been changing considerably since China was admitted as a member of the World Trade Organization (WTO) in 2001. In the competitive market, instead of relying on the government's allocation, maximizing profits, achieving the projected annual output and increasing market share have become the top three major concerns of decision-makers from the construction firms (Wang et al. 2006). However, many local firms are still plagued by low levels of profitability (Cheah et al. 2007; Zeng et al. 2005). Due to the fierce competition in the contracting market, excessive reduction of contract prices imposed by the client on the contractors during the contract negotiation stage has been normal (Wang et al. 2006). Other abnormal behaviour of the client, including asking contractors to finance projects during construction, delays in payment, and others, have caused severe financial difficulties for Chinese construction firms (Wang et al. 2006). This has resulted in the profit being squeezed so much so that the majority of the profitable construction firms are only marginally profitable (Wang et al. 2006). For example, according to Wang (2012) who reported that the NBSC's Shenzhen

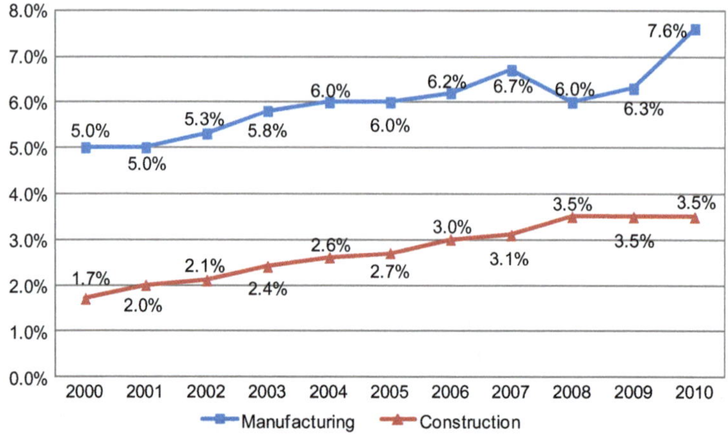

Fig. 6.2 Comparison of profitability level of manufacturing and construction sectors in China, 2000–2010 [*Source*: adapted from NBSC (2011)]

team—which investigated 15 local construction firms in Shenzhen, China—declining profit margins are one of the problems that all these firms are currently facing. With fierce competition, the local construction firms must to squeeze their margin down further in order to win projects. Such competition is not healthy, and results in shrinking margins and increasing potential operating risks, as well as increasing labour costs. It also shows that workers' salaries in Shenzhen have gone up by 8.9 %, compared to the previous year (Wang 2012).

Figure 6.2 shows the difference in the ratio of profit to gross output (a measurement of profitability commonly used in the Chinese construction industry) generated in the manufacturing and construction industry in China.

The data revealed that the profitability level in the Chinese construction industry is lower than in the Chinese manufacturing industry. The gaps between the two trends in Fig. 6.2 are notable. Although the ratio of profits to output value has increased gradually since 2000, up until the slight decline in 2008, the ratio is lower than the benchmarked profit rates suggested by the government of 3–6 % (Xiong 2007). According to Li (2012),[1] the profitability of SOEs in Chinese construction industry is only around 2.5 %, which seems low, but is nevertheless much higher than that of regional private builders, whose profit margin on projects is usually less than 1 %. In line with MOHURD (2009a), Fig. 6.2 infers that the overall operational efficiency of the construction sector is still low. This seems to reflect the fact that budgets in the construction business were being pushed down, and company's profit margins were further squeezed. These added difficulties to construction business operations. However, Li (2012) also warned that caution should be exercised when viewing a company's financial reports so as to understand the company's operations and financial condition. It can be difficult to obtain accurate information, as project

[1] Deputy Director of the Chinese Construction Association.

profits are split across different stakeholders. Only a small portion would be reported at the company level. Cheah et al. (2007) explained that one of the possible reasons for this was the lack of long-term strategies for survival and growth in the Chinese construction industry. Hence, under such circumstances, it may appear difficult for firms to adopt lean principles because the tangible benefits cannot be reaped so readily in the short term.

6.3.4 Technology Development

Xu et al. (2005) have noted that Chinese construction firms have for a long time neglected the use of advanced technology and new equipment. Inevitably, much work undertaken on construction sites is manual, rather than machinery-based, while advanced equipment is only used for large or major construction projects (Xu et al. 2005). Since the 1990s, there has been the development of technology improvement in China's construction industry, as measured by value of machine per person (see Fig. 6.3). In 1991, the value of machines per person was only 2,527 RMB/person. In 2003, this amounted to 9,957 RMB/person, an increase of nearly four times. However, this metric has declined from 2003 to 2006, and saw a slight improvement during 2008–2009. Although the data in Fig. 6.3 points to an increasing trend in terms of the value of machinery and equipment over the past decade, the efficiency and effectiveness of construction equipment is less utilized in China and appears to be a major factor contributing to the difference in productivity between China and the USA (Shen et al. 2011).

Zheng Yi-jun, President of the China Construction Association (Zheng 2008) noted that China was still lagging behind the technologically advanced countries in terms of the construction technology they used. Some areas identified as being particularly weak in China were: (1) technology innovation, (2) level of technology and equipment in use, (3) construction mechanization and (4) prefabrication technology. At present, the level of construction technology and equipment in use in China is approximately a quarter less than those in developed countries; a considerable number of companies are still largely relying on manual labour and on-site operations (Zheng 2008). For example, the strength of steel and concrete used in China is 1–2 grades lower than those in developed countries and the consumption of steel per m^2 floor area is 10–25 % greater than that in developed countries.

Nevertheless, at the company or project level, there are arguably several leading construction firms in China have recognized the role of technology management in their construction business, with incorporating the concept of "self-design" and "self-development" (Li-Hua and Khalil 2006). For example, Shanghai Construction Group (SCG)'s strong IT and R&D capability have helped SCG achieved high performance in a wide range of projects such as high-rise intelligent buildings (Cheah and Chew 2005). In addition, according to the National Audit Office of China (NAOC 2012), the Beijing-Shanghai High-Speed Railway Project has obtained encouraging results in technological improvement in several key areas, such as deep-water long-span bridge construction technology, foundation

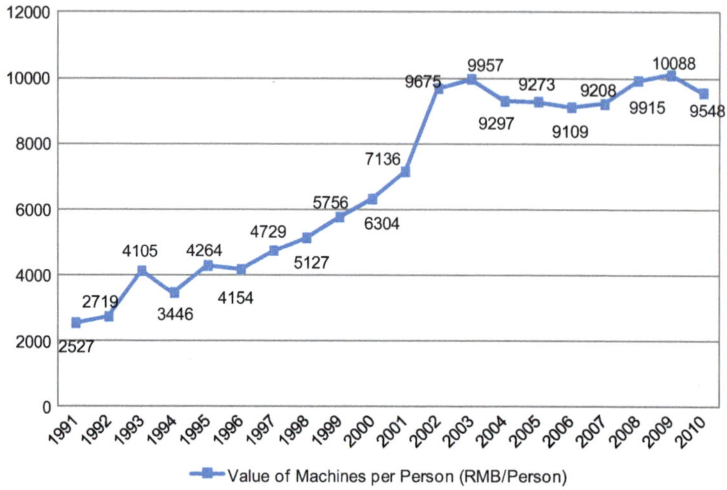

Fig. 6.3 Value of machines per person from 1991 to 2010 in China's construction industry [*Source*: adapted from NBSC (2011)]

settlement control technology for deep spongy soil and so on. These technologies were introduced into project practices and increased the efficiency of construction. However, this is rare and does not represent the whole industry. At present, these large Chinese construction firms are more enthusiastic to use cutting-edge technologies, given they are essential to support the competitiveness of companies (Li-Hua and Khalil 2006). Toyota's philosophy positively offers certain insights for Chinese construction firms on how they can adopt suitable technology.

6.3.5 Project Management

The management systems in the Chinese construction industry have changed towards a commercial approach (Chen et al. 2009). One example is that competitive bidding has become the common practice for Chinese construction firms to obtain contracts (Wang et al. 2006). Wang et al., (2006) also noted that a majority of Chinese construction firms have changed from satisfying only the state target to prioritizing firms' commercial objectives, such as profit maximisation. Chen et al. (2009) added that one major reform was the introduction of western project management concepts and practices. In the past, the World Bank was a key facilitator in this process by its introduction of competitive bidding and international contractors for the first time in the Chinese construction industry, in one of the Bank's early projects, Lubuge (鲁布革) (Chen et al. 2009). This project was investigated later in great detail because it was completed to a high standard of quality and with the project schedule shortened by 5 months. Thus the "Lubuge impact" triggered a top-down approach to reforming the management systems in the Chinese construction industry (Chen et al. 2009). Chen et al. (2009, p.1017) commented that "*Project management concepts and practices, after being piloted and proved as effective, and have been widely pursued in the Chinese construction industry*".

However, Chen et al. (2009) argue that research on project management in the Chinese construction industry has usually been inadequate and single-faceted. They hence examined the project management practices in China's construction industry from a holistic perspective, and noted that the Chinese construction organizations have put and are still putting efforts in to improve their project management skills and capabilities. Observations such as "lack of project management skills in China" have repeatedly been highlighted in the book "Building Modern China", which presented 21 extraordinary individuals' views of the state of the Chinese construction industry (CIOB 2009).

6.4 The Need for Change

The literature reviewed in the previous chapters has reported that the application of lean principles in construction works is relevant in China, although it may be in its infancy stage. Based on the evidence presented thus far, lean practices can help construction firms across the globe to address competitiveness issues such as quality improvement, improved productivity, cost reduction and so on (see Sect. 3.4.5). To remain competitive, several world-leading construction firms have already embarked on lean construction and shared the benefits claimed on their websites. These include Turner Construction Company (4th in the ENR ranking), whose project team working at a project in Maryland, USA, used lean construction methods to reduce waste and rework and to improve operation efficiency (Turner Construction Company 2012). Another large construction company, Erhardt Construction (2012), based in the USA has highlighted that "through Lean's reliable work-flow, projects are delivered better, faster, and at a lower cost to our clients". In Europe, one high profile contractor, Skanska, (7th in the ENR ranking) reported on a residential project (a block of 18 apartments) in Finland which used lean construction methods such as standardized prefabricated elements and Building Information Modeling (BIM) and was built in 6 months—a substantial time reduction (Skanska 2009). Moreover, Thomassen et al. (2003) studied the largest contacting firm (MT Højgaard) in Denmark, which has been implementing lean construction for years, aiming to create competitive advantage. Thomassen et al. (2003) discovered that lean construction projects undertaken by MT Højgaard enjoyed 25 % more profits compared with non-lean construction projects.

Although China's construction industry over the past three decades has achieved much progress in areas of quality, productivity, technology, overseas market shares and so on, there is room for further improvement. The disturbing trends that exist and discussed earlier have driven large Chinese contractors to adopt initiatives to enhance their competitiveness and change for the better. Lu et al. (2008) investigated the critical success factors for the competitiveness of contractors in the context of the Chinese construction industry and found that the sustainable

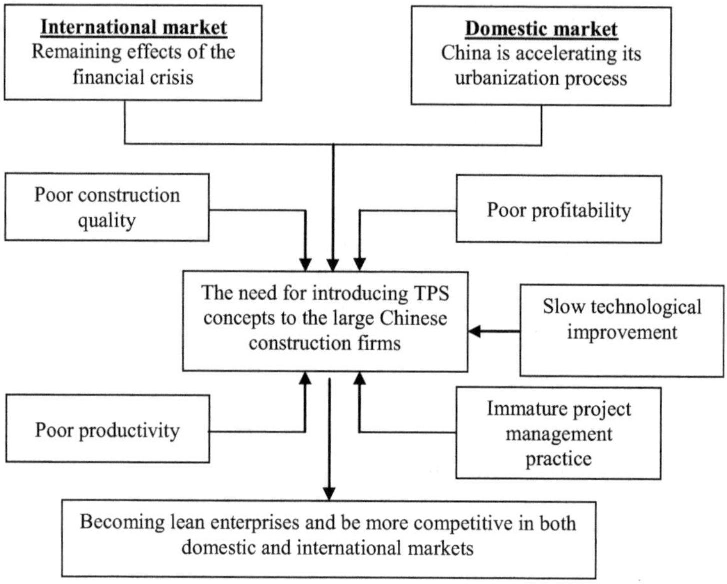

Fig. 6.4 Driving forces behind the need for introducing lean principles in Chinese construction firms

development of human resources, quality management, and site management are among the most highly ranked success factors. The Toyota Way principles are closely relevant to these areas, and are thus considered to be important factors that need to be addressed. Figure 6.4 highlights the forces behind the potential for implementing lean principles in the Chinese construction industry.

6.5 Lean Implementation in China

6.5.1 Introduction

Lean implementation has been exercised in China for more than a decade, but the practice can only be seen in manufacturing joint venture organizations, as well as their ancillary companies in the supply chain (Zhao 2007). Since China became a world-class manufacturing hub, foreign supply chain executives see the opportunities that opening a manufacturing facility in China would bring with it, including lower labour costs and increased profit margins. However, they also understand that certain cultural practice can make a lean initiative more difficult to implement. In recent years, the state-owned and private companies have gradually started to accept the lean approach. So far, according to Zhao (2007), most companies have only focused on the application of lean tools and very few have fully started the whole lean enterprise transformation. Chinese firms held two different attitudes towards lean (Zhao 2007):

1. Many companies claimed they are lean companies only because they have already implemented 5-S activities, or *kanban* activities or etc., but they have failed to appreciate the interrelationships between many other tools. Practicing only one lean tool does not allow the desired results to be fully reaped.
2. Some companies thought that they completed their lean transformation years ago. It is a misconception that these companies failed to acknowledge that TPS/lean is a continuous improvement process of pursuing perfection.

Morris and Lancaster (2005) highlight that management ideas, during their transferability from manufacturing to other sectors, have to be adapted sufficiently to local conditions, which otherwise may generate barriers. Much work has focused on the implementation models of TPS/lean from the manufacturing sector to the construction sector. To generate a better understanding of the TPS/lean framework in the China context, this section will examine the barriers during the transferability process of TPS/lean in China. Paolini et al. (2005) observed that the challenges for lean implementation in Chinese manufacturing firms come mainly from: (1) cultural differences, (2) workforce challenges and (3) supplier challenges.

6.5.2 National Policy in the Chinese Construction Industry

In developed nations such as the UK, Singapore and Australia, there exists "master plans" for the improvement of the construction industry. An examination of several key reports from the UK (Egan 1998), Singapore (Construction 21 Steering Committee 1999) and Australia (Department of Industry, Science and Resources 1999) highlights that the adoption of lean construction or other relevant tools and techniques is one of the initiatives that these countries have recommend for future development to tackle the problems identified in their own construction industries. The Chinese construction industry lacks such a national strategy or policy to highlight the importance of lean and to recommend it for nationwide implementation. MOHURD (2010b), however, has mentioned that Chinese construction firms need to enhance their management efficiency. In MOHURD's (2010b) report, lean management was mentioned for the first time. MOHURD (2010b, pp. 74–75) did not elaborate what lean management or its tools and techniques are in-depth, but did report how lean management can be implemented in four cases. These are:

1. Company A restructured its organizational structure and project structure in a lean way. Efforts were made to reduce project costs by centralizing the purchase of key materials (63.2 billion RMB) and equipment (2.45 billion RMB).
2. Company B revamped its organizational structure, optimized its construction methods, and layout plans, along with the resource plan. In company B, centralized purchasing was adopted, and project finance and budget were closely monitored by a new unit.
3. Company C reconstructed its organizational structures in a way with multiple centres (a technical centre, a finance centre and a procurement centre).

4. Company D published a series of internal documents, such as "enterprises guidelines", "operation manual", "company/project procedures", etc., aiming at establishing norms for company governance and process standardization.

It seems that their attempts to implement lean were focused on organizational restructuring, centralized purchasing, management standardization and so on. These endeavours are indeed part of lean implementation in the Chinese context, but are far from the Toyota Way model, which is more comprehensive and prescriptive. Moreover, it appears that these changes are meant to occur at the company level, while the chief target of lean is to take the non-value-adding activities out of the process, which implies a connection with shop-floor process improvement. The Toyota Way model has one layer dealing with lean tools which can be utilized to improve performance at either the operation or project level, which has a place in the Chinese construction industry.

6.5.3 Chinese Cultural Influence on Lean Implementation

Many researchers have investigated the cultural constraints of lean implementation (Atkinson 2010; Lewis 2000; Paolini et al. 2005; Sim and Rogers 2009). Lewis (2000) noted that lean implementation is more than technical adjustment but is a culture change. Atkinson (2010) argued that lean is all about cultural issues. Given that the Toyota Way has strong cultural roots in Japanese management (Liker 2004; Low and Gao 2011a; Marksberry 2011), this section briefly discusses the cultural influence on lean implementation in the Chinese context. It should be noted that this discussion is in the context of scant literature available on the relationship between Chinese cultural influence and lean implementation.

As far as culture is concerned, it is commonly agreed that the concept has many definitions. At the national level, Hofstede and Hofstede (2005) categorized national culture into power distance, individualism vs. collectivism, masculinity vs. femininity, uncertainty avoidance, and long vs. short-term orientation. Moreover, Pun (2001), based on the research work done by Hofstede and Bond (1988) and Martinsons (1994), listed a number of cultural elements that distinguished Chinese culture from the western culture. This includes intuitive, holistic thinking, family-oriented, high power distance, collectivism, implicit communication, relationship-oriented expression, personal trust and top-down information system. Moreover, Burrill and Ledolter (1999, p.298) regard culture as "*a set of values and patterns of behaviour that focus on customers, quality, and individuals of the organization*". Discussions on organizational culture seem to be more appropriate as lean initiatives are usually undertaken at the organization level, where changes are expected to occur away from those of traditional management. The research undertaken by Paolini et al. (2005), one of the few, cautioned that the cultural difference must be acknowledged and managed for lean concepts to be successful in China. Three cultural elements were discussed in Paolini et al.'s (2005) study, including (1) concept of *guanxi*, the Chinese practice of building relationships, (2) perceptions of personal empowerment and (3) a tolerance for an untidy or disorganized workplace. These three cultural elements are relevant to workforce characteristics, which could be shaped by organizational culture. It is worth mentioning that these workforce behavioural features are viewed as barriers to lean implementation in China (see Paolini et al. 2005; Aoki 2008; Taj 2005). For example, at Toyota, operators are encouraged to expose problems as much as possible for continuous improvement, and where stopping the production line is often seen to occur. In a similar situation with *guanxi*, managers may adopt a quick solution that does not address the problem's root cause.

In addition, Ling et al. (2007) examined how foreign architectural, engineering and construction (AEC) professionals manage cross-cultural encounters in China. Several cultural elements examined by Ling et al. (2007) bear implications for lean implementation in China's construction industry, including:

- Lack of team spirit: the Toyota Way approach places an emphasis on teams and team-working (Liker 2004). If the Chinese building staff are calculative, uncooperative, holding their own views, and reluctant to implement what the team had decided (Ling et al. 2007), it is difficult to bring them together, to work together, for a common objective.
- Procedure driven and obey many rules and regulations: the standard work at Toyota not only demands employees to follow the SOP closely but also encourages employees to be innovative and to improve the existing SOP with *kaizen* thinking (Liker 2004). If the Chinese culture at the workplace promotes obedience to rules and regulations without continuous improvement, it may slow down the adoption of lean practices.
- Lack of initiatives in solving problems and unwilling to take additional responsibilities: in the context of lean, employees are good problem-solvers and they see problems as opportunities. Given that uncertainty (i.e. risks and problems) avoidance is a feature of Chinese culture (Fan 2000; Singh et al. 2003), the Chinese workers are unlikely to take the lead to try to solve the problems, as they are uncomfortable with problems or unstructured situations (Zuo et al. 2009). Instead, they will wait for the supervisor to discover and then act. If the supervisor has other priorities, nothing gets done. Moreover, the culture of "avoid taking additional responsibilities" would hinder the practice of *genchi genbutsu*, as it requires a high level of commitment and responsibility.
- Culture of distrust: lean workplaces usually show higher level of trust (Olivella et al. 2008; Forza 1996). Companies, like Toyota, trust their employees' capabilities, and thus treat them as valuable assets. Ang and Ofori's (2001) study found that trust is one of the most important cultural values of the Chinese. However, according to Ling et al. (2007) and Zuo et al. (2009), project managers in China do not fully enjoy a high level of trust by the stakeholders whom they were engaged. Thus, this seems to suggest that lean is more likely to fail in these Chinese construction firms, particularly in the area of empowerment, collaborative planning and partnership where mutual trust is required.
- Strong networking culture: the Chinese culture values networking and appreciates long-term partnership (Wang and Huang 2006; Zuo et al. 2009) in construction business. This cultural element is also a key facilitator to lean implementation in the area of partnership (Liker 2004; Liker and Meier 2006).

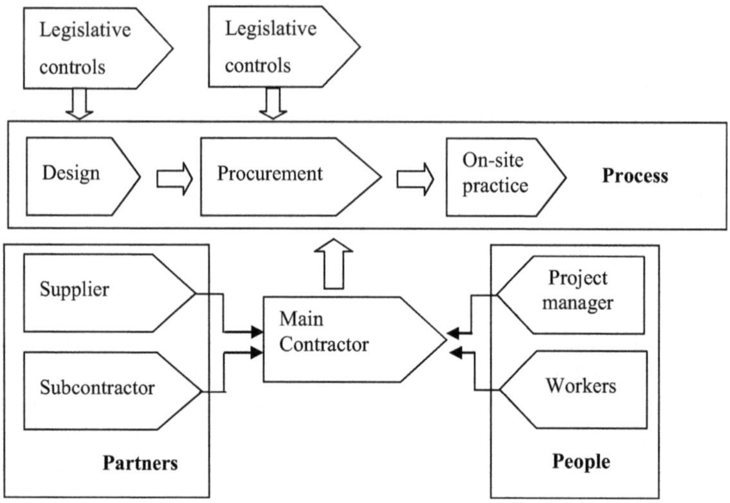

Fig. 6.5 Integrating the process, people, and partners in China's construction industry

6.6 Challenges for Lean Implementation in China's Construction Industry

This section describes Fig. 6.5 to examine the potential challenges faced in the implementation of lean in China's construction industry.

6.6.1 Design

6.6.1.1 Design with Low Buildability

China has a very well-established system of design institutes (Low and Jiang 2003), with a large number of professionals who are equipped with the necessary qualifications and know-how relating to construction technology, economics and management (Xu et al. 2005). Traditionally, as Bajaj and Zhang (2003) observe, China's construction industry was influenced by the planned economy system, with design institutes and construction firms in most cases allocated different tasks. The very limited interaction between the designers and the contractors, as well as the designers placing more emphasis on building appearances, have resulted in the low level of buildability in China (Liu and Low 2007). Liu and Low (2007) suggested that the Chinese construction industry could learn from Singapore's experience of incorporating buildability into the designs, processes, construction techniques, products, and materials, in the hope of enhancing efficiency and standardization in the construction industry. As presented earlier, standardization is one of the crucial principles of the Toyota Way. Encouraging appropriate repetition and standardization in the Chinese construction industry will therefore help to lay the foundation for a lean environment.

6.6.2 Procurement

6.6.2.1 Limited Use of Design and Build Procurement Mode

Ballard and Howell (1998b) argue that the traditional design-bid-build system parallels mass production's wasteful sequential method, making it virtually impossible to achieve improvement and to avoid suboptimization. Currently, the most widely adopted procurement system in the Chinese construction industry is the design-bid-build approach (Chen, et al. 2010). It was implemented by the Ministry of Construction (MOC), and means that domestic design institutes and contractors are responsible only for their own work. Hence they fail to fully foster collaboration prior to the completion of the drawings (Liu and Low 2007). According to Xu and Greenwood (2006), the MOC recommended the "Conditions of Contract for Works of Building Construction" as a template for competitive (design-bid-build) tendering. This form of contract was prevalent because it can be modified to suit individual projects (Xu and Greenwood 2006), since China does not yet have standard conditions for design-and-build contracts. Only few construction companies have adopted this mode. A survey by the MOHURD (2008) to investigate the use of design and build (D&B) procurement in China showed that only 18 out of 37 large construction firms surveyed were experienced in D&B. In addition, the number of small and medium-sized enterprises (SMEs) experienced in D&B was far less than the proportion of the large firms. Again, the limited use of D&B procurement in China has directly discouraged buildability. Thus, the Chinese construction industry seems to be far from an ideal situation for applying lean techniques, as the traditional design-bid-build procurement form does not facilitate lean approaches to project planning and execution. This will ultimately and indirectly hinder the practice of standardization on site as well.

6.6.3 On-site Practices

6.6.3.1 Poor Material Management and Construction Waste

China is an intensive user of raw materials. Chen (1998) highlighted that China' s construction industry consumed 20–30 % of the country's total steel production, 70 % of cement, 40 % of timber, 70 % of glass, 50 % of paint and 25 % of plastic products each year. It was believed that the increasing demand for the raw materials was due to China's high demand for infrastructure construction in recent years. According to Deputy Minister of Construction Qiu Baoxin,[2] China has the most new buildings in the world in each year; the newly constructed area in each year is about two billion square metres, equivalent to 40 % of the world consumption of cement and steer. However, the construction lasts only 25–30 years on average

[2] Qiu Baoxing is Deputy Minister of Housing and Urban-Rural Development. The quotation was part of his speech delivered at the 6th International Green Building and Energy Conservation General Assembly.

(China Daily 2010b). In addition, rising oil prices, raw material prices, fuel prices and transportation costs will inevitably and continually have a direct impact on the building materials industry (MOHURD 2008). To take advantage of the discounts on prices, Chinese contractors usually order large amounts of materials and place early purchase orders from their suppliers. This was identified as a normal practice in developing countries by Polat and Arditi (2005). Moreover, poor material management in Chinese construction sites appears to be a serious problem, leading to the following undesirable consequences:

1. Bulk materials are often stored on site, taking up valuable space in often confined sites. Stored materials are also susceptible to damage.
2. Poor estimation and know-how in fabricating materials (e.g. cutting reinforcement bars) on site, thus also causing a large amount of waste and air pollution.
3. Poor material procurement schedule and congested traffic conditions, especially in large cities, usually lead to frequent delays. Suppliers' inability to deliver materials on time has been identified by Zou et al. (2007) as one of the risks in China's construction industry. This is usually caused by the unskilled construction workers and poor project management skills.

6.6.3.2 Limited use of Prefabricated Components

According to Lu (2002), the prefabrication industry in China experienced rapid development between 1970 and 1990, before falling into a period of recession. Tens of thousands of prefabrication plants of different sizes were set up in urban and rural areas during the prefabrication boom period, more than 90 % of which were small-sized rural companies. Lu (2002) identified the deterrent factors that set back the precast concrete industry below:

1. The price mechanism of the prefabricated components in the planned economy largely restricted the development of companies in the area of technological innovations.
2. Low level of product quality standards.
3. The industry placed undue emphasis on saving construction materials as a strategy in manufacturing prefabricated components in the 1970s, which resulted in poor quality products that jeopardized the industry's reputation. Moreover, the existing bias misled the industry, suggesting that traditional site in-situ construction can provide a better solution than prefabrication. It seems that only in-situ concrete technology was considered a modern technology at that time. Consequently, this has badly affected the development of the prefabrication industry in China.

Recently, The China Daily (2010b) reported that a six-storey energy efficient Expo pavilion was built in less than a day during the Shanghai Expo construction period. The Chinese construction industry has begun to recognize the potentials of prefabrication and to acknowledge that this may become a prevalent approach for the construction of future residential communities, offices and hotels (China Daily 2010b).

6.6.4 People and Partners

6.6.4.1 Poor Labour Skills and Insufficient Training (Workers)

It is important to note that the Chinese construction industry is a highly labour-intensive industry, and a large number of construction workers are peasants and unemployed workers. Most of the people at the worker level are unskilled and many are not round-the-year construction workers (Chan et al. 1999). The construction workers can be recruited easily as a result of the relatively low requirements for skills in construction and massive urban development needs in China. It is common to see many construction workers beginning work without any professional and/or vocational training, thus leading to problems. On average, the educational level of the construction employees is rather low. According to Xu et al. (2005), 97 % of the educational level of the construction employees is below that of a diploma. Lu and Fox (2001) also noted that 50 % of the 600,000 migrant workers working in construction sites in Beijing have received no more than primary education and over 10 % are illiterate. The official statistics from the Ministry of Construction revealed that only 10 % of the 32 million farmers who became construction workers had basic training in their new career, compared with more than 70 % in developed countries (Xinhua 2009b). Ling et al. (2005) noted that the level of professional work was so low that workers required very detailed drawings for them to operate. It is recognized that training is very important in the Toyota Way. Thus poor labour skills and insufficient training will therefore pose a large hurdle for the implementation of TPS/lean concepts.

6.6.4.2 Lack of Project Management Practice (Project Managers)

Researchers have highlighted that project management skills in China are lagging behind those of developed countries. Limited management skills have prevented work from proceeding efficiently. Inadequate project management skills were identified by Zhao and Shen (2008) as the most significant weakness of the Chinese contractors in the international market. About half of the Chinese construction firms have not established an effective project management system (Hu 2003). It has increasingly been recognized that it would be difficult to systematically and effectively manage a project without a sound project management system. Consequently, due to the lag in project management knowledge and techniques, it is likely that there is little awareness of TPS or lean concepts in the Chinese construction industry.

6.6.4.3 Poor Organizational Structure

As Zeng et al. (2003) note, most Chinese construction firms have three or four layers of hierarchy in their organizational structure. There is, however, a vague division of work and economic relationships between these layers. Each level sets its own objectives, liabilities, and targets, and carries out its individual tasks. This perceptibly poor organizational structure results in project managers having to bear the risk in making decisions and solving problems. This also explained why the Chinese manager is poor at taking on responsibilities (Flanagan and Li 1997).

Flanagan and Li (1997, p.154) have also noted that *"They avoid making decisions and prefer to report to their supervisors to get approval. None seems to want to take the risk and responsibility and too many people are involved in decision making"*.

Chan et al. (1999) also noticed that the Chinese construction managers lacked the motivation to carry out the work in the most cost-effective way because they were not given enough authority to make decisions and were not responsible for profits and losses. Furthermore, Chen and Partington (2004) compared the cultural differences between Chinese and Western project managers' way of handling construction projects and highlighted that management processes in China have been heavily influenced by the relationship culture, which emphasized hierarchy and the need to maintain harmony, as well as valuing long-term cooperation for mutual benefit. Nevertheless, it was observed that the construction industry in China provides little or no opportunity to construction practitioners, particularly the lower level workers, for personal enhancement, which in turn badly affects workers' commitment.

6.6.4.4 Supplier Relationships

The bargaining power of local building material suppliers in China has gradually been eroded over the last decade (Lan and Jackson 2002). The risk relating to suppliers with low bargaining power is that they have to bear the extra costs of delivering the materials in small quantities (Harber et al. 1990) if the JIT delivery system is to be implemented. Lan and Jackson (2002) explained the reasons for this:

1. Conventional building materials were in excess supply due to deregulation. This ensures the supply of materials within a short time, so loss due to late arrival of goods at a job location is a low risk event (Fang et al. 2004).
2. Low concentration of production in building materials manufacturing. There were more than 200,000 building material suppliers throughout China in the mid-1990s, with no key player in the industry. In most cases, long-term partnership between a single source supplier and the main contractors is rarely seen.
3. *Guan xi* (relationship) is still a critical factor in determining who the suppliers are.

Moreover, Lu and Yan (2007) highlight that the current strategic partnering applications in China are only observed in the tender preparation stage. Further applications at the project level, for instance the strategic relationship between contractors and suppliers, are rarely seen because most Chinese construction companies do not understand formal partnering approaches clearly, and cannot perceive the project-based benefits of partnering. Therefore, establishing long-term relationships with suppliers should be encouraged in China in order to reduce risks, such as those relating to "poor quality materials".

6.6.5 Legislative Controls

Another potential obstacle to the successful implementation of TPS/lean concepts is the intervention of regulatory authorities and the tedious approval process required in the construction industry (Koskela 1992, p. 48). In China, intervention by governmental authorities exists throughout the whole construction process, which imposes difficulties in managing construction projects (Chan et al. 1999). The risk of government intervention in construction was ranked very high by Fang et al. (2004) from the Chinese contractors' perspective. For example, in the pre-construction period, clients are required to submit very detailed business and economic information of their projects to the government for feasibility assessment. In the tendering stage, because the assessment of a tender has to be carried out with the joint efforts of the government tender administration department and the client, the contract may not be awarded to a competent contractor who has worked with the client for a long time. Moreover, during the construction phase, it is compulsory to engage the government quality inspection office to monitor the major construction activities, which thus duplicates the works of the supervision unit (Chan et al. 1999). These stringent requirements demand approvals from many governmental bureaus (e.g. Construction Bureau, Fire Bureau, Environmental Bureau, etc.) and are a result of the multi-layered and fragmented nature of the governance structure in the Chinese construction industry (Chan et al. 1999; Cheah and Chew 2005). For most construction and design work, this will potentially delay the commencement of construction works. The associated inefficiency would have a knock-on effect on hampering a construction firm's planning and coordinating with other stakeholders (e.g. suppliers and subcontractors). The complex and time-consuming processes to obtain building approvals in China therefore require further revamping and enhancement.

6.7 Summary

The previous chapters have shown that construction companies have adopted the lean production philosophy (lean construction) or the Toyota Way principles in order to reap benefits similar to those that many manufacturing firms have reaped earlier. It appears that an increasing number of studies have been conducted on the transferability of this operational system beyond manufacturing to a non-manufacturing sector. Nevertheless, little has been studied on its applicability to the Chinese construction industry. This current study addresses that gap. In this chapter, after examining the status quo of the Chinese construction industry, the study shows that the Toyota Way principles have the potential to improve the performance in the following five aspects: quality, health and safety, productivity, profitability and management. Furthermore, this chapter follows the road map of the process, involves people and the partner's chain in China's construction industry and identifies a number of challenges that may potentially impinge on the promotion of lean practices in construction projects.

Research Methodology 7

7.1 Introduction

This chapter explains the choice of research methodology for this study. It begins with an overview of the general approach and methods used in the research. As the focus of the research is on investigating Toyota Way-styled practices within large Chinese construction firms, the methods commonly applied to lean construction studies are discussed. The choice of the mixed research method is justified as an appropriate research strategy. The determination of the study sample and the techniques of data collection are also described.

7.2 Fundamental Concepts

According to Lincoln and Guba (2000), research is always constructed on the basis of people's ontological, epistemological, and axiological beliefs about the world. Klenke (2008) noted that the researcher's philosophical assumptions about these three aspects are critical in framing the research process, and require transparency.

Ontological belief concerns the nature of knowledge or the nature of reality (Lincoln and Guba 2000; Fellows and Liu 2008). It is about whether the research views the world from an objective or a subjective perspective. On the ontological level, this research has adopted a realist position, as implementation of lean and Toyota Way may exist in China, but practitioners sometimes do not recognize or follow a structured procedure or processes that would make the implementation effective. The ontological stance of this study is to explore whether Toyota Way-styled principles have been implemented within LCCFs.

On the other hand, epistemology concerns the question of what is (or what should be) regarded as acceptable knowledge in a discipline (Bryman 2004). It is about "how we know", and the methods through which knowledge is acquired (Klenke 2008). Klenke (2008) emphasized the importance for every researcher of bringing some set of epistemological assumptions into the research process, and

DOI 10.1007/978-981-287-014-8_7,

that these assumptions influence how the data are understood and interpreted. Epistemologically, there are two ways of viewing the world—positivism or interpretivism. The former advocates the application of the method of natural science to the study of social reality and more. It is of the belief that the world conforms to fixed laws of cause and effect, and that complex issues can be tackled using simplified or fundamental approach. It is therefore possible for the researcher to be objective from the detached position of the research situation. This research holds to the positivistic position in the sense that it identifies the relationship between the level of Toyota Way implementation and the performance of various projects. Interpretivism advocates the absence of a universal truth and places more emphasis on understanding the meaning of actions from actors' perspectives. This research also maintains that the Toyota Way model, as a comprehensive and complex model, can be proposed as a way for implementation. In this research, inquiries are made to consider the meaning and possible implementation opportunity for each Toyota Way principle within large Chinese construction firms.

Axiology deals with the question of what is valuable in research. According to Klenke (2008), values are part of the "basic beliefs" that undergird and affect the entire research process: research problems, guiding paradigm, framework, data collection methods, analysis strategy, and others. It is worth mentioning that values play a significant role in the study of lean or Toyota Way, especially as evidenced, for example, in the fact that the primary aim of lean is to maximize value to the customer. Yet the principles of Toyota Way reflect Toyota's endeavours, not only in managing production well by adding value to the process, but also by adding value to people, the company, and society at large.

7.3 Research Design, Approaches and Methods

Kumar (2005) describes a research design as a procedural plan that is adopted by the researcher to answer questions validly, objectively, accurately and economically. Yin (2008, p.26) sees research design as a "blueprint" for researchers. According to Kumar (2005), the following two objectives need to be fulfilled in a research design:

1. To conceptualize an operational plan and to undertake various procedures and tasks required to complete the study.
2. To ensure that these procedures are adequate to obtain answers to the research questions.

Research strategies can be broadly categorized as either quantitative or qualitative. Quantitative research is objective in nature (Johnson and Harris 2002; White 2000). It usually requires respondents to record their attitudes, opinions, or beliefs on a five- or seven-point scale measured with numbers. The scores are then analysed using statistical procedures to test the hypothesis (Creswell 1994). Three main approaches were suggested to collect the data: asking questions of respondents by means of questionnaires and interviews, undertaking experiments, and performing extensive reviews of the relevant literature (Johnson and Harris

2002; Fellows and Liu 2008). On the other hand, qualitative research is subjective in nature (Naoum 2007). It relies on observing people in their own environment, communicating with them in their own language, and on their terms, with an equal relationship between the researcher and the participants. Hence, the data are gathered primarily in the form of words and observations, as opposed to numbers, and are then analysed to discover the unifying concepts and patterns that give meaning to the data (Johnson and Harris 2002; White 2000). Fieldwork and case studies are the major types of qualitative study. In addition, a research study using both qualitative and quantitative approaches can be called a "mixed-methods" approach or "methodological triangulation" (Creswell 2003). The mixed method is based on the premise that an effective body of research on a topic should include more than one research approach.

7.4 Research Methodology Employed in This Study

7.4.1 Selected Research Approach

The nature of each objective (see Chap. 1) implies that one overarching approach would not be appropriate for this research. This research concerns both qualitative and quantitative characteristics, which are explained below:

1. Quantitative aspects of this research:
   - Objective 2 is concerned with the implementation of Toyota Way principles (identified as measurables) within LCCFs, as well as the perceived importance of the attributes to the company.
   - The potential barriers to the implementation of Toyota Way principles by Chinese construction firms are inherently relevant to a number of quantitative variables, which were reviewed and identified in the literature and pilot study.
2. Quantitative aspects of this research:
   - Investigation into the status quo of LCCFs' site management practice, HRM, problem-solving behaviours, and other similarities and differences (gaps) with the Toyota Way standard. These are featured with qualitative elements. Interviews and case studies appear to be more appropriate for capturing these relevant information.
   - Clarification of how the implementation of Toyota Way principles can be fulfilled by multiple case studies of construction projects. This again depends on an overall understanding of the Toyota Way approach in the whole process of a project.
   - The development of the implementation framework also depends on the understanding, interpretation, and summary of the research findings by the researcher.

Despite the reported growing presence of lean initiatives in construction, there has been limited research conducted into China's construction context addressing the attitudes and actions towards Toyota Way-styled practices. Therefore, mixed-methods research seems to be appropriate and is helpful in gaining a better

Table 7.1 Qualitative and quantitative characteristics of this research

Research objectives	Research process	Qualitative or quantitative
Objective 1: To breakdown the Toyota Way principles into measurable or quantifiable parameters	Preliminary work	Literature review
Objective 2: To investigate the status quo of the Toyota Way principles having been implemented by large Chinese construction firms (LCCFs)	Phase I	Quantitative
Objective 3: To investigate how the Toyota Way principles can be (better) practised by large Chinese construction firms	Phase II	Qualitative
Objective 4: To establish the Toyota Way model implementation framework for Chinese construction firms	Discussion	This can be concluded on the basis of the findings arising from Phases I and II

understanding through an in-depth study of the new phenomenon. As can be seen from the research aims and objectives outlined in the earlier chapter, this research concerns both qualitative and quantitative characteristics, which are explained in Table 7.1.

7.4.2 Selected Research Methods

The approach to data collection should be based primarily on the nature of the investigation and on the type of data and information that are required (Naoum 2007). So that the aim of the study—to establish an implementation framework of the Toyota Way model within the large Chinese construction firms—could be met, a diverse range of methods to collect applicable data were employed. This included questionnaires, semi-structured interviews, and case studies.

7.4.2.1 Questionnaire Survey

The survey method involves collecting information from a larger sample of the target population, which is selected based on systematic and representative sampling methods, by means of a standardized questionnaire administered identically to all the target respondents in the sample population (Creswell 1994). Among the research conducted into the area of investigation—concerning the application of lean production or lean construction methods—questionnaire surveys appear to be a popular method for assessing the level of lean construction implementation (see Salem et al. 2006; Diekmann et al. 2004; Johansen and Walter 2007). With this in mind, in this study the questionnaire survey was also used to obtain information about the implementation of Toyota Way in a wide range of Chinese construction firms.

7.4.2.2 Interviews

In the event that a questionnaire survey might not easily allow the researcher to probe some themes highlighted in response to a certain question on the survey (Fellows and Liu 2008), interviews were employed to supplement the information required. An interview involves questioning a respondent through discourse on a defined theme or subject area to obtain responses aimed at addressing a research hypothesis (Naoum 2007). In this research, interviews were a suitable data collection method for answering the question of "how" (see Table 7.1) in relation to the development and/or fine-tuning of the Toyota Way principles, in order to guide Chinese construction firms in implementing the lean concept.

7.4.2.3 Case Studies

The next level of this research employs case study analyses. Yin (1994, p.13) gave a widely accepted definition of a case study as an empirical inquiry that "*investigates a contemporary phenomenon within its real life context, especially where the boundaries between phenomenon and context are not clearly evident*". This definition is highly relevant, because the implementation of the Toyota Way principles in the Chinese construction industry is indeed relatively new. In contrast to the questionnaire survey with its rigid limits, a case study can lead to new and creative insights, the development of new theories, and can have high validity with the practitioners who are the ultimate users of the research findings (Voss et al. 2002).

The objective of case study analysis is twofold. First of all, one of the hypotheses proposed that the Toyota Way principles are applicable in China's construction industry. The evidence used to test this hypothesis in the questionnaire survey may be inconclusive. The projected data will only reveal to *what* extent the Toyota Way principles are applicable to LCCFs, but they do not tell us how. Yin (1994) recommended that for research projects of this nature focusing on "what, why, and how" questions, the case study approach is ideal. The case study, in contrast to the questionnaire, allows the researcher to uncover how in daily practice the Toyota Way principles can affect the work of construction firms. Overall in this research, the case studies aim to give an in-depth understanding of Toyota Way implementation in practice.

7.5 The Research Framework

Figure 7.1 outlines the research steps taken in this study. The structure of this research has three stages: Literature review (planning), two-phased data collection/analysis, and validation.

7.5.1 Survey Sampling

The *population* of a research is defined as the units belonging to the category of interest (Creswell 1994). In this study, the population of interest is LCCFs. Using

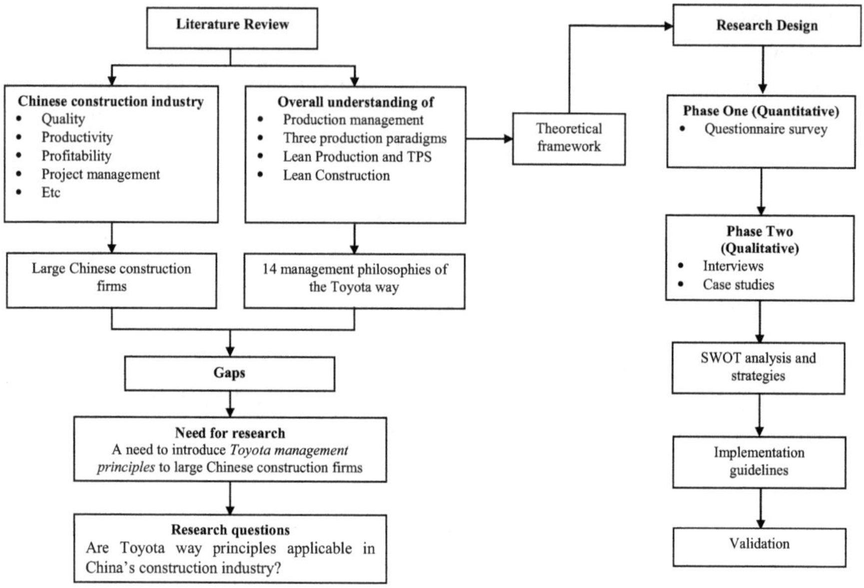

Fig. 7.1 General structure of the research design

two indicators, namely "total revenue" and "value of assets owned", China's National Bureau of Statistics (NBSC 2003) categorized Chinese construction firms into "large", "medium", "small", and "micro" (see Table 7.2).

Referring to classification, all the "premier" and "first" grade construction firms fall in the large category, and thus form the population of this study. The reasons that the study is focused on these large Chinese construction firms are:

1. They play the most important role in China's construction industry, and thus represent a typical business model in terms of management style, site management practices, HRM, and others in China.
2. Some of the leading Chinese construction firms have gained a reputation on the international stage. The Toyota Way model, which is highly promoted in this study, can serve as a ready strategy for them to enhance their competitiveness in the international construction market.
3. LCCFs have abundant resources, which increase their likelihood of embark on the lean/Toyota Way: the implementation of some principles may require investment and managerial effort, which Chinese SMEs may not be capable of doing presently.

7.5.2 Data Collection

It has been acknowledged that it is difficult to achieve a satisfactory response rate for a questionnaire survey in China's construction industry (Liu et al. 2004;

Table 7.2 Classification of Chinese construction firms in terms of size

Indicators	Large ('10,000s RMB)	Medium ('10,000s RMB)	Small ('10,000s RMB)
Total revenue (*R*)	$R \geq 80{,}000$	$6{,}000 \leq R < 80{,}000$	$3{,}000 \leq R < 6{,}000$
Value of assets owned (VA)	$VA \geq 80{,}000$	$5{,}000 \leq VA < 80{,}000$	$300 \leq VA < 5{,}000$

Note: 10,000 RMB = US$1,585. $R \geq 125$ million (in US$) is for large-scale firm
Source: NBSC (2003)

Kang 2006). To avoid the constraints regarding direct access to construction companies in China, questionnaires were actually sent out by the researcher's personal contact in China. Basically, the sample frame for this study is the list of LCCFs (of "premier" and "first" grade) registered with two organizations, namely the Chinese Construction Association (CCA) and Beijing VENCI consulting.[1] In this way, 93 completed questionnaires were obtained from 400 firms contacted, representing a 26 % response rate.

7.5.3 Data Analysis

The questionnaire was designed using a Likert scale, which is ordinal in nature. Concerns over ordinal data are not new and have been the subject of considerable debate in construction-management literature (Fellows and Liu 2008). Several researchers in the domain of construction management suggest that ordinal data can be appropriately analysed with procedures that require rank-order information, such as non-parametric methods (see Fellows and Liu 2008; Naoum 2007). However, in the studies of assessing lean production or lean construction implementation, one of the primary statistical methods employed was based on the average mean value calculated. For example, Salem et al. (2006) used a five-point Likert scale to measure a number of lean construction tools, which were adopted in case projects selected. Panizzolo (1998) assessed the degree of lean production adoption by looking at the percentage of plants that rated them as "adopted" or "fully adopted". Panizzolo (1998) also employed Likert scales, with average mean values to investigate whether each lean programme was pervasive or significant. Doolen and Hacker (2005) used mean values to measure the implementation level of lean practices, though with non-parametric data. Similarly, the use of mean values as the chief form of statistical analysis is also found in assessing other new management philosophies, such as TQM (see Valmohammadi 2011; Yusuf et al. 2007; Zhang et al. 2000) and JIT (see Fullerton and McWatters 2001; White et al. 1999). Hence, in this research, in order to make the statistical analysis more robust and rigorous, mean values is adopted, but only for the purpose of outlining the overall picture of

[1] Beijing VENCI has partnered with the CIOB (Chartered Institute of Building) China to set up the VENCI-CIOB learning centre for Chinese building professionals.

the status quo of large Chinese construction firms in implementing the Toyota Way-styled practices. More information can be referred to the in-depth discussion in Chap. 8.

7.5.4 Phase II: Interviews and Case Studies

7.5.4.1 Interviews

The design of the semi-structured interviews was based mainly on the research objectives, especially for the one concerning "how to better implement the Toyota Way principles within Chinese construction firms". Prior to the interviews, their contents were tested with two Chinese academic experts. Minor alterations were made as a result of this pre-test.

During the fieldwork, the interviews were conducted with a selected range of participants who had agreed to further take part in the study. These consisted of companies' deputy managers and production managers at the company level, as well as project managers, chief engineers, and foremen from the projects. For some companies, site visits were allowed in addition to interviews. This was the best opportunity to confirm the findings from Phase I—whether Toyota Way-styled practices had been actually implemented, and to what extent. The interviewees were chosen by a snowball sampling method. The first point of contact at the company was usually a project manager or a chief engineer. Once the research was explained and the first contact was requested to connect the researcher with people in various roles, they pointed the researcher to colleagues within their firms or counterparts outside the firms. The interviews usually took 1 h or more for each participant. The interviews focused on various aspects of Toyota Way's 4P model. Additionally, the status quo of the construction-management approach was first investigated, and from there suitable Toyota Way practices were introduced. In summary, in each interview, the interviewees were asked a subset of the following questions:

1. How do you implement this particular Toyota Way principle in your current project?
2. What are the gaps between your current practice and this particular Toyota Way-styled practice?
3. What are the main issues that hinder the implementation of this particular Toyota Way principle?

Interviewing a number of Chinese building professionals assisted in understanding the present status of their company philosophy, site management and practices, people management, and problem-solving behaviours. In addition, the results of the interviews also helped to identify the gap between their current practice and the Toyota way principle indicators. In carrying out interviews, advices from the interviewees were also given, which were actually quite helpful in contributing to the Toyota Way model, as they concerned very practical issues.

7.5.4.2 Case Studies: Determination of the Number

For a given set of available resources, Voss et al. (2002) highlight that the greater the number of case studies, the greater the opportunities for in-depth observations. This is because, as they highlighted, two limitations exist with a single case: the first is the limit on the generalization of the conclusions drawn, and the other is the presence of biases, such as misjudging the level of representativeness of a single event or exaggerating the importance of easily available data. Furthermore, it can be anticipated that a single case study will not supply enough evidence to validate all the principles of the Toyota Way and its corresponding approaches in the study. Multiple case studies look at several different construction projects, in order to reach more general conclusions than those provided in a single case. Although multiple cases may reduce the depth of the study when resources are constrained, it can both augment external validity and help guard against observer bias (Voss et al. 2002). Hence, a multiple case study approach appears to be most appropriate for this research, because it offers the following three important advantages (Yin 2006, p.115):

1. It shows the audience that the complete cycle of case study research can be practiced with more than a single case, reducing the suspicion that the researcher's skills are limited to a single case that may also be personally special to the researcher in some way.
2. It better responds to a common criticism of single case studies: that the case in question is somehow unique and idiosyncratic.
3. It gives modestly greater amount of comparative data.

The case study method is prevalent in the context of lean construction research. Santos (1999) employed six cases to investigate the implementation of flow principles in the construction industry. In the USA, Kim (2002) selected eight cases to assess the implementation of lean construction. In this study, the units of analysis chosen are two projects from Company A and one project from company B. Companies A and B are private and state-owned, respectively. This choice was made partly because A and B presented a convenient sample and were available. Moreover, it also represents a diverse choice for the case studies, as companies of different natures are covered. The purpose was to ensure the data gathered enable the sufficient generalizations of the findings.

7.5.4.3 Selection of Case Studies

Creswell (2003) wrote that a case for study needs to be (1) revelatory (when an investigator has an opportunity to observe and analyse a phenomena previously inaccessible to scientific investigation), (2) unique, and (3) critical to testing the theory. In addition, case study choices should provide opportunity to replicate and generalize the study. The selection of the case studies in this study observed the theoretical considerations and practical considerations, as described below:

1. The relationship between the case study firm and the author had already been developed. This assisted in obtaining access to the companies for detailed information.

2. The case study firms should reflect or present the current practice of site management, people management, as well as the problem-solving practices of LCCFs.

The case study research firms are presented in Chap. 10. With respect to data collection, Jankowicz (2000) stated that semi-structured interviews are a powerful data collection technique when used in the context of a case study research method. This is because in organizations such as construction companies, project managers are always very busy on site, due to the pressure of an aggressive schedule; they are unlikely to allow research access to their organizations unless they can see some commercial or personal advantage that can be derived from it. Therefore, this issue should be taken into consideration before an interview technique is selected. A particular type of respondents in case projects is focused on: the high level project leader/manager who generally has the most influence over the degree of adoption of new management philosophy, such as Toyota Way-styled practices.

7.5.5 Validation

The validation of the framework will be conducted by interviewing a number (5–10) of Chinese building professionals in Singapore. The purpose is to see if they agree or disagree with the implementation framework based on the Toyota Way model, as well as with the strategies proposed for improving implementation, especially those that address the need to modify several Toyota Way-styled practices in the Chinese context. It is worth noting that these 5–10 Chinese building professionals will be different people from those who participated in the questionnaire survey phase. The design of the validation effort is the result of professional judgment. With respect to the numbers of participants for the validation process, researchers such as O'Keefe et al. (1986) and Bryman and Bell (2003) suggested that for a group of professionals, less than 10 would be appropriate for validating the results.

7.6 Summary of the Research Process

This chapter justifies and explains the research approaches and methods adopted in this research. Acknowledging the characteristics of the major research approaches and methods, methodological triangulation is designed by incorporating both a qualitative and quantitative approach into a two-stage research process. A questionnaire survey, the most common method for collecting quantitative data from a large sample, was conducted to allow more rigorous testing of the hypotheses in this research. It also provides an overall picture of the extent to which the Toyota Way principles that Chinese construction firms may have adopted.

The methodology adopted in this study includes desk research, questionnaire survey, and case studies. Firstly, literature review is undertaken to obtain an overall understanding of the Toyota Way model, including its background and history, its sub-principles, techniques and tools, the typical approaches to TPS in construction (lean construction), and the challenges faced when applying the Toyota Way principles in construction. Secondly, once the quantitative and qualitative aspects of this study have been discussed, a three-phased research design is proposed. The questionnaire survey and follow-up interviews will be used to explore the current state of the application of the Toyota Way in the Chinese construction industry, as well as the challenges the Toyota Way practice faces in China. Cases studies will be used to answer the question of *how* the Toyota Way model can be applied to the large construction firms in China. Lastly, the framework for implementing the Toyota Way model in the Chinese construction industry will be established based on the three-phase research findings as well as on references from the Toyota Way guidance notes (Liker 2004; Liker and Meier 2006).

Survey Data Analysis 8

8.1 Overview

This chapter is structured according to research objectives two and three, and presents the findings and results of the survey, in particular concerning the extent to which Toyota Way-style practices are implemented by LCCFs. The survey questions are provided in Appendix A. Each of the 14 principles of the Toyota Way is denoted by the abbreviations Principle 1 (P1) to Principle 14 (P14). A distinct scale, containing 5–9 actionable attributes or practices, was included in the survey for each Toyota Way principle.

8.2 Data Analysis Procedures

The collected data were analysed in four stages, using the statistical package SPSS version 17.0. In the first stage, the degree of implementation, as well as perceived importance of each attribute under various Toyota Way principles, was assessed. The assessment was made purely on the basis of the respective mean values. Following this, tests of significance were carried out in order to investigate the difference between the levels of implementation of the Toyota Way and their perceived importance. In the third stage, a correlation analysis was performed to determine the relationship between the firm's (project) performance and the level to which each Toyota Way principle was implemented. In the final stage, the potential hindrances encountered by large Chinese construction firms in adopting the Toyota Way were investigated. It is worth mentioning that a Kolmogorov–Smirnov test was used to determine whether the data were normal. It was found that, for a large majority of the variables, the distributions were significantly different from normal ($p < 0.05$). Hence, it is not possible to conduct a parametric analysis.

S. Gao and S.P. Low, *Lean Construction Management*,
DOI 10.1007/978-981-287-014-8_8, © Springer Science+Business Media Singapore 2014

8.3 Sample Characteristics

A total of 400 copies of the questionnaire were sent to potential respondents in February, 2011. By the end of May 2011, 93 completed copies had been returned, representing a valid response rate of 24 %. Table 8.1 presents the characteristics of the respondents. Among the respondents, a large majority are highly experienced, with 75 % and 33.3 % having more than 10 years and 20 years work experience, respectively. There were 10 missing entries in the work experience, which accounted for the remaining 13.1 %. Moreover, a majority of the respondents were managerial personnel working in Chinese construction firms. This included 34 general (deputy) managers (36.6 %), 23 project managers (24.7 %) and 15 engineers (16.1 %). Respondents from these three groups have a good understanding of construction works, and can thus provide reliable answers to the questionnaire. With respect to the respondents' firms (see Table 8.2), three-quarters of the responding firms were general contractors, and the remaining one-quarter were qualified to conduct business as professional contractors (subcontractors). There were 55 (57.4 %) firms registered in the "Premier" category of firms' qualification, which outnumbered the first-grade (41.7 %) firms. In terms of the ownership of firms, the sample comprised 62 stated-owned firms and 31 private firms.

8.4 Reliability Tests

Reliability refers to the extent to which there is a consistency in responses on repeated application of the same measurement tool (Blythe and Tripodi 1989). Such measures are necessary in order to ensure that the same results will be consistently reproduced in subsequent administrations of the instrument. Cronbach's alpha coefficient is used as the reliability indicator. The higher the coefficient (e.g. 0.8 or 0.9), the stronger the linear relationship of the items is correlated and the higher the internal consistency. Overall, as shown in Table 8.3, the scale reliability is high. All the adopted Toyota Way principles exceeded the usual recommendation of alpha = 0.70 (Black 1999) for establishing the internal consistency of the scale.

8.5 The Toyota Way Principles: Implementation and Perceived Importance

This section presents and briefly discusses the extent to which the Toyota Way principles are adopted; it then assesses their importance for firms or projects from the perspective of the respondents. In addition, the Wilcoxon Signed-Rank Test is performed for each Toyota Way attribute in order to ascertain whether any significant difference exists between the current implementation of Toyota Way-style practices and the level of perceived importance. The practices have been grouped into four different models: philosophical model, process model, people and partner

Table 8.1 Characteristics of the respondents

Description	Number	Percentage
Position (*N* = 93)		
General (deputy) Manager	34	36.6
Project Manager	23	24.7
Engineer	15	16.1
Contract Manager	5	5.4
Quality Manager/technician	4	4.3
Regional Manager	3	3.2
Not stated	9	9.7
Working experience in China (*N* = 93)		
1–5 years	17	18.3
6–10 years	19	20.4
11–15 years	8	8.6
15–20 years	11	11.8
Above 20 years	28	30.1
Not stated	10	10.8

Table 8.2 Profiles of the responding firms

Description	Number	Percentage
Type of firm (*N* = 93)		
General contractor	71	76.4
Professional contractor (subcontractor)	18	19.3
Both	4	4.3
Type of ownership (*N* = 93)		
State-owned	62	64.9
Collective	0	0
Private	31	33
Grades (*N* = 93)		
Premier	54	57.4
First	39	41.7

model, and problem-solving model. The empirical findings are presented in a series of tables (see Tables 8.4, 8.5, 8.6 and 8.7).

8.5.1 The Toyota Way Philosophy Model

8.5.1.1 Long-Term Philosophy (P1)

The Philosophy model refers to practices relevant to the guiding principles of the company. Referring to the bottom layer of the Toyota Way model (Liker 2004), the Philosophy model has one principle which consists of nine attributes. This is shown in Table 8.4.

On the whole, the mean values of the practices under Principle 1 were rated highly by the responding firms in terms of the degree of their implementation and

Table 8.3 Scale reliability: Cronbach's alpha of the data collected

Toyota Way principles	Number of items in the scale	Cronbach's alpha Level of practice	Cronbach's alpha Level of importance
P1 Long-term philosophy	9	.882	.845
P2 One-piece flow	6	.841	.830
P3 Pull *kanban* system	5	.793	.792
P4 Level out the workload	5	.791	.794
P5 *Jidoka*	7	.878	.878
P6 Standardized tasks	5	.859	.860
P7 Visual management	6	.874	.897
P8 Using reliable technology	4	.905	.907
P9 Grow leaders and leadership	7	.895	.882
P10 People management	7	.920	.930
P11 Supplier relationships	9	.918	.914
P12 *Genchi genbutsu*	5	.801	.698
P13 Decision-making	6	.870	.874
P14 *Kaizen*	8	.909	.939

their perceived importance. In the case of implementation, as depicted in Table 8.4, four attributes have been "moderately" (average mean ≥ 4) practiced by the responding firms. Among these, P1.8, "*be able to rapidly respond to meet the changing requirements of the customers*" stood out with highest score received ($m = 4.23$). Following were "*sustain a constant purpose*" (P1.1), "*understanding customer's requirement is priority work*" (P1.7) and "*a clear view of its core competency*" (P1.5). The top-ranked attribute (P1.8) reveals that the responding firms placed greater emphasis on customers' requirements and were able to respond quickly. This is parallel to the main target of lean practice, which is to create value for customer.

With regard to the perceived importance, it is evident that the respondents placed a higher degree of importance on all the attributes than their implementation. Strongly supporting attributes such as "*sustain a constant purpose*" (P1.1), "*have a clear view of the company's core competency and endeavour to become an expert in this area*" (P1.5) and "*be able to rapidly respond to meet the changing requirements of the customers*" (P1.8) were important attributes among the respondents. The importance of "*constant purpose*" has been highlighted by Deming (1986), who believed that it should be management's number one priority and obligation. On the contrary, it was found that "*employees/suppliers are treated as internal customers*" (P1.9) was rated as of least implemented and least importance. This indicates that the responding firms were not aware of "*employees/suppliers are internal customers*" as a philosophical concept. Something that might contribute to this low awareness is probably the high level of worker mobilization. Therefore, for this attribute to be adopted and to a larger extent, the workforce should be trained to understand that people working in the downstream work-flow are actually their customers. Moreover, attribute P1.4, "*short-term losses*

Table 8.4 Descriptive statistics of Principle 1 practices in terms of the implementation and importance

Principles and attributes of the Toyota Way Philosophy model	Implementation			Importance			
	Mean	S.D.	Rank	Mean	S.D.	Rank	*p*-Value
P1 Long-term philosophy	3.81	.91	–	4.18	.82	–	
P1.1 Sustain a constant purpose (company vision, mission and values)	4.07	.779	2	4.48	.714	1	.000
P1.2 Have a high purpose or mission which generates value towards employees, society and customers	3.73	.997	5	4.12	.926	6	.000
P1.3 Formulate a plan towards the realization of company's long-term vision	3.66	1.048	7	4.27	.809	4	.000
P1.4 Short-term losses affect decision-making, but are less important than pursuing long-term goals	3.49	.880	8	3.89	.902	8	.000
P1.5 Have a clear view of the firm's core competency and endeavour to become an expert in this area	4.00	.897	4	4.39	.781	2	.000
P1.6 Be responsible for products, employees and society	3.68	1.039	6	4.14	.875	7	.000
P1.7 Understanding the customer's requirement is priority work	4.01	.886	3	4.21	.760	5	.006
P1.8 Be able to rapidly respond to meet the changing requirement of the customers (e.g. design change)	4.23	.725	1	4.32	.691	3	.299
P1.9 Treat employees/suppliers as internal customers	3.45	.935	9	3.83	.912	9	.000

affect decision making" was rated as second last in implementation and importance. This reflects the way the Chinese construction industry is plagued by short-sighted behaviour that contributed to the industry's poor performance. Aiming at short-term profit is likely to kill long-term constancy, and hence prevents the continuous improvement advocated in the overall philosophy.

8.5.2 The Toyota Way Process Model

The Process model of the Toyota Way refers to various well-known lean tools or practices that have been widely adopted on the shop floor. As with the approach used in the analysis of the philosophical practices employed by the responding firms, the mean value and standard deviation have been calculated for each of the attributes identified under the Process model in terms of their implementation and importance. The importance values are generally rated comparably highly by the respondents. As can be seen from Table 8.5, the mean value of the implementation

Table 8.5 Descriptive statistics of Principle 2–8 practices in terms of the implementation and perceived importance

Principles and attributes of the Toyota Way Process model	Implementation			Importance			*p*-Value
	Mean	S.D.	Rank	Mean	S.D.	Rank	
P2 One-piece flow	3.45	.844	–	4.06	.787	–	
P2.1 Employee is concerned with waste elimination	3.39	.765	25	4.13	.779	12	.000
P2.2 Material flow is adhered to consistently throughout the daily work activities	3.28	.822	29	3.90	.835	26	.000
P2.3 Material, equipment and other resources are provided in a "just-in-time" manner when needed	3.56	.890	14	4.21	.746	7	.000
P2.4 Site layout is organized to enhance material flow, employee movement, etc. to minimize wastes due to movement, motion, travel, etc.	3.73	.792	6	4.19	.708	9	.000
P2.5 Strive to cut back to zero the amount of time any work is sitting idle or waiting for someone to work on it	3.29	.850	28	4.00	.762	20	.000
P2.6 Make flow evident through organizational culture	3.46	.947	20	3.94	.890	25	.000
P3 Pull "kanban" system	3.44	.952	–	3.91	.901	–	
P3.1 Materials are ordered as close as possible to exact needs	3.90	.843	4	4.21	.760	7	.000
P3.2 Strive for possible low level of (even stockless) material inventory in construction site	3.23	.999	32	3.57	1.000	36	.000
P3.3 Use simple signals—cards, empty bins, etc. to monitor the level of inventory and to order the needed material/component	3.11	.967	36	3.66	.957	34	.000
P3.4 Monitor the quantity of material/component/equipment that the teams actually take away	3.55	1.043	17	4.23	.860	5	.000
P3.5 Clear job contents, work time, material requirements, among other information are prepared before releasing a work task to a crew	3.43	.910	23	3.88	.926	27	.000
P4 Heijunka (level out the workload)	3.55	.852	–	3.99	.804	–	
P4.1 Project manager plans the work with input from other parties including subcontractors, clients, suppliers, etc.	3.73	.918	6	4.11	.836	14	.000
P4.2 Daily work activities are planned to balance material availability, manpower, machine availability and workload between operations	3.79	.760	5	4.18	.775	10	.000
P4.3 Foremen (Last Planners) make commitments on what the crews will do each week based on what is ready to be done	3.56	.824	14	4.00	.776	20	.000

(continued)

Table 8.5 (continued)

Principles and attributes of the Toyota Way Process model	Implementation			Importance			p-Value
	Mean	S.D.	Rank	Mean	S.D.	Rank	
P4.4 Weekly/daily work assignments are completed in accordance with the weekly/daily schedule	3.46	.851	20	4.12	.774	13	.000
P4.5 Levelling the daily work activities without overburdening workers and machinery	3.19	.907	33	3.57	.861	36	.000
P5 Built-in Quality	3.70	.963	–	4.23	.776	–	
P5.1 Employees are dedicated to provide “built-in” quality into every aspect of operations	3.67	1.020	9	4.32	.779	3	.000
P5.2 Preventing defective or “no inspection” assignments from entering the next process	3.70	1.014	8	4.43	.695	2	.000
P5.3 Rejecting defective materials, components and equipment	3.99	.910	2	4.46	.650	1	.000
P5.4 Employees are encouraged to seek support from their supervisors when something goes wrong at work	3.92	.797	3	4.22	.735	6	.000
P5.5 Employees are empowered to be responsible for quality	3.67	.920	9	4.18	.829	10	.000
P5.6 Employees who work in the same team meet on a regular basis to discuss quality problems and lessons learned	3.41	1.082	24	3.98	.867	22	.000
P5.7 Feedback about quality is routinely given by the employees	3.57	1.000	12	4.04	.879	17	.000
P6 Standardization	3.34	.995	–	3.77	.916	–	
P6.1 Established standard operating procedures (SOPs) (e.g. work processes) are practised by employees for each major operation/process	3.57	.861	12	4.07	.806	16	.000
P6.2 Employees play a key role in creating the SOPs	3.34	.911	27	3.67	.897	33	.000
P6.3 Employees are encouraged to improve the existing SOPs based on their own practical experience	3.35	.981	26	3.77	.955	31	.000
P6.4 Incorporate employee’s creative improvement of the standard into new SOPs	3.26	1.077	30	3.78	.941	30	.000
P6.5 Using standardized prefabricated components from off-site shops	3.18	1.145	34	3.55	.980	38	.000
P7 Visual management	3.32	1.036	–	3.87	.934	–	
P7.1 Visual aids are adopted to make wastes, problems and abnormal conditions readily apparent to employees	2.86	1.053	38	3.62	1.059	35	.000
P7.2 The posted information in terms of job status, schedule, quality, safety, etc. is in place that most workers can see it on a daily basis, and it is up-to-date	3.24	.991	31	3.80	.957	28	.000

(continued)

Table 8.5 (continued)

Principles and attributes of the Toyota Way Process model	Implementation			Importance			*p*-Value
	Mean	S.D.	Rank	Mean	S.D.	Rank	
P7.3 Appropriate signages are used to identify layouts, traffic, safety concerns, etc.	4.04	1.004	1	4.27	.792	4	.003
P7.4 The construction site is kept clean at all times	3.56	.957	14	4.02	.842	19	.000
P7.5 Employees take pride in keeping the construction site organized and clean	3.18	1.077	34	3.80	.957	28	.000
P7.6 The workplace follows the principles of 5-S	3.03	1.131	37	3.68	.997	32	.000
P8 Use of reliable technology	3.53	.946	–	4.01	.911	–	
P8.1 New technology must support the company's values	3.55	.957	17	4.10	.928	15	.000
P8.2 New technology must demonstrate its potential to enhance processes	3.51	.925	19	3.95	.872	24	.000
P8.3 New technology must be specific solution oriented	3.59	.921	11	4.03	.921	18	.000
P8.4 New technology must be thoroughly tested and proven to provide long-term benefits	3.45	.980	22	3.97	.921	23	.000

ranges from 2.86 to 4.04, whereas the mean value for the importance ranges from 3.55 to 4.46. In the case of the implementation, all the practices scored less than 4.0, except for "*usage of signage to identify layouts, etc*". (P7.3), which was given the highest mean value ($m = 4.04$) among the 38 attributes.

8.5.2.1 One-Piece Flow (P2)

In this category, only P2.4 ($m = 3.73$) was rated highly in terms of the degree of their implementation. It is believed that the site layout is arranged by the main contractor and this highly rated attribute indicates that efforts have been put into the planning of site layout by the responding firms to enhance the flow of material, manpower, etc. To achieve this, visual management tools are helpful and that might explain why attribute P7.3, "*usage of site signage*" was given the highest mean value amongst all the attributes in the Process model. The remaining attributes under Principle 2 were centred below the average level of the ranking table.

8.5.2.2 Pull "*Kanban*" System (P3)

There are five attributes identified in Principle 3 which collectively assess the extent to which the pull principle and its associated tools have been implemented. According to Table 8.5, it was found that P3.2, "*strive for possible low level of material inventory*" and P3.3, "*using simple signals to monitor the level of inventory and to order the needed material*" were rated as of the least implemented as well as least important attributes in this category. This infers that the pull system is not highly appreciated by the respondents. Apart from the technical difficulties, it

Table 8.6 Descriptive statistics of Principle 9–11 practices in terms of the implementation and perceived importance

Principles and attributes of the People and Partner model	Implementation			Importance			
	Mean	S.D.	Rank	Mean	S.D.	Rank	p-Value
P9 Leaders and leadership	3.68	.886	–	4.15	.805	–	
P9.1 Leaders are motivated to inspire people to achieve goals	3.49	1.02	15	4.31	.790	1	.000
P9.2 Leaders must have in-depth job knowledge	3.87	.806	1	4.31	.748	1	.000
P9.3 Leaders possess teaching ability and are able to pass their knowledge on to others	3.45	.887	17	3.95	.872	18	.000
P9.4 Leaders must support the employees doing their work	3.85	.789	2	4.13	.797	7	.001
P9.5 Leaders will take time to understand problems and root causes before acting	3.76	.826	4	4.15	.765	5	.000
P9.6 Leaders strongly encourage employees to develop "continuous improvement" in thinking and action	3.71	.980	5	4.11	.823	8	.000
P9.7 Leaders must understand the company policy and procedures, and communicate these to their team	3.63	.892	9	4.06	.840	11	.000
P10 Develop people and promote teamwork	3.49	.997	–	4.05	.899	–	
P10.1 Select the best person for a given job	3.64	.866	8	4.28	.754	3	.000
P10.2 Training is provided to equip the employees with the required skills before they are assigned to work	3.59	.966	11	4.14	.887	6	.000
P10.3 On-the-job training is provided to further develop employee's exceptional skills	3.49	.959	15	4.04	.903	13	.000
P10.4 Employees are cross-trained to perform additional functions	3.13	1.04	23	3.79	.971	19	.000
P10.5 Training materials are standardized	3.50	1.18	14	4.01	.989	16	.000
P10.6 Employees are encouraged to cooperate with others to complete the whole task	3.62	.963	10	3.99	.898	17	.000
P10.7 Daily work activities are organized into team function	3.59	.966	11	4.11	.836	8	.000
P10.8 Internal motivation methods	3.43	.967	20	4.02	.916	15	.000
P10.9 External motivation methods	3.45	1.07	17	4.07	.942	10	.000
P11 Respect partners relationships	3.43	.997	–	3.81	.940	–	
P11.1 Respect partners' capabilities	3.65	.839	7	4.06	.865	11	.000
P11.2 Challenge the partners by setting collaborative targets	3.34	1.02	21	3.72	.977	22	.000
P11.3 Take part in partners' production process	3.69	.916	6	4.03	.861	14	.000
P11.4 Work with the partners to improve project effectiveness	3.09	1.22	24	3.63	1.040	23	.000

(continued)

Table 8.6 (continued)

Principles and attributes of the People and Partner model	Implementation			Importance			
	Mean	S.D.	Rank	Mean	S.D.	Rank	*p*-Value
P11.5 Work with the partners in various areas to develop their technical capabilities	3.09	1.05	24	3.54	.958	25	.000
P11.6 Share information with partners in a structured manner	3.19	1.04	22	3.62	.963	24	.000
P11.7 Conduct joint improvement activities with partners to solve problems	3.44	.99	19	3.79	.938	19	.000
P11.8 Strive to establish a long-term relationship with reliable partners	3.82	.915	3	4.19	.846	4	.000
P11.9 Limit the number of suppliers	3.53	.985	13	3.74	1.015	21	.008

also infers that the responding firms used the traditional way of procuring materials in which orders in big batches were generated. This might explain the reason why P3.2 was poorly practised. The results are consistent with the findings of researchers such as Arbulu et al. (2003), who highlighted that the "big batches" mindset is one of the biggest constraints for the development of *kanban* strategy at a construction site. Moreover, P3.1, "*materials are ordered as close as possible to exact needs*" was ranked among the top five most implemented practices in the Process model. Logically, if materials are ordered based on site needs, it would yield lower level of inventories. This seems to contradict attribute P3.2, "*strive for low level of inventory in site*" as it was ranked at the bottom of the ranking table. One possible explanation here could be that the materials in P3.1 might refer to the order-to-construction materials (e.g. ready-mixed concrete), which is usually placed immediately upon arrival at the jobsite rather than being stockpiled.

8.5.2.3 *Heijunka* (Level Out the Workload) (P4)

The principle of *heijunka* in this research consists of five identified attributes that pertain to project scheduling. Efforts have been made to link the LPS to the *heijunka* principle in the earlier chapter as both aim at achieve stability and reliability in the work-flow. According to Table 8.5, two attributes, namely P4.1 and P4.2 were found as being implemented to a larger extent with mean values of 3.73 and 3.79, respectively. The former implies that the Chinese project manager would somehow consult with other parties in terms of key resources including people, equipment, work space, etc. The latter describes that the responding firms are adhering to the rationale of planning for daily activities to balance the different available resources. Despite the fact that attribute P4.2 was rated highly, it does not mean that reliable construction work-flow has been achieved by the responding firms. Liker (2004) introduced other objectives of *heijunka* which include *muri* elimination—overburden or strenuous work. Clearly, as Table 8.5 indicates, attribute P4.5, "*levelling the daily work activities without overburdening workers and machinery*" was rated more poorly than other attributes in terms of implementation and importance. This infers that the responding firms have seemingly made

Table 8.7 Descriptive statistics of Principle 12–14 practices in terms of the implementation and perceived importance

Principles and attributes of the Toyota Way Problem-solving model	Implementation			Importance			*p*-Value
	Mean	S.D.	Rank	Mean	S.D.	Rank	
P12 Genchi genbutsu	3.78	.901		4.07	.804		
P12.1 Solve problem by going to the places where problems are discovered	4.01	.898	1	4.22	.819	1	.005
P12.2 Analysing and thoroughly understand the situation before making decisions	3.81	.883	3	4.30	.669	2	.000
P12.3 Making decisions based on the verified data	3.77	.955	4	4.20	.742	4	.000
P12.4 Making decisions based on management team's past experiences	3.65	.839	8	3.63	.961	18	.928
P12.5 *Genchi genbutsu* has become part of the company culture	3.67	.932	7	3.98	.829	13	.000
P13 Consensus decision-making	3.43	.944		3.86	.835		
P13.1 Using appropriate problem-solving methodologies (e.g. 5 Whys) to determine the root causes of problems	3.56	1.01	12	3.99	.874	12	.000
P13.2 Possible experiments are conducted to test the potential cause of a problem	2.99	1.10	19	3.46	.980	19	.000
P13.3 Broadly consider alternative solutions	3.59	.835	11	3.97	.714	14	.000
P13.4 Valuing the process through which the decision was reached	3.49	.913	15	3.84	.780	17	.000
P13.5 Building consensus within the team, including employees and outside partners	3.41	.921	17	3.85	.829	16	.000
P13.6 Addressing the root causes of problems via effective communication vehicle	3.55	.887	13	4.07	.833	10	.000
P14 Kaizen	3.56	.953		4.10	.882		
P14.1 Reflection on mistakes (e.g. defects, rework, safety issues, etc.) on a regular basis	3.65	.947	8	4.14	.850	6	.000
P14.2 Management treats problems as development opportunities for employees	3.69	.939	5	4.14	.911	7	.000
P14.3 *Kaizen* activities are conducted in your workplace	3.47	.991	16	4.07	.871	11	.000
P14.4 Management supports the *kaizen* activities	3.68	.964	6	4.12	.949	8	.000
P14.5 The improvement will be codified into documents and/or policies used by organization	3.06	.925	18	3.89	.873	15	.000
P14.6 Each hierarchy of the organization develops measurable objectives as well as actions to support the executive-level goals	3.85	.939	2	4.23	.822	3	.000
P14.7 Managers are keen on measuring the objectives and give feedback	3.63	.892	10	4.11	.861	9	.000
P14.8 PDCA methodology is used to solve problems	3.52	1.03	14	4.15	.915	5	.000

progress on levelling the work by considering the balance of various resources, but issues of alleviating overburden of workers and machines are presently not taken into account.

8.5.2.4 Built-In Quality (P5)

From Table 8.5, it can be seen that 5 of the top 10 most implemented attributes in the Process model are actually from Principle 5, which has made it the most implemented principle compared to the remaining process-focused principles. Some highly implemented attributes include:

- P5.3, "*rejecting defective materials, components, and equipments*" ($m = 3.99$)
- P5.4, "*employees are encouraged to report problems occurred*" ($m = 3.92$)
- P5.2, "*preventing defective or 'no inspection' assignments from entering the next process*" ($m = 3.70$)

These three attributes were given mean values of 3.99, 3.92, and 3.70, respectively. It implies that the quality management practices have been in place within the responding firms. Meanwhile, these attributes have all, in the respondents' opinions, been crucial for improvements in firms' performance. This is reflected by the mean value given to the degree of their perceived importance too (see Table 8.5). However, under Principle 5, two relatively less implemented attributes include P5.6, "*quality problems discussion and lesson learnt*" and P5.7, "*feedback about quality is routinely given by employees*". Both highlight that the quality culture in the responding firms has not been fully established, where the employees/frontline workers seem less proactive in discussing quality problems and are lacking a "*kaizen*" mindset to engage in meaningful quality improvement activities for continuous improvement. In addition to that, the relatively low mean values given to these two were partly due to the pressure of tight schedule that employees' time was compromised and thus left them with limited time for quality control activities.

8.5.2.5 Standardization (P6)

Overall, the principle of standardized work (P6) was the second least-implemented principle ranked by the respondents, with an average mean value of 3.34. Under this category, the most frequently exercised attribute was P6.1, "*established standard operating procedures* (SOPs)". Given that the nature of individual construction projects is very different, it is comforting to see that some Chinese construction firms have defined SOPs to some extent upon processes that may be repeated. Apart from P6.1, all of the remaining four attributes were given relatively low scores for implementation and importance. Judging from the relatively poor ratings of attributes P6.2, P6.3 and P6.4, it seems that employees' involvement in creating and implementing SOPs was limited. This is in line with the standardization of work methods that was practised in Taylor's time, or what Adler and Borys (1996) termed *coercive standardization*. This worked in the way that the new SOP was decided by management, and the production employees were required to follow it. Worse, employee involvement and creation was simply ignored. Lastly, of all the less commonly adopted attributes, "*using standardized components*" (P6.5) was perceived as the least implemented and undervalued. This is understandable, since industrialized construction was still in the infancy stage in China, and so the standardized components were not typically used. All the findings seem to suggest that none of the remaining key elements of standardized work (with the exception of SOPs), i.e. bringing in employees in the course of creating and improving standardization, and using extensive standardized components are commonly practised by the responding firms. The consequences of non-standardized operations have been well documented (Liker 2004; Monden 1998). Under such work conditions, not only would variations be introduced, that could affect product quality, but in the long-term, the development of *kaizen* activities might be severely affected. As Imai (1997) has highlighted, standardized tasks are the basis for continuous improvement. Hence, it is reasonable to assume that the slow adoption of standardized work may affect continuous improvement activities conducted among the responding firms.

8.5.2.6 Visual Management (P7)

In terms of the implementation, Table 8.5 indicates that the responding firms have dedicated least amount of attention to visual management practices except using appropriate signage to identify layouts, traffic, safety concerns, etc. (P7.3). This practice was singled out as the most practised individual attribute in the Process model. Extensive use of signage is a good practice of visual management on the site. A possible explanation for this is on account of the construction related bureau's efforts in passing laws to make it mandatory for firms to provide visual signages (e.g. site banners, corporation image, etc.) on the site. In contrast, the remaining attributes identified in P7 were given relatively low means. These include:

- P7.5, "*take pride in keeping the site organized and clean*" ($m = 3.18$)
- P7.6, "*5-S practice*" ($m = 3.03$)
- P7.1, "*visual aids*" ($m = 2.86$)

This suggests that visual management has not been pervasive among responding firms. In fact, expect "*appropriate site signages*" (P7.3) were rated highly because of the mandatory norms; alternative visual aids (P7.1) such as using mistake proofing tool to identify abnormal conditions, established boards with critical information at the point of need, etc., were rated as being infrequently practised. This infers that visual management is used at a superficial level, and it is poorly practised compared to what the manufacturing firms have benchmarked. To reiterate, visual management and 5-S need to go hand in hand. The ratings of P7.5 and P7.6 highlight that the responding firms put limited efforts in practising 5-S. Again, it is probably due to the frontline workers who do not take pride in keeping the construction site organized and clean (P7.5).

8.5.2.7 Use of Reliable Technology (P8)

Four attributes identified in this principle were given mean values close to 4 in terms of its importance. Under this principle, "*New technology must support company's value*" (P8.1) was rated first in importance, followed by principles such as "*whether it is specific solution oriented*" (P8.3), "*whether it has long-term benefits*" (P8.4) and "*whether it has potential to enhance the process*" (P8.2). Moreover, P8.3 received the highest mean value for implementation. These are clear indicators that the Chinese construction firms held a moderately pragmatic attitude towards the acceptance of new technology. Their primary concern is whether the new technology could help solve the existing problem or supports the firm's value rather than pursue its long-term value.

8.5.3 The Toyota Way People and Partner Model

The People and Partner model involves issues in leadership, training individuals, teamwork, motivational strategies, supplier relationships, etc. that may influence the process conditions of the working environment, and therefore improve organizational performance. Table 8.6 shows the mean values for attributes listed in the People and Partner model. The overall mean values of P9, P10 and P11 for implementation are 3.68, 3.49 and 3.43, respectively.

8.5.3.1 Leaders and Leadership (P9)

As Table 8.6 indicates, overall, the attributes listed in Principle 9 have mean values below 4 ("moderate" level). Four attributes were placed at the very top ranks in the People and Partner model. P9.2 ($m = 3.87$) is in the first place, then P9.4 ($m = 3.85$) in second place, followed by P9.5 ($m = 3.76$) and P9.6 ($m = 3.71$), in the order of implementation. This shows that the leaders from the responding firms, possess "*in-depth job knowledge*" (P9.2) to a moderate extent, intend to "*provide support to the employees while they are doing their work*" (P9.4), intend to "*take time to understand problems and root cause*" (P9.5) and "*encourage employees to develop a 'kaizen' mindset in thinking and action*" (P9.6). This is reflected in the fact that the Chinese building professionals who are in leadership positions do exhibit some of the good qualities and abilities that are in line with the Toyota Way-style leadership features. With respect to perceived importance, both P9.1 and P9.2, which concern leaders capability to "*inspire people to achieve goals*" and the extent to which leaders must "*possess in-depth job knowledge*" have been assigned the maximum points ($m = 4.31$), putting them in the first place on the importance scale in the People and Partner model. Furthermore, according to Table 8.6, of all the attributes identified in Principle 9, attribute P9.3 "*possess teaching ability and pass knowledge on to others*" which is one key element of leadership, was given a relatively low rating in terms of implementation and perceived importance. One interpretation of this low rating is that passing in-depth knowledge to followers is not a strength of Chinese leaders, even though they are described as technically knowledgeable (P9.1). Together with P9.1, "*leaders are motivated to inspire people*", which was also poorly rated, the inference seems to be that Chinese construction professionals may not view themselves as strong in soft skills such as inspiration to lead, coaching, motivating employees, etc. Conversely, visible leadership characteristics such as providing support to employees (P9.4) and taking time to understand problems (P9.5) were more highly appreciated and practised than the less tangible behaviours just mentioned. Slattery and Sumner (2011) explained that this could be due to the dynamics of managing projects within the construction industry, in which managers take responsibility to lead the team to achieve tangible results.

8.5.3.2 People Management (P10)

In contrast to the other two principles of the People and Partner model, P10 was notably practised poorly, with a mean value of 3.49 for implementation. Under this principle, four key practices were assessed, namely personnel selection (P10.1), various forms of training (P10.2–P10.5), teamwork (P10.6–P10.7) and motivational strategies (P10.8–P10.9). Only the first attribute (P10.1), pertaining to personnel selection, was given a relatively high mean value ($m = 3.64$) as it was agreed upon by a majority of the respondents that its importance ought to be stressed. The mean values for training, teamwork and motivational strategies were not high. This seems to suggest that the responding firms only focus on the first step of human resource management—personnel selection, whereas disproportionally small efforts were made in the areas of training, teamwork, and use of appropriate motivational strategies. Moreover, all the training opportunities, crossing training (P10.4) was rated the least practised and least important attribute, compared to the other two pre-job training and on-the-job training (P10.2 and P10.3). One possible explanation for this low mean value might be the resistance to multi-skill training of most respondents who are in senior management positions and have also specialized in certain functions. For this reason, they were less interested in any change that might result from multi-skill training. Additionally, multi-skill training demands more resources and a greater budget to develop the training plan, and this may discourage implementation too, as resources for such investments are normally limited.

8.5.3.3 Partner Relationships (P11)

The concept of partnership has long been applied among Japanese manufacturing firms and has also been promoted through Latham's (1994) report in the construction context. The results pertaining to P11 in Table 8.6 shows that P11.8, "*establish long term relationships with suppliers*" emerged as the top-ranked attribute both in terms of implementation and perceived importance. However, it is reasonable to doubt that this practice is in fact well implemented by the responding firms. It seems that their efforts are limited to describing the importance of long-term relationships, and do not focus on effectively changing behaviour or fostering long-term relationships. The conclusion can be drawn since a number of attributes that are associated with long-term relationships were not highly rated, and are ranked at the bottom. These include:

- P11.9, "*limit the number of suppliers*" ($m = 3.53$)
- P11.6, "*share information with partners*" ($m = 3.19$)
- P11.4, "*collaborate with partners to improve project effectiveness*" ($m = 3.09$)
- P11.5, "*develop partners' technical capacities*" ($m = 3.09$)

One possible explanation for this behaviour could be the differences between manufacturing and construction in terms of supply chain characteristics.

The Toyota Way integrated its tried and tested suppliers into its so-called extended enterprise (Liker 2004), in which the suppliers are given the opportunity to learn Toyota's philosophy and lean production skills. Without these perquisites, this principle can hardly be implemented in the construction industry. Apart from very large construction firms which do have their own material producing subsidiaries, most construction firms need to establish relationships with construction material companies or with vendors who only process the construction firm's order and arrange payment to the manufacturer. As project locations change, it is likely that the relationships with local vendors simply terminate. Without establishing sound relationships with manufacturing firms, we can expect it to be less likely that construction firms are involved in partners' (e.g. in manufacturing firms) production processes, never mind offering assistance to develop their technical capacities. Overall, the poor implementation of principle 11 reflects the fact that this could be the structural problem of China's construction industry, where severe competitions prevent partnerships in which suppliers or subcontractors are not treated as internal customers (refer to P1.9), but as competitors.

8.5.4 The Toyota Way Problem-Solving Model

The Problem-solving model of the Toyota Way consists of three underlying principles: *genchi genbutsu* (P12), decision-making (P13) and *kaizen* and *hansei* (P14). These three principles represent four different stages of the Plan, Do, Check, Act (PDCA) philosophy for problem-solving. Table 8.7 shows the mean values and standard deviations for the problem-solving practices of the large Chinese construction firms surveyed. The overall mean values of P12, P13 and P14 for implementation are 3.78, 3.43, and 3.56, respectively.

8.5.4.1 *Genchi genbutsu* (P12)

Table 8.7 shows a much higher level of certain *genchi genbutsu* practices than the remaining principles in the problem-solving model. These include:

- P12.1, "going to the place where problems are discovered" ($m = 4.01$)
- P12.2, "analysing the situation before making decisions" ($m = 3.81$)
- P12.3, "relied on the verified data" ($m = 3.77$)

Moreover, these three attributes were ranked at the top in terms of their perceived importance. The practice of *genchi genbutsu* is considered to be a valuable quality and ability of a leader. This also corresponds to one of the highly rated attributes identified in P9, measuring a leader's attitude to problem-solving (P9.4). It reinforces the idea that Chinese building professionals have a serious attitude to deal with problems by seeing them at first hand. Furthermore, "*making decisions simply depends on past experience*" (P12.4) is technically not lean behaviour. Although it was not highly rated, it can be inferred that mixed approaches were adopted by managerial people in the course of making decisions, in which reliable data was also consulted.

8.5.4.2 Consensus Decision-Making (P13)

Table 8.7 reveals that practices concerned with decision-making were relatively poorly implemented in practice. For example, "*possible experiments are conducted to test the potential cause of a problem*" (P13.2) was rated among the least practiced, with a mean value of 2.99. It is understandable that there are few circumstances in which experiments need to be performed at the project level. The most common one that can be seen at construction sites is conducting routine quality checks on the incoming materials. It is particularly true that when a problem is discovered at site, where the situation might become so chaotic that time is in short supply, and the likelihood of carrying out experiments may be lowered. In addition, other key attributes identified in Principle 13 such as "*using the 5 whys to determine the root cause of problems*" (P13.1) and "*consider alternative solutions*" (P13.3) were only moderately adopted by the responding firms. Moreover, it is noteworthy that "*building consensus within the team*" (P13.5) was ranked the second lowest among all the attributes under Principle 13. This reflects how consensus has not been widely adopted, as in some cases employees (e.g. frontline workers) were not allowed to participate in the decision-making process. Rather, the project manager and other managerial personnel were in favour of the authoritative method, whereby the decision was made in a top-down manner. Similarly, another low-ranking attribute, P13.4, "*valuing the process through which the decision was reached*" highlighted the way that the entire decision-making process is not undertaken with sufficient attention.

8.5.4.3 *Kaizen* or Continuous Improvement (P14)

The last principle of the Toyota Way contains several management practices that are unique to the Japanese culture, such as reflection (or *Hansei* in Japanese) (P14.1), *kaizen* (P14.2–P14.5), policy deployment (P14.6–P14.7) and the PDCA method (P14.8). According to Table 8.7, key element of policy deployment, such as "*measurable objectives as well as means to support the executive-level goals*" (P14.6), and *kaizen*-related practice, such as management "*treats problems as opportunities*" (P14.2) and "*support the kaizen activities*" (P14.4), were most frequently practised by the respondents. Both the highly rated attributes P14.2 and P14.4 pertain to management's role in *kaizen*, whereas P14.3 and P14.5 which require employees' efforts and participation in *kaizen* were given low scores. This seems to suggest that *kaizen* is sufficiently recognized among top management, that they stressed its importance and put effort into promoting it. However, when it comes to the operational level relating to employee participation, the results suggest that the penetration was low and the *kaizen* activities were not effectively conducted in the site. Clearly, this is a gap which needs to be filled. It is important to secure management commitment in *kaizen* actives (Imai 1997; Liker 2004), and the success of *kaizen* additionally depends on employees' skills, mindset and motivation. These are the areas that LCCFs need to enhance. Furthermore, it is worth mentioning that using the PDCA approach in problem-solving (P14.8) was perceived as an important management philosophy, but the result suggests that in the actual workplace, the use of PDCA to resolve problems was still immature.

8.5.5 The Implementation–Importance Gap

The above data reveals the status quo at large Chinese construction firms in terms of adopting Toyota Way-styled practices. With respect to the perceived importance, a one-sample *t*-test comparing the attribute scores with the mid-point (3 = "neutral") was performed, and 99 % confidence was selected. The *t*-test showed that respondents had positive view towards all of the Toyota Way attributes, and agreed that they are all significantly important ($p = .000$). Moreover, in comparison with the measures of the implementation dimension, it is interesting to note that all attribute measures in the "perceived importance" dimension are rated comparably higher, with the exception of P12.4 (*making decisions based on management team's past experiences*), which showed a higher mean score on the implementation dimension ($m = 3.65$) than on the "perceived importance" dimension ($m = 3.63$). In order to ascertain whether there was any significant difference between implementation levels and perceived importance, the Wilcoxon Signed-Rank two-tailed test was used. This non-parametric test for significance between attributes is the most appropriate, as the attributes were measured on a Likert scale (ordinal). One hypothesis was formulated, and a 5 % level of significance used. To test for a significant difference between the practice (P) and the perceived importance (I) for the sample, the following hypotheses were formulated:

1. H_0: $\mu_P - \mu_I = 0$, i.e. there is no difference between the implementation level and perceived importance
2. H_1: $\mu_P - \mu_I \neq 0$, i.e. there is a significant difference between the implementation level and perceived importance

The calculated *p*-values are shown in the extreme right columns of Tables 8.4, 8.5, 8.6 and 8.7. As noted, if the significance value is less than .05, there is a significant difference. If the significance value is greater than .05, there is no significant difference. With the exception of two of the attributes (P12.4 and P1.8), statistically significant differences were found between the actual implementation level and the perceived importance. This implies that the responding firms were aware of the importance of Toyota Way practices, but were not yet fully ready to implement them. One possible explanation for this is that since many of the responding firms are in the early stages of implementing such a management philosophy in their work, they are not aware of the whole spectrum of its implementation. Their lack of understanding of what is needed for the implementation of the lean or Toyota Way philosophy might have affected their focus and thus their implementation level. Hence, the overall statistical description in this section shows that more efforts need to be focused on promoting all aspects of the Toyota Way practices within Chinese construction firms. As mentioned earlier, two attributes, namely P12.4, "*decision-making depending on experiences*" ($p = .928$) and P1.8, "*quick response to clients*" ($p = .299$) were implemented to the same extent as they were perceived as important ($p > 0.05$).

Furthermore, if one normalizes the scales and plots the aggregated responses using the median response as the point where the two scales cross, one can examine the relative ranking of each principle (see Fig. 8.1).

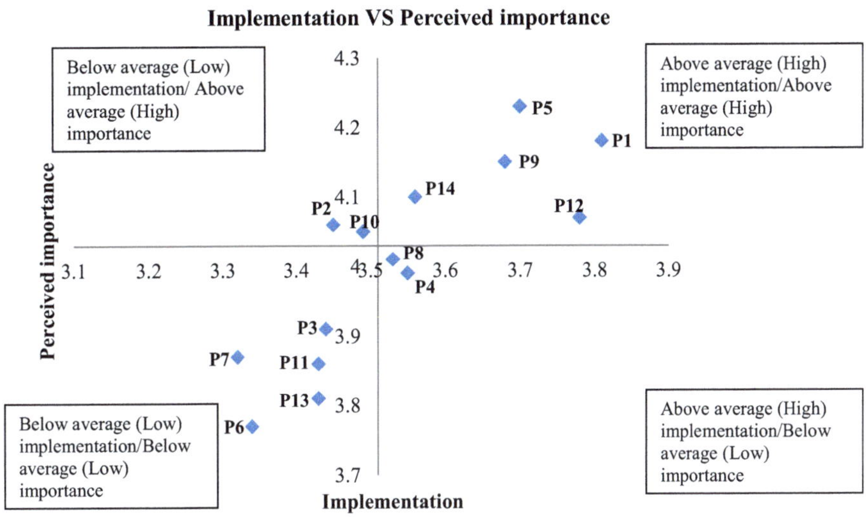

Fig. 8.1 Implementation–importance gap (normalized)

From a somewhat linear relationship, where principles that are ranked high in terms of implementation are also ranked high in terms of importance. In most cases, the Toyota Way principles are centred on two categories of extremes, namely High-High and Low-Low. In a few cases, these means cross over into contradiction (High-Low) for cases (P2, P10, P8 and P4) that are close to the borderlines. Following the line from P7 to P1, it appears that more attention and resources are put into Toyota Way P1, P5, and P9, whereas several other principles (such as P6, P7 and P13) are areas of less importance and thus less effort is put into their implementation.

Following the discussion above, it is a further challenge that priority needs to be given to certain attributes, on account of their low implementation level. To reiterate, since the majority of attributes were perceived to be of importance, the limited resources and efforts need to be given to those least implemented. Hence, a one-sample *t*-test was also conducted to compare the mean of all the Toyota Way-styled attributes, in terms of their implementation level, with the mid-point 3 (3 = "somewhat"). It was shown in Table 8.8 that of 91 attributes, 15 attributes were found *not* to be significantly implemented, or insignificant efforts were put into implementing these attributes ($p > 0.05$ or $p > 0.01$).

8.6 Relating the Toyota Way Practices and Project Performance

Project performance has been assessed according to different indicators (see Table 8.9). In general, the responding firms evaluated their results higher than the performance of their competitors. The highest scores are allocated to clients'

Table 8.8 Descriptive statistics and one-sample *t*-test of Toyota way attributes in terms of implementation

Category	Item	Description	Mean	S.D.	*t*-Value (*p*)
Process model	P3.2	Strive for low level (even stockless) of material inventory at construction site	3.23	.999	2.27* (.025)
	P3.3	Use simple signals to monitor the level of inventory and to order the needed material/ components	3.11	.967	1.067** (.289)
	P4.5	Levelling daily work activities without overburdening workers and machinery	3.19	.907	2.046* (.044)
	P6.4	Incorporate employee's creative improvement of the standard into new SOPs	3.26	1.077	2.298* (.024)
	P6.5	Using standardized prefabricated components from off-site shops	3.18	1.145	1.531** (.129)
	P7.1	Visual aids are adopted to make waste, problems and abnormal conditions readily apparent to employees	2.86	1.053	−1.273** (.206)
	P7.2	The posted information in terms of job status, schedule, quality, safety, etc. is in a place that most workers can see it on a daily basis, and it is up-to-date	3.24	.991	2.393* (.019)
	P7.5	Employees take pride in keeping the construction site organized and clean	3.18	1.077	1.628** (.107)
	P7.6	The workplace follows the principles of 5-S	3.03	1.131	.274** (.785)
People and partner model	P10.4	Employees are cross-trained to perform additional functions	3.13	1.04	1.191** (.237)
	P11.4	Work with partners to improve project effectiveness	3.09	1.22	.679** (.499)
	P11.5	Work with partners in various areas to develop their technical capabilities	3.09	1.05	.783** (.436)
	P11.6	Share information with partners in a structured manner	3.19	1.04	1.785** (.077)
Problem-solving model	P13.2	Possible experiments are conducted to test the potential cause of a problem	2.99	1.10	−.094** (.926)
	P14.5	The improvement will be codified into documents and/or policies used by organization	3.06	.925	.669** (.505)

$*p > 0.01$, $**p > 0.05$

Table 8.9 Descriptive statistics of performance indicators

Qualitative performance measures	Mean	S.D.
Profitability	3.59	.811
Productivity	3.63	.762
Quality	3.87	.765
Delivery time	3.72	.739
Client satisfaction	3.98	.703

satisfaction ($m = 3.98$), followed by project quality ($m = 3.87$) and project delivery time ($m = 3.72$). On the contrary, profitability ($m = 3.59$) and productivity ($m = 3.63$) are rated the lowest. All these indicators are what the Toyota Way model endeavours to improve at the shop floor. It implies that there is much room for improvement in these areas if Toyota Way principles could be implemented.

The performance indicators used in this study were chosen on the basis that in the literature, it has been reported that the application of lean could enhance performance in these areas (see Sect. 3.4.5). Admittedly, the measurement of these performance indicators is, however, not as strong in a Likert scale, which has been criticized as being subjective. Nevertheless, as Whitelaw (1969, p.51) has noted, "*it is better to have a less-than-perfect standard, the shortcomings of which are known, than to have no standards at all*". Investigating the relationship between the Toyota Way practices and project performance can at least provide a snapshot of what the relationships are between the two domains in the Chinese construction firms. To assess the relationship between various practises of the Toyota Way and project performance (e.g. quality, productivity, profitability, and so on), Spearman's rho correlation analysis was performed. Again, due to the ordinal nature of data, Spearman's rho correlation is regarded as the most appropriate to measuring the strength and direction of the linear relationship between the pair of variables. The level of significance was set at $p < 0.01$. When $p < 0.01$, the conclusion is that there is a significant correlation. When $p > 0.01$, there is no significant correlation between the two variables. The results of the correlation analysis are shown in Table 8.10. As expected, all correlation coefficients are positive. This indicates that better performance is associated with more extensive use of the Toyota Way principles. Most coefficients are highly significant at $p < 0.01$ (with **). It is worth mentioning that the correlation coefficients between the Toyota Way principles and productivity and quality measures are relatively higher than the other correlation coefficients.

By examining the individual correlations from Table 8.10, one can observe that 6 out of possible 70 correlations between the Toyota Way principles and the performance measures are above 0.5. These are denoted as high correlation (H). Each of these six highest correlations is between quality performance (Q) and one of the following Toyota Way principles:

- P11—supplier management ($r = .584$, $p < 0.01$)
- P7—visual management ($r = .583$, $p < 0.01$)
- P5—built-in quality ($r = .529$, $p < 0.01$)
- P10—employee management ($r = .527$, $p < 0.01$)
- P9—leadership ($r = .509$, $p < 0.01$)
- P14—*Kaizen* ($r = .501$, $p < 0.01$)

Additionally, correlation coefficients between 0.3 and 0.5 are denoted as medium (M), while correlation coefficients below 0.3 are classified as low (L). This classification should help to develop an overall portrayal of the various Toyota principles correlating with the performance measures. In contrast, among the 70 possible correlations, the Pull *kanban* system (P3) is found to be insignificantly correlated with client satisfaction ($r = .189$, $p = .68$), while *genchi genbutsu* (P12)

Table 8.10 Correlations between the use of the Toyota Way practices and project performance measures

	Profitability	Productivity	Quality	Delivery time	Client satisfactory	Average correlation
The Toyota Way Philosophy model						
P1 Long-term philosophy	.320** (M)	.434** (M)	.496** (M)	.322** (M)	.485** (M)	.41
The Toyota Way Process model						
P2 One-piece flow	.314** (M)	.451** (M)	.482** (M)	.398** (M)	.365** (M)	.40
P3 Pull "*kanban*" system	.211* (L)	.317** (M)	.361** (M)	.238* (L)	.189 ($p = .068$)	.26
P4 *Heijunka*	.287** (L)	.350** (M)	.499** (M)	.376** (M)	.422** (M)	.39
P5 Built-in quality	.295** (L)	.296** (L)	.529** (H)	.337** (M)	.321** (M)	.36
P6 Standardization	.385** (M)	.338** (M)	.474** (M)	.357** (M)	.367** (M)	.38
P7 Visual control	.371** (M)	.403** (M)	.583** (H)	.387** (M)	.483** (M)	.45
P8 The use of reliable technology	.285** (L)	.360** (M)	.472** (M)	.224* (L)	.262* (L)	.32
The Toyota Way People and Partner model						
P9 Leaders and Leadership	.296** (L)	.464** (M)	.509** (H)	.342** (M)	.297* (L)	.38
P10 People management	.310** (M)	.396** (M)	.527** (H)	.333** (M)	.357** (M)	.38
P11 Partners relationships	.337** (M)	.396** (M)	.584** (H)	.320** (M)	.376** (M)	.40
The Toyota Way Problem-solving model						
P12 *Genchi Genbutsu*	.150 ($p = .152$)	.377** (M)	.437** (M)	.296** (L)	.238* (L)	.30
P13 Decision-making	.395** (M)	.388** (M)	.494** (M)	.350** (M)	.385** (M)	.40
P14 *Kaizen* or continuous improvement	.365** (M)	.452** (M)	.501** (H)	.349** (M)	.360** (M)	.41

*Correlation is significant at the 0.05 level, **Correlation is significant at the 0.01 level

exhibits an insignificant association with profitability ($r = .15, p = .152$). It is worth noting that these are the only two correlation coefficients below 0.2 whose p-values are greater than 0.05.

8.6.1 The Toyota Way Philosophy Model

The effects of the implementation of Principle 1 on all the performance indicators are statistically significant. Of all the associated performance indicators, the most significant correlations are with quality ($r = .496$, $p < 0.01$), followed by client satisfaction ($r = .485$, $p < 0.01$). A closer examination of the correlations of nine attributes and performance indicators (see Appendix C) shows that P1.4, "*short-term losses affect decision-making, but are less important than pursuing long-term goals*" was found not to be significantly correlated with client satisfaction. It was understandable that decisions that potentially cause short-term financial losses would affect client satisfaction. Overall, each of these significant correlations indicates that firms which put more effort into Principle 1 practices were more likely to achieve better performance.

8.6.2 The Toyota Way Process Model

As is evident from Table 8.10, all the seven principles that underlie the Toyota Way Process model are statistically significant in their correlations with all six performance measures, expect for P3 and client satisfaction. This correlation of several Toyota Way Process principles and (project) performance measures contains a number of surprises. First, it seems that the magnitude of the correlation coefficient of both the Pull *kanban* system (P3) and the adoption of reliable technology (P8) is small in association with most performance measures.

In the case of P3, there is a significant correlation coefficient with delivery time ($r = .238$) and with profitability ($r = .211$), even though its magnitude is classified as low. Moreover, P3 does not have a significant correlation with client satisfaction ($r = .189, p > 0.05$). This suggests that putting more effort in pull system in terms of managing materials can increase various aspects of project performance, but has virtually no impact on client satisfaction. Greater understanding can be gained if the relationship between operationalized *pull* attributes and performance measures are further analysed (see Appendix C). For example, one operationalized pull practice, "*materials are ordered as close as possible to exact needs*" (P3.1) is found to be statistically insignificant with all the performance indicators but profitability ($r = .274$). Another pull practice, "*achieve as low a level of inventory as possible*" (P3.2) is found not to be significantly correlated with the increase in all the performance measures. Theoretically, a pull system is often credited with facilitating a faster work-flow and faster project completion, through using JIT material delivery based on actual demand. However, the insignificant correlations in P3 suggest that the effort devoted to using the pull system in minimizing the site inventory (P3.2) does not contribute to higher client satisfaction, faster delivery time, or better profit in the Chinese construction context. This result may also suggest that the survey participants have limited knowledge of using *kanban* to control material. A possible explanation is that high levels of commitment to implementing the pull system with the hope of reducing on-site inventory may suffer from potential risks, including late delivery by the suppliers, materials going out of stock, and so on. All of these will result in unnecessary project delay and client unhappiness. Moreover, given the increasing logistical costs, as well as the fluctuating material market, it is entirely understandable that the pull practices such as P3.2, P3.3 and P3.4 would put construction firms in a position where they suffer from financial loss if the materials are purchased at the time when the price is at its peaks.

Moreover, P8 is another principle that has relatively low correlation coefficient to all performance measures. Similar to the pull *kanban* system (P3), P8 is found to have a small correlation coefficient but is significantly associated with client satisfaction ($r = .262$), project delivery time ($r = .224$) and profitability ($r = .285$). This suggests that, for specific performance measures such as client satisfaction, project delivery, etc., the influence of adopting reliable new technology may operate in an indirect manner. One possible explanation is that clients may not be satisfied to see their own projects employing new technology, which may require high initial investments, even if it is shown to be reliable. In most cases, the client is the one who needs to pay for all the purchase of the technology. Furthermore, operationalized attributes under this category, such as (P8.2) "*new technology must demonstrate its potential to enhance processes*" are found to be insignificantly correlated with project delivery time ($r = .149$, $p > 0.05$) and client satisfaction ($r = .262$, $p > 0.05$). This may suggest that Chinese clients, who tend to be more results-oriented, usually lack the patience to see the potentials of technology that would add value to project. Moreover, the adoption of new technology may affect project delivery as it demands a necessary learning process for people to master the utilization of the new technology.

Additionally, both P6 and P7, which are rated the least implemented Toyota Way principles, and thus fall in the Low-Low (low implementation, low importance) quadrant in Fig. 8.1, are surprisingly found to be strongly and significantly associated with various performance measures. Profitability ($r = .385$) has the largest correlation coefficient with P6, whereas quality ($r = .583$) and client satisfaction ($r = .483$) have the largest correlation coefficient with P7. It is worth noting that visual management (P7) is a multi-variable that contains various useful visual aids (P7.1) which in many areas can be used to identify abnormal features and defects. For example, thanks to visual tools like *andon*, *poka-yoke*, etc., which are regarded as the primary approaches to exposing non-conformance to operatives (Monden 1998; Liker 2004), it seems that quality defects could be prevented at an early stage. Apart from the use of *andon* and *poka-yoke*, 5-S practice (P7.4–P7.6) is found to be positively and significantly associated with all the performance measures. This suggests that by keeping the workplace clean and tidy, the client can be more satisfied.

The remaining Toyota Way practices in the Process model also offer insights into project performance. The principle of the employees building-in quality (P5), for example, is less significantly correlated with all performance measures except for quality performance ($r = .529$). The correlation coefficients for all the operationalized practices listed in P5 are positive and statistically significant for quality performance. P5.3, "*rejecting defective materials, components and equipment*" has the largest coefficient ($r = .504$) and is significant. This is followed by P5.1 ($r = .497$) and P5.2 ($r = .450$) in that order. These indicate that proactive quality programmes, processes and procedures do increase quality. In other words, these efforts build quality in. Conversely, except for quality performance, the potential influence of P5 appears to have moderate or weak associations with the remaining performance measures, such as productivity and profitability, in the short-run. More specifically, it was found that P5.7, "*feedback about quality given by the employees*" and P5.6, "*employees meet regularly to discuss quality problems*" are not significantly correlated with delivery time and with client satisfaction. It seems logical then that if defects and rework due to quality problems become extensive, delivery time and client satisfaction would then be adversely impacted.

8.6.3 The Toyota Way People and Partner Model

All the three underlying principles in the People and Partner model are significantly and positively associated with all the performance measures, in particular, quality, which has the largest correlation coefficient with P9, P10 and P11.

In the case of leaders and leadership (P9), the results indicate that project quality and productivity are significantly correlated positively with all the features of leaders' behaviour and leadership skills. One possible explanation for this is that when employees are working in a team where everyone is motivated to think and act with "*kaizen*" thinking (P9.6), the leaders are helpful in interpreting the policy and procedures from the firm level (P9.7), etc. They could improve their abilities to become more quality-oriented and to take on more responsibilities. All these may in turn improve project quality and productivity. Moreover, even though "*leaders possess teaching ability and are able to pass their knowledge to others*" (P9.6) was rated as a poorly implemented practice, the results indicate that a significant positive correlation can be found between P9.6 and all the performance measures. This might be explained by the fact that this practice is perhaps the most effective way to enhance employees' abilities, which in turn benefits the projects themselves in various ways.

Similarly, as noted from the results, quality performance is found to have significant and positive correlations with all the features and practices identified in employee management (P10). It is understandable that when individuals are selected through high-standard and quality-selection processes, and are giving training that goes far beyond simple teaching, but also includes problem-solving techniques, problem analysis, and work within a team environment, that all of these elements could act to improve the employee's skills in building-in quality when performing their jobs. Moreover, there is a positive but insignificant correlation between profitability and a number of employee management practices, including P10.2, P10.3, P10.4, P10.5 and P10.7. This seems to suggest that various kinds of training do not actually improve project profitability. A possible explanation for this might be that current housing methods are so mature that training could be regarded as ineffective.

Lastly, as can be seen from the results, it is worth mentioning the relative impact that these operationalized practices have on project performance measures. All the supplier-relationship practices are positively correlated with almost all the performance measures, particularly with quality. In the meantime, the correlations of P11.7 ($r = .201$) and P11.8 ($r = .149$) with delivery time are found to be positive but not significant ($p > 0.05$). One possible explanation is that joint-improvement approaches to problem-solving involving partners may enhance quality while also affecting the delivery time negatively, if consensus is not quickly reached. Likewise, P11.1–P11.4 are found to be significantly correlated with quality performance ($r > 0.5$). This suggests that project quality is likely to improve when firms show more respect for their partners' capabilities (P11.1), when more challenges are given to suppliers in terms of collaborative targets (P11.2), when more active participation is taken in their working process (P11.3), and when more opportunities are given to collaboratively work together (P11.4) arise.

8.6.4 The Toyota Way Problem-Solving Model

In Table 8.10, project performance measures are examined in their relationships to three specific principles underlying the Toyota Way problem-solving model, namely P12, P13 and P14. For all the correlation measures listed, only one correlation coefficient between quality performance and *kaizen* (P14) is found to be greater than .05. When firms are dedicated in *kaizen* activities, Imai (1997) found that there is an improvement in the quality of work. In contrast, the principle of *genchi genbutsu* (P12) is found to be insignificantly associated with profitability ($r = .15, p > 0.05$). First of all, P12.4, "*decision making based on management past experience*", has a negative coefficient ($r = -.043$) and is insignificant ($p > 0.05$). Secondly, P12.1 and P12.3 have positive but insignificant coefficients. This suggests that several *genchi genbutsu* practices do not impact profitability with any degree of statistical significance.

8.7 Hindrances to Implementing Toyota Way Practices

Certain barriers exist to the successful implementation of the Toyota Way in construction. Understanding this and the use of appropriate strategies to overcome these barriers should increase the chances of the Toyota Way being successfully implemented. A number of these barriers have been reported in the literature review, but may not be limited to the area of lean construction. This section examines the extent to which this list of factors can be considered as the hindrances to implementation. Table 8.11 presents responses to the question concerning hindrances between "premier" and "first-grade" Chinese construction firms. The higher the mean, the greater the importance of the hindrance. It can be seen from

Table 8.11 Ranking of hindrances to implementing the Toyota Way Principles in China

Hindrances	Overall		Premier firm		First-grade firm		H^b
	Mean[a]	Ranking	Mean[a]	Ranking	Mean[a]	Ranking	value
H1: Lack of a long-term philosophy	3.84	1	3.81	2	3.84	1	.202
H2: Absence of a "lean" culture in the organization	3.81	2	3.85	1	3.81	2	.127
H3: Limited use of design and build procurement mode	3.21	23	3.26	20	3.21	22	.063
H4: Construction firm's limited involvement in the design stage	3.21	23	3.20	22	3.21	22	.001
H5: Foremen's (last planner) insufficient knowledge on project planning	3.59	11	3.33	19	3.59	10	10.72*
H6: Multi-layers subcontracting	3.81	2	3.78	3	3.81	2	.891
H7: Limited use of off-site construction techniques (e.g. prefabrication)	3.23	22	3.22	21	3.19	24	.015
H8: Lack of project management skills (e.g. leadership skills, problem-solving skills, etc.)	3.77	4	3.65	6	3.69	5	.284
H9: Lack of support from the top management	3.75	5	3.65	6	3.71	4	.967
H10: Frequent turnover of workforce	3.66	9	3.65	6	3.62	9	.089
H11: Insufficient training	3.67	8	3.67	5	3.63	8	.037
H12: Employee's resistance to change	3.31	21	3.35	18	3.28	21	.381
H13: Management's resistance to change	3.68	6	3.78	4	3.64	6	1.34
H14: Employee's tolerance for an untidy or disorganized workplace	3.40	18	3.20	22	3.36	18	4.03*
H15: Absence of a "lean" culture in the extended network of partners	3.47	16	3.43	13	3.39	16	.231
H16: Unhealthy competition among suppliers	3.52	14	3.41	14	3.48	14	1.32
H17: Inadequate delivery performance	3.60	10	3.50	11	3.56	11	1.55
H18: Hierarchies in the organizational structure	3.33	20	3.17	24	3.30	20	3.98*
H19: Financial constraints	3.68	6	3.63	9	3.64	6	.096
H20: Less personal empowerment	3.39	19	3.37	17	3.35	19	.002
H21: Avoid making decisions and take responsibility	3.54	13	3.44	12	3.50	13	1.12

(continued)

Table 8.11 (continued)

Hindrances	Overall Mean[a]	Overall Ranking	Premier firm Mean[a]	Premier firm Ranking	First-grade firm Mean[a]	First-grade firm Ranking	H^{b} value
H22: Using "*guan xi*" or relationships to conceal mistakes/errors	3.56	12	3.56	10	3.52	12	.011
H23: Stringent requirements and approvals	3.47	16	3.41	14	3.39	16	.204
H24: Lack of support from the government	3.49	15	3.41	14	3.46	15	.582

*$p < 0.05$

[a]All the listed hindrance was scored from 1 to 5, with 1 indicating not a hindrance, to 5 a very important hindrance

[b]Kruskal–Wallis H-values have been reported for all hindrance

Table 8.11 that all listed hindrances in both groups received a mean value greater than 3.0, which implies that these items are somewhat hindering them from implementing Toyota Way-style practices.

No matter what group the respondents belong to, the most significant barriers that can be ascertained from the rank are H1, "*Lack of a long-term philosophy*"; H2, "*absence of a 'lean' culture in the organization*" and H6, "*multi-layers subcontracting*". Clearly, these three items appear to be of central concern to Chinese building professionals and need to be addressed at an organizational level. Establishing a long-term philosophy and forming a lean culture are of paramount importance for firms embarking on the lean journey. These might seem abstract, but they serve as powerful guiding principles for firms in this changing world. As the hierarchical Toyota Way model implies, the cornerstone of this is possession of a long-term philosophy. Moreover, without a stable lean culture, initiatives such as *genchi genbutsu*, built-in quality, and others will remain empty promises. In the context of China's construction industry, the ranking infers that the two groups of respondents recognized the importance of these two philosophical elements but yet were still found to be lacking.

Furthermore, "*multi-layers subcontracting*" (H3) was among the top three most significant barriers. Given that multi-layered subcontracting is not uncommon in China's construction industry, the problem is that multi-layered subcontractors may have different company culture and different business philosophies. This suggests that it might be a challenge to implement the Toyota Way or lean principles, unless the subcontractors align their firm culture to the general contractor. Moreover, given their relatively high rankings, the barriers "*Lack of support from the top management*" (H9) and "*Management's resistance to change*" (H13) were also found to be significant obstacles for Chinese construction firms. On the other hand, it is surprising to note that the two groups of respondents rated "*limited use of off-site construction technique*" (H7) and "*limited use of design and build procurement*" (H3) as insignificant hindrances. Arguably, lean practices such as just-in-time, built-in quality (*jidoka*), one-piece flow, etc. could be more adaptive to the prefabricated environment, which shares much similarity with the manufacturing setting. In countries like Singapore and Japan, as well as in the Nordic countries where tremendous efforts have been made in promoting greater use of off-site fabrication, construction sites have increasingly become places where the various parts of buildings are assembled. However, the low-ranking of this hindrance implies that the majority of construction projects in China still operate in a conventional way, where off-site fabrication techniques have not yet been commonly adopted. Furthermore, with respect to design and build procurement, Johansen and Walter (2007) outlined that the lack of integration between design and construction is a sign of the early stage of lean construction. They argued that traditional procurement forms do not facilitate lean approaches to project planning and execution. In contrast, this poorly ranked H7 seems to suggest that design and build procurement is not commonly adopted in China's construction industry.

Since the data are ordinal and the responses may not naturally distribute, non-parametric tests have been used in the analysis. Table 8.11 also has a bearing on the last research objective, concerning the extent to which premier and first-grade construction firms (two groups) in China perceive hindrances. As the Kruskal–Wallis *H*-value indicates, for most factors listed in Table 8.11, there were statistically insignificant differences between the two groups. However the difference between the two groups' perception of H5, H14 and H18 was statistically significant (at $\alpha = .05$ level of significance). In the case where the foreman was insufficiently capable in planning (H5), the premier construction firms believed that their foreman's planning ability was relatively better, and so they see this factor as a less significant hindrance. Other outstanding disparities can be found in the way that respondents viewed organizational structure (H17) as well as employees' tolerance for an untidy workplace (H14). The literature review highlighted how hierarchical structure, along with a top-down leadership style, is one of the many culture barriers that causes lean initiatives to fail. It was surprising to see that premier construction firms did not see this factor as a significant hindrance to the implementation of the Toyota Way, and have ranked it at the bottom. This seems to imply that, even though the organization structures in the responding firms were hierarchical in nature, management practises "servant" leadership that is intended to facilitate lean initiatives. In the case of H14, it is implied that the premier construction firms perceived their employees as having less tolerance for an untidy workplace compared to their counterparts from the first-grade firms, thus they will be more willing to practice housekeeping on the site.

8.8 Summary

This chapter presents the results of the questionnaire survey. The questionnaire survey investigated the responses of building professionals in large Chinese construction firms to questions relating to the importance of attributes derived from the Toyota Way model, as well as to the current status of implementation in their firms. The results indicated that there is positive acknowledgement among the respondents as to the importance of the various attributes. The results also highlighted that large Chinese construction firms seem to have implemented Toyota Way practices rather unevenly, which implies that they have both strong and weak capacities in adopting different practices of the Toyota Way. The result of this is that significant differences were found between the actual implementation level and their corresponding perceived importance. The reasons behind this will be explored in the following interview stage. In addition, with respect to the impact of the implementation of Toyota Way practices on project performance, the survey pointed to strong positive correlations between implementation of the Toyota Way principles and performance measures, except that the pull system was found to be insignificantly correlated with client satisfaction, and *genchi genbutsu* exhibited an insignificant association with profitability. The last part of the questionnaire survey is concerned with possible obstacles hindering implementation of the Toyota Way, and its findings have several important implications for managers. For example, they confirmed the widely held view that successful implementation of the Toyota Way in the Chinese context is found among firms possessing a long-term philosophy and a "lean" culture. Without a culture shift to embrace a long-term philosophical statement, and without a culture of commitment, little is likely to be achieved by Toyota Way implementation.

Interview Findings

9

9.1 Introduction

This chapter presents the interview findings to explore in more depth the survey questionnaire responses that concern the extent to which the Toyota Way practices have been practised. Interviews allow a detailed investigation of each interviewee's perspective to gain an in-depth understanding, for example, of their perceptions on the adoption of the Toyota Way-style practices in China's construction industry. For some practices not yet implemented, the interviews also seek to understand the barriers to their implementation. It should be noted that the purpose of the interviews was not to validate the framework but to supplement the survey findings present in the preceding chapter.

9.2 Data Collection

The data collection method involved multiple interviews using a questionnaire. Multiple interviews of key participants were conducted in 16 firms over a period of 2 months (from March to May, 2011). It hoped to provide rich source of data to determine whether there is a practical approach that the Toyota Way principles could be implemented. The interviewees were selected from the earlier participants who showed keen interests in the questionnaire survey. The interviews, which took approximately 1½ h, were conducted at two main venues, namely interviewees' site offices and the head offices. The interview coverage is summarized in Table 9.1. It comprises 17 site staff (e.g. project managers and engineers) and 10 management staff (managing directors, deputy managers, and vice-president).

All the questions were read out by the researcher from a written form (see Appendix D) and their replies were recorded. After some background questions about the firm, their working experience and their overall understandings of "lean", the topic of the Toyota Way model was then introduced. The interviewees were first asked, based on their recent projects or past experiences, how each underlying

S. Gao and S.P. Low, *Lean Construction Management*,
DOI 10.1007/978-981-287-014-8_9, © Springer Science+Business Media Singapore 2014

Table 9.1 Profile of the interviewees and their companies

Code	Designation (years of working experience)	Grade	Ownership	Location
A[a]	1 Project Manager (8)	Premier	SOE	Beijing
B[a]	1 Engineer-in-Charge (5)	Premier	SOE	Beijing
C[a]	1 Engineer-in-Charge (7)	Premier	SOE	Shanghai
	1 Site Engineer (5)			
	1 Commercial Manager (10)			
D	1 Project Manager[a] (10)	Premier	SOE	Wuhan
	1 Managing Director[b] (22)			
	1 Contract Manager[b] (14)			
E	1 Project Director[a] (15)	Premier	SOE	Hangzhou
	1 Deputy manager[b] (26)			
F	1 Manager[b] (12)	Premier	SOE	Shanghai
	1 Head[b] of Engineering Management Department (16)			
	1 Project Manager[a] (6)			
G[b]	1 Vice-President (20)	Premier	SOE	Beijing
H[b]	1 Manager (16)	One	SOE	Beijing
I[b]	1 Regional Manager (12)	Premier	SOE	Beijing
J[a]	1 Project Manager (13)	One	Private	Shanghai
K[b]	1 Vice-President (26)	Premier	SOE	Nanjing
L[a]	1 Site Engineer (5)	Premier	SOE	Wuhan
M[a]	1 Project Manager (8)	Premier	Private	Zhejiang
N[a]	2 Project Manager (15|7)	Premier	Private	Zhejiang
	2 Site Engineers (4|2)			
O[a]	1 Site Engineer (4)	Premier	SOE	Zhejiang
	1 Quality Engineer (6)			
P[a]	1 Project Manager (20)	Premier	SOE	Zhejiang

Note: 27 interviewees in total
[a]The interviews were conducted at project site
[b]The interviews were conducted at firm office

principles of the Toyota Way can be implemented in the context of China's construction industry. Of particular interest was to understand if any barriers or opportunities can affect or facilitate the implementation process.

9.3 Interview Results

9.3.1 Understanding Lean or Toyota Way-Style Practices

In brief, the questionnaire survey found that operationalized Toyota Way-styled practices have not yet been effectively implemented by the large Chinese construction firms (see Chap. 8). Even though there were only a small number of interviewees who stated that they have heard of the term "lean" or components of the lean approach, such as just-in-time (JIT), quality control, etc., a consensus can nevertheless be found on the term "lean management" ("精细化管理" or "*Jing xi hua guan li*") which has the opposite meaning of "extensive management". Lean management, highlighted by several interviewees, was seen as a management approach that requires efforts in planning details, processes, groundwork, execution, performance (e.g. quality, cost, time) and continuous improvement. Not surprisingly, these have usually been ignored by most practitioners. To be more specific, several responding firms stated that attempts were previously made at introducing several new initiatives related to lean principles, including:

1. Tight control and meticulous planning on work plans (e.g. Firms F, G, K)
2. Making reliable work schedules through diligent coordination between trades (e.g. Firms C and I)
3. Enhancing the coordination between design and construction in the early stages (e.g. Firm D)
4. Just-in-time delivery for certain materials (e.g. Firms D and L)

It can be concluded that the term "lean" was interpreted and implemented differently by responding firms. This seems to be implied when several senior interviewees have some knowledge about lean processes, but these tend to be process-focused initiatives rather than a holistic philosophy. Their perceptions of "leanness" fell into the first type of lean model identified by Green and May (2005), in which the elimination of inefficiencies is set as a priority and which involves various lean production tools. Only a few interviewees (e.g. from Firms F and P) mentioned the role of people in the implementation of lean approaches. One project manager from Firm P spoke about the missing link in implementing any new initiatives, including lean. He stated that:

> They (construction workers) are the most critical element in successful implementation of lean or Toyota Way-styled practices. However, Chinese construction workers are extremely undervalued in terms of their social status, their pays, etc. Their turnover is high. In circumstances like this, it is difficult to implement lean.

In addition to the limited focus on people, industry-specific challenges, such as the lack of a supporting environment, also make the task of lean implementation more complex and challenging. One project engineer from Firm C added that:

> It is almost impossible if only one party (e.g. the construction firm) determines to embark on the lean journey. That construction firm will soon give up. This is because lean is about working in a different way that all the involved parties need to understand and agree on accordingly. In another word, all the parties (e.g. clients, subcontractors, etc) involved in one project should collaboratively implement lean, or the desired benefits can hardly be achieved.

Furthermore, respondents from different type of firms hold different perceptions of lean or Toyota Way practices. Those from state-owned enterprises (SOEs) appeared to have more knowledge of lean practices, whereas their counterparts from the private sector were equipped with less knowledge.

9.3.2 Long-Term Philosophy

The earlier questionnaire survey relating to the assessment of long-term philosophy implementation revealed that Chinese construction firms appreciate these values and philosophies, and had implemented some of these practices to significant extent. The interview findings reinforced these survey findings, including that these philosophical guidelines are not new to large Chinese construction firms.

9.3.2.1 Sense of Constant Purpose

Interviewees noted that, in most cases, constant purpose pertains to their firms' visions, missions, and values, all of which were explicitly highlighted on their firms' websites. One deputy manager from Firm F spoke about how the historical events that the firm had experienced would in turn have influenced the firm. He stated that:

> Constant purpose concerns who we are. Our firm was founded with a military background and this has been around for a long time. The high quality of military norms and conducts were maintained and encouraged by our leadership to apply in our daily work. I believe the firm, as well as the employees, have benefited from such norms and codes, which later became our constant purpose.

9.3.2.2 Be Self-Reliant and Responsible

In the same way that Toyota takes responsibility for car owners, society, the environment, etc., the responding firms that were interviewed also uphold similar responsibilities in many areas. To be responsible for the construction quality, for decisions made in the bidding, and for employees' health and safety, to name a few, are all considered important. For example, Firm B is one of the few firms that addresses the issue of being responsible in terms of what it promised in the bidding proposal. One engineer from Firm B highlighted that:

> Since the bidding process is very competitive in China, some bidders deliberately squeeze the timeline to win the project. We assure the client that the project will be completed on time, based on what we promised in the bidding stage. This shows our responsibility.

With respect to the issue of self-reliance, a large number of responding firms know what their core competencies are and what key technologies they possess. For example, Firm D specializes in airport work nationwide, now actively sought opportunities in the international airport-construction market. Its deputy manager explained:

> We have built several airports both in the domestic market and outside China. I am confident that we are the leaders in this area.

Additionally, Firms I, K and L are construction engineering firms with a strong engineering focus, and specializing in steel-structure buildings, nuclear power plant projects and infrastructure projects, respectively. Each firm has its own research and development institutions to support their businesses and to enhance their technical know-how. Because of their rich experiences accumulated and cutting-edge technologies acquired from their in-house research and development department, they have become very successful in China.

9.3.2.3 Long-Term Perspective

In the interviews, most firms acknowledged that it is not easy to do business in China's construction industry with complete focus on the long-term perspective. This is unlike business in the Toyota Way which is not about making a single profitable deal, but is about building long-term relationships with business partners, customers and employees (Liker 2004). The Chinese construction industry has many short-sighted players, so much so that some firms have to follow similar practices. This was reflected by several interviewees that money is important and perhaps the most important aspect of managing a firm or a project. For example, one project manager from a large private construction firm (Firm J) revealed that it is common among private firms to tend to be more short-sighted in this regard. This is because private firms in China's construction industry are cost sensitive, and therefore place greater emphasis on cost control. With such a mindset, it is less likely that they can see a big picture from a long-term perspective. This attitude was viewed as one of the major problems that affected the Chinese construction firms in developing their employees, fostering relationships with suppliers and others.

9.3.2.4 Client (Customer) Focus

A majority of interviewees agreed that client focus was one of their chief priorities in the project. Here client focus was interpreted to mean meeting the client's requirement by adding value to the project. This perhaps explains the reason why P1.8 "*be able to rapidly respond to clients*" emerged as a high-order attribute in the questionnaire survey (see Chap. 8). From the collective replies, several means were mentioned that could be adopted to add possible values to the project as well as to enhance clients' satisfaction. These include:

1. Cost minimization: this can be applied during various phases of a project. When applied earlier, such as at the bidding stage, tools including value engineering, process optimization and others can be used to identify opportunities that may be associated with cost deduction. Responsible contractors will not propose cost-cutting ideas that they cannot promise. When applied during project execution, one regional manager from Firm I spoke about standing in the client's shoes, as a way to look at opportunities to minimize cost. For example:

    > We understand the fact that the client is sensitive to cost, and therefore always look at the cost issue from their perspectives. That encourages us to advise the client various ways of how to reduce project costs. For instance, we can advise the client which materials could be replaced by others of equivalent quality but much cheaper.

2. Speedy completion: there seems to be an emerging trend that timely completion is becoming clients' primary concern. To assure the clients that their projects can be delivered on time, some interviewees mentioned that their project team should be more responsive when the projects were slipping behind schedule. Responsiveness is measured by the speed with which the construction firm could

get the project back on track, as well as the extent to which the construction firms were willing to invest in more resources.

3. Quality improvement: a majority of the firms interviewed focused on the quality of project. For example, one project manager from Firm A highlighted how his firm was in the middle of a transition as the emphasis was becoming less "product-oriented" and more "process-oriented". One reason for the introduction of the "process-oriented" philosophy was that the leadership realized that only the right process can yield the quality product and that can further enhance the client satisfaction.

9.3.3 Process-Related Practices

Open-ended questions during interviews were asked to understand practices which the firms interviewed could adopt in order to implement the underlying principles of the Toyota Way Process model from their experience or relevant projects. Correspondingly, questions in this section covered the areas of uninterrupted work-flow, material planning system, (built-in) quality control, planning and scheduling, standardized work, visual management and use of new technology.

9.3.3.1 One-Piece Flow (P2)

One-piece flow is concerned with eliminating the waste or non-value-adding activities that would result in disrupting the flow or efficiency loss. Based on the interview findings, it appears that an uninterrupted work-flow can be achieved by three possible ways that were adopted by a majority of the responding firms, including:

1. to maintain a balanced volume of work through good construction plans,
2. to ensure that the materials supply is uninterrupted, and
3. to ensure adequate workforce.

Maintain a Balanced Volume of Work

A large number of interviewees replied that an uninterrupted work-flow is achievable in the context of China's construction industry. They highlighted that idle crews were uncommon in their respective projects. One department head from Firm F pointed out that:

> Even at the peak of the construction when the number of our frontline workers amounts to more than 2000, there are still gaps which show signs that more manpower is needed and that there is sufficient work space to accommodate them. So, whoever comes to my project, I will assign jobs to them easily. Hence, crew idleness is not an issue we are concerned with.

This seems to suggest that creating more workface availability and maintaining a balanced volume of workface is one effective means of ensuring that the crews can work on a continuous flow of work. Some volume of work might be a plan buffer, which refers to a step, or some tasks that are part of a plan, but are not yet scheduled, or can be rescheduled (Tommelein 1998). This, however, increases the likelihood of space conflicts, because more workers would be working on the site since more workfaces are released. To alleviate the space tensions between different trades or subcontractors, one of the interviewees—a project manager from Firm A with 10 years of experience—pointed out that, in large construction firms like Firm A, a manual or handbook on "flow construction" was used which provided "how-to" guidelines for the project manager in designing the process of flow construction involving plot-planning, path planning and sequencing within and between different trades and teams.

Uninterrupted Materials Supply

Ideally, in the one-piece flow environment, materials are to be delivered only when they are needed (Liker 2004). It is difficult to apply the Toyota Way method of materials management to China's construction industry, as the interview results indicated that the current practice of materials management is to stockpile materials on site despite the site being faced with a constraint. With an adequate inventory on site, not only is it more likely that rising costs of construction materials [e.g. reinforcement bars (rebar)] can be overcome, but that it also allows uninterrupted work-flow caused by failures in the delivery of the materials needed (Schmenner and Swink 1998).

Role of Workforce in Creating Uninterrupted Work-flow

Workers also play a pivotal part in creating uninterrupted work-flow. This is because one important tenet of one-piece flow is that workers should be capable of identifying and eliminating the non-value-adding activities that would affect their productivity (Liker 2004). Not surprisingly, the interview results revealed that presently, Chinese frontline workers are not regarded as productive compared with their Japanese counterparts. Most respondents indicated that a majority of the frontline workers are low-skilled migrant workers, and most of them cannot be expected to have a clear picture of what non-value-adding activities or *muda* are, let alone to improve the process by eliminating them. The way they learn and perform their work is simply through observing experienced co-workers or following direct instructions from supervisors, and trying things out by themselves. The majority might not bother to think about how to improve the present process by eliminating non-value-adding activities. Rather, what concerns them most is how much they would be paid at the end of the day. Because of the inability of the frontline workers, it is therefore required of the project manager and his or her project team to work harder and to be dedicated from the outset of the project to removing the uncertainties and creating an uninterrupted working environment for the workers. Unlike the frontline workers, most interviewees understood the consequences of various types of *muda* in the construction context. Even though some felt that, due to the unpredictable nature of the construction site, the non-value-adding activities such as double handling, inventory, and defects were not easy to eliminate. A number of areas were repeatedly mentioned that were worthy of the project team's attention, including:

1. Optimizing construction plan and site layout design,
2. Optimizing site logistics to reduce double handling,
3. Enhancing on-site quality management to eliminate defects,
4. Closely monitoring the status of frontline workers to quickly identify idle time, and
5. If possible, providing training for frontline workers in the areas of waste elimination.

9.3.3.2 Pull *Kanban* System (P3)

The discussion on the implementation of Principle 3 mainly focuses on materials planning and management. In the construction industry, material management consists of materials procurement, delivery (transportation), storage (inventory), etc. (Thomas et al. 2005).

Pull and Push: A Mixed Method

Even though the pull or push system terminology was unfamiliar to the interviewees, the interview results suggest that the industry predominantly used the push or just-in-case method to managing the commonly used materials. Table 9.2 illustrates these mixed characteristics in three areas, namely material ordering, delivery frequency and inventory.

1. Materials ordering from suppliers

   Through qualitative analysis of the interview results concerning firm's ordering system, it was found that presently, a majority of the firms interviewed have quite sophisticated software to calculate the total quantities of various materials used for bidding purposes. However, when it comes to the project level, most interviewees stated that a simple ordering policy was used. Basically, the project office gives the head office (or regional office) its demand figures, usually in a weekly or monthly form which is an estimate calculated in accordance with the project's progress. Other factors, such as level of inventory and labour resource, were also taken into account. The head office then reviewed the materials plan and processed the orders accordingly. Generally speaking, there were three approaches for different materials:

   (a) *small items (make-to-order)*: these materials or items are typically ordered or picked from merchants for immediate use such as nuts and bolts. This only requires a short cycle time (1–2 days or even within a few hours) for delivery. These items are usually kept in the store room at the project site.
   (b) *commonly used materials*: these include steel rebar, bricks, etc. Several interviewees from SOEs revealed that the project team needed to submit its weekly or monthly materials requirement plan a few days in advance to the purchasing department in the head office for review and approval, whereas private firms decentralized materials procurement at the project level. To request for quotations, price requests were sent to potential suppliers, whose particulars were listed in the firms' own databases.
   (c) *special orders*: special orders apply to engineer-to-order parts, which are unique for construction projects and typically have long lead times prior to

Table 9.2 Characteristics of materials planning and management in China's construction industry

	Practices of Chinese construction firms	The pull method
Materials ordering	• Largely determined by the forecasted price of the material (push) • Purchase-to-stock if the material price is predicted to be low (push)	• Based on actual needs to subscribe needed materials (pull) (Monden 1983)
Delivery frequency	• Materials such as ready-mixed concrete and customized components were ordered using pull system • Most commonly used materials are delivered once or twice every week (push)	• Multiple delivery (e.g. several times within a day) from suppliers (pull)
Inventory	• Relatively high level of inventory • The stock can last for several days (push)	• Low level of inventory (pull)

their deliveries to the construction site. As a result, these parts are ordered on a monthly basis.

2. Delivery
   In the JIT system, suppliers are expected to deliver small lot sizes of materials frequently. The lot size is determined by the actual consumption of materials, rather than by forecast demand. According to the interviewees, it is hard to adopt JIT delivery when considering how the supply chain works in China's construction industry. Given that one supplier usually has many business partners, the daily priority is to satisfy the high or medium volume orders. Because an order of a construction project for one piece of material is regarded as small, it is therefore harder to demand the supplier to make commitments to multiple deliveries. This is particularly true when frequent deliveries are required and logistic costs increase. Accordingly, in order to reduce the logistic costs, agreements were made between the two parties to the effect that the materials are delivered in large quantities, despite causing unnecessary inventory. Several interviewees mentioned that ready-mixed concrete was one exception. This type of material could be delivered to the site using JIT method, but only if the delivery rate from the concrete firms is compatible with the progress of a concrete job in the field.
3. Inventory
   In contrast to Toyota's pull system, which results in limited inventory at the shop floor, almost all the firms interviewed preferred to keep an accepted level of safety stock in the site. The level of safety stock varied, but it was generally agreed by most responding firms that it is necessary to maintain an inventory of at least 5 days' worth of material, or up to 2 weeks for site use. One project manager from Firm A pointed out that 5 days' worth was the smallest amount of time that his site could achieve. Gypsum boards were used as an example by this interviewee:

If the gypsum board was ordered one week in advance, we are able to get a price of about 20 RMB per m2. However, the vendor will charge an extra 0.3 RMB per m2 if they are asked to delivery and given only one day notice in advance.

Not surprisingly, most firms did not have a computerized system to electronically monitor the level of inventory at the project level. Therefore, the site personnel has to conduct daily, weekly, and monthly inventory checks and inform the project manager of the inventory level. In general, the item is re-ordered if its number drops below a specified re-order quantity. This practice indicates that, unlike the case of manufacturing firms using "*kanban*" as a vehicle to communicate materials information, the Chinese construction firms manually monitor the inventory level and file a re-order on a regular basis, whether they need it or not. This is evidence that a push strategy is presently applied in most of the firms interviewed.

Causes of High Level of Inventory

According to the interview results, the implementation of a pull *kanban* system with the hope of minimizing inventory is often impeded by the external business environment—by the unstable price of construction materials, as well as by employees' traditional mindsets.

1. Unstable price of construction materials

The upward spiralling costs of important construction materials have put great pressure on almost all the firms interviewed. The Chinese construction industry has seen a steady rise in the prices of steel, cement, bricks and other materials. Material prices updated in several online agencies showed that the prices vary even from day to day. One commercial manager from Firm C highlighted that:

I am very concerned about the costs of materials going up. The biggest issue is the rate of the increase. In the past couple of years, it has escalated significantly.

Taking steel—the most demanded construction material—as an example, one project manager from Firm P outlined that:

Steel for construction use has seen its price increase by as much as 35 per cent from 4,000 RMB/ton (equivalent to US$700) in 2009 to current 5,500RMB/ton (equivalent to US$900) in 2011 March.

Unstable prices are perhaps one major factor that discourages a "pull" system from being implemented in China's construction industry. A majority of the firms interviewed have a full-time estimator to keep an eye on the marketplace, and to update the budget accordingly. To shield themselves from rising costs, almost all the responding firms adopted the strategy of stockpiling, by appropriately and economically securing a considerable amount of construction materials (e.g. steel) at a fixed price in advance. The commercial manager added:

We have to stockpile materials. We also use one estimator to closely monitor the price movement. The main purpose is not to protect against unreliable delivery of materials, but to manage escalating price.

2. Traditional mindset
   Several interviewees highlighted that, apart from unstable prices, keeping safety stocks allows production to continue. Such an attitude may be embedded in a firm's culture. Most interviewees were convinced of the superiority of keeping safety inventory on site. This agrees with Polat and Arditi's (2005, p.710) study, which concluded that: *one possible benefit of keeping inventories on site is shielding downstream activities from upstream activities*. One interviewee from Firm N responded that:

   > The fact is that these materials are only temporarily stored here. Clearly, all of them will be consumed eventually when the project progresses. At the end of the project, we are very cautious to procure materials as we need them and to carefully monitor the inventory. Overall, the wastage of material is under control.

   Moreover, the interview results showed that keeping safety stocks is also strongly associated with the client's intentions. One interviewee mentioned that, from the client's perspective, a sufficient inventory symbolized that the overall progress of the project was good. Excessive inventory on site gave a strong signal that the project was ready to release more workfaces for workers. One experienced site engineer added that:

   > Sometimes, Jianli (known as a site supervisory company) will not allow the job to be performed if the material is just the exact amount needed. For example, a concrete in-situ job is scheduled to commence in the afternoon and the planned amount of cement is 20tons. Jianli will not authorize the construction firm to perform the job if there are only 20T of prepared cement on site. This is because the wastage of cement must be taken into account.

9.3.3.3 *Heijunka*: Level Out the Workload (P4)

Heijunka is a production planning method, which evenly distributes the production volume and production variety over the available production time. To translate it in the construction context, this study connects the principle of *heijunka* (P4) with the Last Planner System's four levels of planning phases to find common grounds (see Sect. 4.7.2.1). To supplement the survey findings, questions were raised during the interviews concerning the followings:

1. What levels of project planning are adopted?
2. Who is the Last Planner at the project?
3. How reliability is project planning (e.g. weekly plan)?

Level of Project Planning

The interviewees were asked whether their project plans were designed with hierarchical levels (known as master plan, phase plan, look-ahead plan and weekly plan). Table 9.3 illustrates the levels of details in terms of project planning by the responding firms.

1. The master/phase plan
   Table 9.3 indicates that the master plan along with the phase plan is available in nearly all the firms interviewed. The master plan was a mandatory requirement in the contract which was the contractual agreement between the client and the

Table 9.3 Adoption of LPS or *Heijunka* principles

Firm	Master plan/ phase plan	Look-ahead plan	Weekly plan	Daily huddle meeting	Critical roles in the planning process	Reliability of plan (do they use a matrix; yes/no?)
A	√	X	√	√	• Project Manager (PM) • PM needs to constantly consult with foreman as well as subcontractor	• Fairly good
B	√	Monthly plan	√	√	• PM	• Fairly good, but sometimes delays occur because of design variations
C	√	Monthly plan (seldom update)	√	√	• PM and Project Engineer (PE)	• Fairly good
D	√	Monthly plan	√	√	• PM	• Fairly good, as milestones can be achieved, but the course to achieving it may be different to what is in the planned schedule
E	√	Monthly plan	√	√	• PM makes request to subcontractors to review their own weekly plans	• Fairly good. If a delay occurs, PM makes a request to the subcontractor to engage more resources in order to make up the lost time
F	√	Monthly plan	√	√	• PM has a critical role in producing the plan, but it requires the parent company (e.g. Planning/ Estimating Department) for approval and documentation • Subcontractors submit their weekly plans for PM's approval and reference	• Reliability is good. However, constant minor adjustments are needed but acceptable

(continued)

Table 9.3 (continued)

Firm	Master plan/ phase plan	Look-ahead plan	Weekly plan	Daily huddle meeting	Critical roles in the planning process	Reliability of plan (do they use a matrix; yes/no?)
G	√	Monthly plan	√	X	• PM	• Fairly good
H	√	Monthly plan	√	X	• PM is in charge, and consults with subcontractors who need to feedback information on resource, weather, etc.	• Fairly good
I	√	5 weeks ahead + current week	√ (issued on every Wednesday)	√	• PM	• Good
J	√ (aim for 1–2 months compression)	Monthly plan	√	Weekly huddle meeting	• PM	• Not very certain about the reliability of weekly plan. Sometimes it is ahead, while sometimes it is behind the schedule
K	√	Monthly plan	√	Weekly huddle meeting	• PM is in charge, and consults with foreman of each trades	• Fairly good
L	√	Monthly plan	√	X	• Full-time staff from head-quarter • Planning/ Estimating Department is stationed at the project site	• Relatively good. Sometimes it will get ahead of schedule by completing unassigned jobs
M	√ (no phase plan)	Monthly plan	Bi-weekly plan	X	• PM	• OK. Often the schedule is affected by material and labour shortage
N	√	Monthly plan	Bi-weekly plan	X	• PM needs subcontractor's agreement on the work contents and target set in the given week	• Labour shortage at the moment, so there is a slight delay

(continued)

Table 9.3 (continued)

Firm	Master plan/ phase plan	Look-ahead plan	Weekly plan	Daily huddle meeting	Critical roles in the planning process	Reliability of plan (do they use a matrix; yes/no?)
O	√	Monthly plan	√	Weekly huddle meeting	• PE is in charge and is required by PM to submit the plan for his final approval	• Generally good. If delay occurs, PM will request subcontractors to engage more resources to make up the lost time
P	√	Monthly plan	√	X	• PM • The ultimate purpose of a weekly plan is to inform not only the project team, but the frontline workers	• Generally good

"√" denotes that the respondents have implemented
"X" means "not implemented"

main contractor prior to the award of the contract. One Project manager from Firm D added that:

> The master plan incorporates inputs from client's team in which a few key deadlines (milestones) are clearly set. However, the master plan is of little value in guiding the daily operations because it only defines tasks at an abstract level. In most times, the master plan is more useful to be used in claims for delay against the client.

In addition, the phase plan is derived from the master plan and is used to guide the work of phase package. There are financial incentives agreed by each party that by successful completion of the work package in the phase plan, the financial awards will be given.

2. The look-ahead plan
Unlike the master plan, a look-ahead plan was found to be absent in almost all the firms interviewed. Technically speaking, Firm I is the only one that has real-life experience in making a look-ahead plan, but this was undertaken in an international project. As revealed by its manager, the look-ahead plan was actually requested by the client, when the firm was involved in an overseas project in Singapore. As noted, the look-ahead plan aims to establish a plan that contains weeks of workable backlog. Similarly, a look-ahead plan has an equivalent term and is more prevalent among the responding firms in the form of a "monthly plan". A monthly plan comprises 4 weeks' work plan and provides good look-ahead views of tasks that need to be completed in the coming month. However, compared to look-ahead plans, monthly plans turn out to be less

dynamic and less rigorous. Some responding firms revealed that if a monthly plan is produced on day 1, no further attempts were usually made to update, even if changes do occur.

3. Weekly plans and weekly meetings
   A weekly plan was adopted by most interviewees at the project level. Only two responding firms stated that they used a bi-weekly plan instead of weekly plan, on account of clients' desire to be updated on the project's progress every 2 weeks. According to the interviewees, before the weekly plan is finalized, it is necessary for the planning team to consult with the foreman and subcontractors about the availability of resources, the constraints of their current work, and others. As a result of these considerations, a weekly plan is created and is announced during the weekly meeting, which all relevant parties attend. According to the interviewees, a weekly plan generally consists of three parts:
   (a) Overview of jobs/tasks that have been completed in the past week.
   (b) Overview of jobs/tasks that need to be completed in the coming week.
   (c) Analysis of roots causes for the schedule delay and associated countermeasures.

   As some interviewees revealed, some levels of consensus on these three parts would be reached among all the parties. However, this form of weekly plan suffered from the lack of details on assigned job, compared to the weekly plan designed in the Last Planner System. Also missing were information on the "detailed contents of daily tasks", "who is in charge", "potential hurdles that might affect reliability", and so on.
4. Daily huddle meeting
   A few interviewees outlined that there were short meetings held by the project team at the start of a workday. The purpose is to convey the daily goals established during the weekly meeting to the entire workforce. Issues relating to OHS, drawing variation and others will also be addressed only. A few interviewees stated that a similar type of meeting was held at the end of each workday to review what had actually been completed and to discuss the potential constraints that the team might encounter the next day.

The Last Planner

As shown in Table 9.3, 14 out of 16 responding firms pointed out that their project manager (PM) was expected to play a "critical role" in making a project plan, with some assistance from site engineers. Clearly, project planning cannot be done as a one-man effort. About half of the interviewees highlighted that it was essential to consult the subcontractors or the foreman (concerning, e.g., the availability of resources, site constraints, etc.) and integrate the work plans with the project team's plan. One engineer pointed out that sometimes the project manager was so busy that he or she might not really be involved in the planning process, but was expected to review and approve the schedule prepared by his direct subordinates (e.g. engineers). This implies that the traditional planning approach in which the plan is prepared by the project manager and his team is still predominately used in the Chinese construction industry, although consultation with subcontractors or foremen can be seen on site. In the end, it is the project manager who makes the final decision on what the weekly plan should look like. Moreover, it is rare for the foreman or subcontractor to reject the project manager's proposal—instead, they need to hammer out a solution by all means based on the weekly plan announced—i.e. make necessary modifications to their own plans.

Furthermore, two possible barriers to the implementation of a bottom-up process of project scheduling were uncovered:

1. Unqualified foreman (last planner)

   The first and perhaps the greatest barrier is the poor capability of the trade foreman or the last planner. It was pointed out that the problem was that subcontractors or foremen (of trades) in China's construction industry tended to concern themselves with their own work's scope, and thus failed to see the bigger picture of interacting with other trades or teams. One project engineer from Firm C commented that:

   > Those foreman or trade leaders basically have no idea of where they are now in the flow of work. Therefore, it requires the PM to extensively coordinate between different trades to minimize (space) conflicts, idle times etc. In most cases, it requires us to employ a top-down plan rather than to do it in a bottom-up way.

   On the other hand, some interviewees seemed to show little confidence in what has been promised by the trade foreman in terms of the work that needed to be done by a certain date. One senior manager from Firm F added:

   > …even though the foremen might have made 'realistic commitments' in delivering what they call the 'can-do' work assignments, due to uncertainties and changing site conditions, they would eventually fail to get the work done. I personally have no confidence in what they have committed to.

2. Tight schedule pressure

   The client's demand for timely completion of the construction project was another barrier cited for the adoption of a bottom-up planning technique (e.g. LPS). Several interviewed firms revealed that they were under enormous pressure to produce a project plan using scientific planning principles. For example, the manager of Firm F explained that:

   > A domestic private client requires us to deliver a project of 530,000m2 in 18 months. If we do not take this job, the client will find a replacement easily. In a case like that, if we adopt the LPS as it allowed and encouraged the foreman to make commitments to completing workable tasks, we would never expect the job to be done on time. It must be a 'top-down' approach to enable the project team to design and to optimize the job sequence, and give trades immovable deadlines, and by all means to provide assistance to help them finish the work within the given duration.

   In addition, the same interviewee introduced a so-called bottom-line management concept in her project to ensure timely completion as required by the client:

   > … for example, a curtain wall package for one specialty subcontractor. What I mean by the bottom line management is that the curtain wall subcontractor is only given "windows", as well as the latest completion time (the bottom line) for the execution of their responsibilities on site. They have no choice but to complete the job before the deadline. If the job is delayed, assistance will be provided accordingly.

Reliability of Work Plans

No matter how well the weekly plan is prepared, most interviewees agreed that a highly reliable weekly plan was not easy to achieve. This is because site conditions are so unpredictable that the weekly plan was sometimes delayed by 1 or 2 days, which are then postponed to the following week. The delay calls for immediate actions by several means such as allocating more workers and other resources, working overtime, etc. to get the project back on schedule as soon as possible. One interviewee from Firm J explained that:

> It is always like that (slipped schedule). As long as the milestones in the critical path are not affected, this sort of schedule slippage is acceptable. We understand it is not uncommon that enormous variations will cause our workers not to complete the job on time. Most of the times, it would be like this: one week we are a bit ahead of the schedule and the next week some tasks have slipped to the week after. Overall, we are always in the middle of re-planning the project, re-estimating the time, and re-adjusting the resources accordingly.

It is understandable that in any project, most activities have some leeway or "float" that permit them to start at a late time. Properly controlled, this float is valuable in regulating the use of workers, materials, and other resources. As part of the lean construction tool, Percent Plan Complete (PPC) is widely used as one key matrix in lean construction for measuring and monitoring variations in the weekly plans (see Ballard 2000). When asked whether PPC or other forms of matrix were employed to monitor schedule reliability, a majority of the firms interviewed stated that these were not adopted. Moreover, the use of PPC was criticized by one of the interviewees, a senior manager of Firm F, who said that:

> In the Chinese construction context, PPC can only be done in the research setting. It is not likely, for my firm at least, to apply this tool even though it sounds theoretically feasible. This is because not only are management efforts always inadequate on site, but also the pressure of tight schedules do not allow us to conduct such matrix work.

Several interviewees stated that although such a matrix was not used at the moment, general constraint analysis was performed on a weekly basis (at the weekly meeting) to analyse the causes of variations that may affect the upcoming weekly plans.

Muri and *Mari*

Apart from LPS discussed under this principle, *muri* and *mari* were also addressed during the interviews. Given that labour is still in high demand, the workforce on site have no choice but to work longer hours to keep the project on track, without which there is only a slim possibility of achieving timely completion. Therefore, it is very common to see workers having to work on the weekend, sometimes having only 2 days off in a month. Even during the holiday period (e.g. Chinese New Year), excessive overtime remains prevalent. As a result, the client is assured that the project progresses well, but *muri*—the uneven work—is frequent and obviously ignored by most responding firms. This seems to suggest that the workforce is undervalued and is often treated as a machine, rather than as the industry's most important asset. A possible explanation could be that the desired deadline for completion is unreasonably short in comparison with a normal schedule based on employees' benchmarked productivity. Several interviewees also attributed this unreasonable pressure from the client in the industry's unregulated environment.

9.3.3.4 Built-In Quality (P5)

Built-in quality (*Jidoka*) is another key element of the Toyota Way process model. The questionnaire survey had assessed the level of built-in quality practices (P5) among Chinese building professionals and found that several practices with *jidok*a thinking were in place. The follow-up interview sought to understand this principle in more depth. Inquiries included:

1. How non-conforming parts are identified?
2. What are the challenges in implementing built-in quality in daily operations?
3. How is the QC implemented at the project level?

Identification of Non-conforming Parts

In response to the question of how "non-conforming parts" were identified, almost all the responding firms indicated that—unlike Toyota's *andon* system or TQM's promotion of "*do it right in the first place*"—Chinese construction firms rely on various inspections. These included the "in-process" inspections and handover inspections. Among multiple participants who are involved in the quality inspection, it is mainly the site engineers, and not frontline workers, who take the credit for identifying non-conformance. It was pointed out that site engineers and foremen spent most of their time on a construction site acting as a source of technical advice and quality control. While on the job, they were very dedicated to keep their eyes open for improper procedures and sloppy work. Site engineers are constantly reminded to adhere strictly to the checklist and encourage to live the belief that "*defects or non-conforming work will not be tolerated*". Once the problems are discovered, these need to be solved by any means immediately. For serious quality problems, a formal rectification order is issued. In their views, it was hoped that this approach would foster quality awareness at the site level as it provides an opportunity for workers to minimize inappropriate performance. Because the workers are surrounded by technical personnel, occasionally, the workers will seek some help to resolve problems. This perhaps explains why one attribute, namely "frontline workers are encouraged to stop the operation and seek assistance from supervisors to solve the problem" (P5.4) was rated highly in the questionnaire survey results.

In summary, simply relying on site engineers for quality assurance is inadequate, as there is the possibility that a site engineer or supervisor might miss some critical checks, with the result that these defects would go unnoticed into the next step. The interviewees also commented that to further boost awareness of the built-in quality philosophy, total involvement of all employees was required, especially to ensure that such ideas would reach the frontline workers, as most construction works are actually done by them.

Challenges in Implementing Built-In Quality

The biggest challenge cited repeatedly by many respondents was that the rapidly expanding construction industry in China takes in a great number of unskilled construction workers who have fairly low levels of awareness and knowledge of quality. As highlighted in the literature, a majority of the frontline workers in China were literally migrant workers who were registered with various labour-only subcontractors. The chief engineer from Firm C revealed that:

> If the offer is low but still accepted by the labour-only subcontractor, it is very likely that they will send the most low-skilled workers to us. It is really hard to manage the quality of such workers. Too often, they kept silent about damage or results of poor workmanship.

This was confirmed by a number of interviewees, such as one manager of Firm H, who put it that:

> Take wall plastering for instance, given the dry climate in northern China, the construction method clearly states that frontline workers must wet the wall prior to the wall roughening treatment. The quality of wall will be affected if 'wet' treatment is not properly done and it will lead to cracks in several months. The problem is that the frontline workers knew that the cracks would not immediately be noticeable by the auditor, and so some irresponsible workers will 'play smart' and skip this essential step, and proceed to the next.

A different perspective was offered by one manager from Firm I, who outlined that:

> ...built-in quality has not been reached at this moment in most Chinese construction firms as far as I am concerned. This is because in construction, each project is different and each drawing is different too. Unless the employees or workers are professional, it is challenging for them to stop and ask for assistance.

Quality-Related Activities

All the 16 firms interviewed had embarked on the quality route using quality circle (QC) for quality improvement. However, in comparison with the Japanese QC, the QC presently conducted in China's construction industry tended to be characterized by some differences, which are highlighted in Table 9.4.

1. Quantities

   According to the interviewees, QC activities are not actively conducted at the project level, even though most responding firms are aware of its importance. Generally, one construction project only aims to foster one case of successful QC. As one Quality Engineer from Firm O explained,

   > It is good enough to have only one QC. This is because if more QC activities are conducted, extra management efforts, as well as more human resources from different departments are needed, and perhaps more budget for it. We really do not have time and resources for this.

   Given that all the firms interviewed are large Chinese construction firms, they are expected to have widely spread sites geographically and projects going on at the same time. Fortunately, by the end of the year, construction firms may achieve multiple cases of successful QC, as well as attain tangible results. In order to recognize the outstanding contributions to project performance and to

Table 9.4 Comparison of Chinese QC and Japanese QC in construction projects

	QC conducted within Chinese construction firms	Japanese QC activities
Quantities	• One or two QC were conducted within one project • No meetings to discuss the quality problems	• A large number of QC activities (e.g. per year) • Frequent meeting to discuss the quality problems
Enabler	• External forces such as to fulfil the requirements and be eligible for the quality competition	• Employees recognize the potentials for improvement (Imai 1997)
Approach	• Top-down approach: project manager determines a QC topic for his team to carry out	• Bottom-up approach: employees voluntarily participate in the QC activities (Lillrank 1995)
Overall goal	• Aim at "ad hoc" improvement if possible	• Continuous and small improvement (Imai 1997)

give the QC participants the satisfaction of achieving their goals, a number of firms interviewed introduced award schemes to recognize excellence in QC achievement in their construction projects. What is more meaningful is that a number of successful QC cases were compiled, published, and circulated within the firm and between the projects, so that all employees can learn from the experience. In addition, employees with outstanding QC cases are encouraged to compete for higher-level competition, i.e. national QC awards.

2. Enabler

   Today, specific requirements such as to conduct effective QC activities in construction projects are indicated in the contract documents. Based on the interview results, the by-product of QC, namely employee motivation, was largely ignored. Conversely, the ultimate drive for them to carry out QC has nothing to do with striving for excellence, but to simply fulfil the contractual obligations, or to use quality awards as the selling point to increase the likelihood of winning future projects.

3. Approach

   As Table 9.4 indicates, QC activities in the Chinese construction industry are predominantly conducted in a top-down fashion. Many of the firms interviewed pointed out that the project manager acts like a moral supporter, rather than a leader of the QC team, whereas the real efforts came from the project engineer who is in charge. Moreover, the interview findings suggest that some QC teams are inappropriate in terms of their structures, in that the frontline workers are often excluded. This implies that the QC activity is limited to among the site personnel and the frontline workers may not even be aware of the existence of QC throughout the project. It is common knowledge that frontline workers and their supervisors know better about the processes and are better able to identify the possible problems in their workplace. One interviewee from Firm P explained that:

> It is close to impossible, at least at this moment, that our workforce could contribute their feedback and be participating in QC activities as their Japanese counterparts do. All in all, the quality of workforce is the biggest difference.

4. Overall goal

   Some respondents stated that they were currently struggling to define an appropriate QC problem to form a QC topic. For a number of projects that the researcher visited, most have not yet identified a QC topic even though the QC team has been formed. Most responding firms outlined that efforts have been made in searching for relevant problems to hope to achieve breakthrough improvements. One interviewee pointed out that:

> It is getting harder for us to choose a topic. Many have been investigated by colleagues or industrial practitioners. So far, our QC team has still not decided on what to work on. This perhaps is the reason why our project has progressed to almost half way but the QC has not yet been formally kicked off.

9.3.3.5 Standardization (P6)

The term "standardization" in the context of lean means established standard operating procedures (SOPs) for workers in performing their job tasks with fewer variations. It has a broader meaning in Toyota's context, which includes empowerment to be given to the frontline operatives to continuously "*kaizen*" the current SOP for the better. Not surprisingly, a majority of the interviewees pointed out that standardization could work better in the repetitive manufacturing environment, rather than in the construction industry where projects are typically one-off and unique in nature. This sentiment helps to explain the challenges the industry faces in the implementation of standardization (P6). However, in their views, standardized work is applicable to some extent and is regarded in a much wider sense. This includes two major levels, namely the standardization at the project operation level and the standard practices at the firm level.

Standardization at the Project Operation Level

Interviews with project managers and engineers revealed the actual use of standardization in the construction site, where three forms of standardization emerge: (1) SOPs or other written standards, (2) technical preparatory meetings and (3) standardized floor (product).

1. Standard operating procedures (SOPs)

   The study hopes to understand how the firms interviewed were involved with SOPs. One of the ways to assess this is to examine whether the firm's written standards were available (e.g. what is the procedure for setting up the formwork?). According to the interviewees, the construction method statement is available at the construction site in which work instructions are clearly stated. It contains the sequence of tasks that need to be carried out in order, as well as the estimated number of workers and materials that are needed. However, the problem was revealed by a majority of the interviewees in that only a few foremen or site engineers would actually read the SOP and other written standards on the site. This implies that the effectiveness of these standards is rather poor. One possible explanation for this is that a majority of the frontline workers do not meet a reasonable standard of literacy and hence they do not understand the contents. Furthermore, there were no attempts from the project team to discuss the standardized work process among the team members and with the foremen. One interviewee from Firm C commented that:

> As long as the workforce can deliver the assigned job on time and within budget, we are not concerned too much about the level of standardization they have achieved. In fact, the majority of workforce should have knowledge on what they are currently doing. They have been doing this for years and this craft work has not changed so much.

When asked whether the workforce will identify new ideas to improve the current processes, the interviewees replied that very few senior workers with positive attitudes contributed ideas and gave constructive feedback for process improvement. One project manager from Firm E revealed that:

> In fact, their workmanship is 'better' than anyone from our project team because they have been doing this for a very long time and are very specialized in it. For some experienced workers, they have the vision of what the outcome will look like, and also understand what the working procedures are. So our duty is to ensure that they are aware of all the standards that they need to comply with before they commence the work.

With respect to their motivation to improve the current SOP, it appears that this was not driven by the kaizen mindset or company culture; it was, however, largely determined by their financial commitments. Basically, they were motivated to work faster and more efficiently in order to increase the volume of their work. This is because increased work volume can bring them extra earnings. A Project Manager from Firm P pointed out that:

> There is actually no need for us to teach them step by step in terms of SOPs, driven by the financial incentives; they will spontaneously speed up their work as they are paid by piece rate and we found this payment strategy is quite effective.

The problem is that the workers may aim to speed up their work but may not necessarily follow the established routines or instructions. Conversely, they may work using their own means and methods that they simply thought was the right thing to do, and by doing that, it can help them to achieve the defined objectives and maximize their earnings.

2. Technical preparatory meetings
   Compared to the written standards that almost no one will consult, technical preparatory meetings were mentioned frequently as a standardized process, where attempt was made to transfer a number of key issues and technical know-how to the foreman and site engineers. For most firms interviewed, the technical preparatory meeting is mandated and is an effective way in which professional engineers deliver a "how-to" guide for relevant processes to the foremen and relevant subcontractors before commencing a particular work. Basically, the construction firm uses relevant codes of practices, SOPs, relevant drawings, and lessons learnt from past experience, to highlight the key issues to

all the participants who attended the meeting. Once the technical preparatory meeting between general contractor and foreman was completed, the latter would accordingly carry out their own internal technical preparatory meetings with their team members to discuss issues in more details.

3. Constructing a standardized floor
   Most interviewees shared their views that standardized work can be possibly achieved through the "repetitive" process in the construction site. Apart from the standardized input (e.g. SOPs) and process (e.g. technical preparatory meetings), constructing a "standardized" floor is frequently referred to as another form of standardization. The standardized floor can be described as the "best practice", which conforms to customer expectations. Using the standardized floor as a reference, it is hoped that the remaining individual floors can be constructed within the same lead time, using the same construction methods, completing the tasks in the same sequence, etc. One project manager from Firm D added that:

> Because the client is keen on project schedule, we come up with a standardized construction method that ensures each floor can be completed within three days. If workers do not follow the established standards step by step, it is highly likely to fail. For example, we strictly demand that the formwork together with the rebar work must be done in two and a half days. By achieving this, it allows another half day, at the timeline between 2-4pm, for the concrete job, and this allows the concrete curing to be carried out in the Day three night.

Standardization at Firm Level

The project management handbook is frequently referred to as a standard approach at the firm level to improve productivity and performance. According to the interview results, it is not uncommon for large Chinese construction firms to develop their own version of project-oriented project management guidelines based on one of the international project management standards (e.g. PMI's PMBOK). The primary intention is for them to reflect on their management experiences to standardize the project management procedures, use and application. This allows the project team to follow the standardized guidelines to manage all kinds of projects undertaken by the firm. These PM guidelines limit the variety of different management approaches and methods and provide the means of support and help to assure the project management quality of each project. It also prevents every project manager from having to struggle in what they should do and how they should do it. For example, PM guidelines introduce the use of standard report forms which allow the collection of information in a uniform way. In practice, the parent firm of the Firm D designed 55 pieces of standardized forms for the project team to use. These forms covered various aspects including material usage, health and safety, risk management, financial report and others. One vice-president from Firm D stated that the current project management handbook has gone through two major modifications, and that this has allowed additional procedures or modifications to be added from collective knowledge sharing and continuous improvement.

Table 9.5 Visual management practices in China's construction projects

Aspects of visual management	Purpose(s)	Example(s)
Visual instructions	• To provide general description of the project	• Corporate image (CI) • Project description
	• To highlight work-related contents	• Schedules (e.g. master plan, monthly plan, etc.) • Work instructions
	• To highlight health and safety-related issues	• OHS posters, banners, etc.
	• To give material/components information	• Material classification boards/labels
	• To give logistics-related instructions	• Routes, signs, etc.
Visual monitoring	• To control site operations • To enhance the security on site	• CCTV
5-S practice	• To beautify the construction sites	• House keeping

9.3.3.6 Visual Management (P7)

The questionnaire survey revealed that visual management practices were not widely adopted in China's construction industry. The interview results correspond with the survey findings that a majority of the interviewees were not familiar with the terms of visual management, but when asked whether its associated tools (e.g. 5-S) could have been adopted in their workplace, some examples for visual management in China's construction industry were mentioned as indicated in Table 9.5. Collectively, three major aspects were identified, including (1) visual instructions, (2) visual monitoring and (3) 5-S practice.

Visual Instructions

(1) General description

All the firms interviewed mentioned CI. Some interviewees pointed out that this was amounted to practising standardization that was promoted by the head office. In most of the cases, CI can easily be seen from outside the construction site. This was mainly for the attention of potential customers and put on display for marketing purposes. In addition, CI also includes the standardized sign board displaying the information to the relevant parties, highlighting the first impression of the project with its objectives in project quality, schedule, health and safety, etc. (see Fig. 9.1).

(2) Work-related contents

In some of the construction firms, construction plans and schedules are simplified and displayed in the site meeting rooms for the workers through visual boards. These boards are created and updated by the project team according to the site plans. In this way, the management and workers can understand the deadlines and their requirements (see Fig. 9.2).

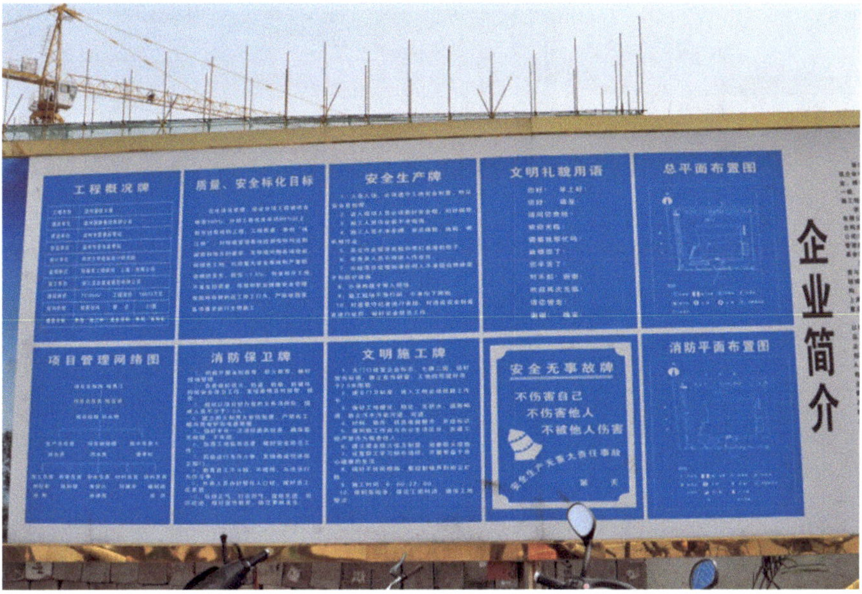

Fig. 9.1 An example of firm's CI sign board

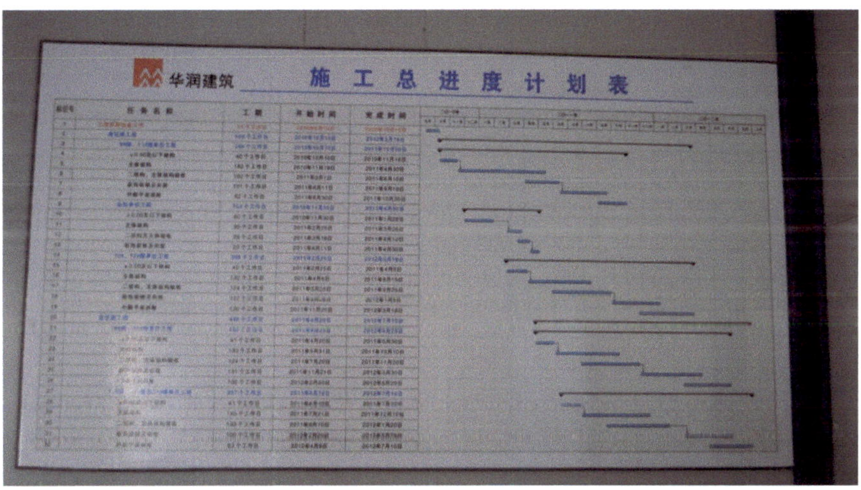

Fig. 9.2 An example of the master plan

Another form of work-related visual tools is that simple work instructions were written in the jobsite. For example, in some of the rebar works, the foreman simply wrote down the instructions for the rebar works (e.g. rebar specifications) on the surface of wooden formwork to inform the workers of

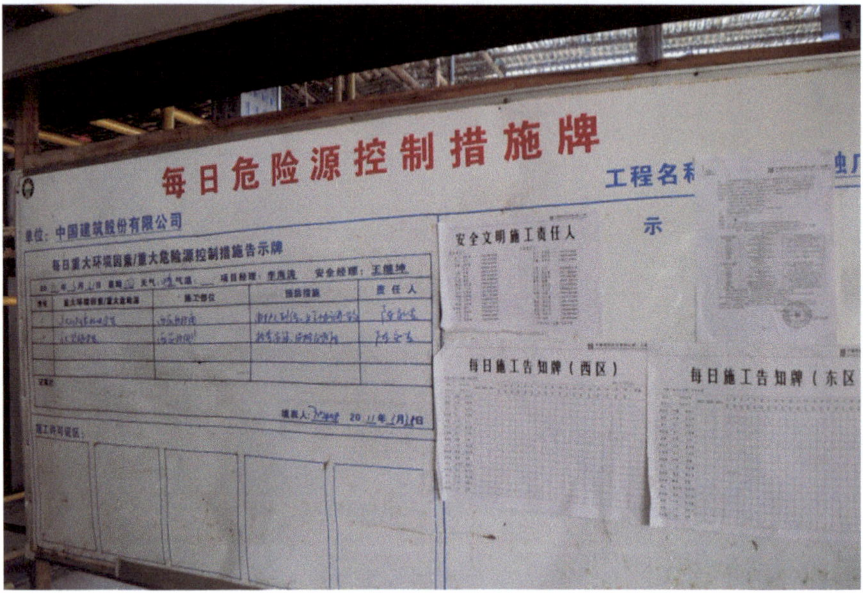

Fig. 9.3 An example of a visual poster relating to hazardous risks

such requirements. According to some site engineers, the more skills the frontline workers gain, the less they would rely on these visual instructions.

(3) Health and safety-related issues

Health and safety instructions were widely displayed at all the projects visited. For example, Fig. 9.3 shows the daily potential hazardous risks identified by the project team that was displayed at site for the workers. It also lists the contact persons who are in charge of the relevant work sections. This serves as timely information to remind the operational workers of the risks they face. Moreover, as Fig. 9.4 illustrates, the safety policies and safety procedures, among other safety information, are displayed on a notice board to raise awareness of workplace health and safety.

(4) Materials management

Another emerging method of visual control tools adopted at the project level includes labels attached to the materials in the material storage areas so that these can be readily distinguished from other materials. Some interviewees pointed out that they used to attach the labels to each material for this purpose, but these were frequently removed by some irresponsible frontline workers. Hence, given the costs of producing these labels as well as ill-disciplined workers, they only used the labels occasionally when someone outside the project comes to visit or to conduct routine checks.

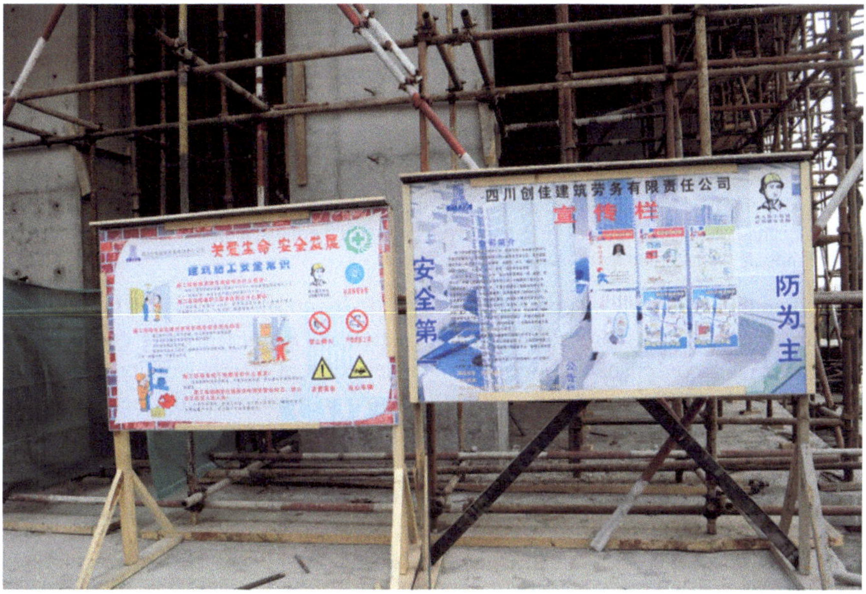

Fig. 9.4 An example of a visual poster relating to health and safety policies

(5) Logistics-related signs

In some of the projects visited, the logistics-related signs were helpful in directing the guests in the site even without any one accompanying them and, who can rely on these logistics-related signs for directions.

Visual Monitoring

CCTV was one of the visual management tools most cited by the respondents. Approximately 40 % of the interviewees reported that CCTV was installed at key entry and several workplace points of their respective construction projects. Effective use of CCTV in some areas helped to identify the problems and track how these were solved. In the night, the CCTV also helps to create a sense of security as it prevents the valuable components and materials (e.g. copper cables, pipes, etc.) from being stolen. In addition, the CCTV cameras were also used to monitor the workers and identify those idling. One project manager from Firm A pointed out that:

> It is actually requested by clients to install CCTV on site although a few workers have sentiments given the fact that they are being monitored.

5-S Practices

5-S was classified as an unfamiliar practice by most interviewees. One project manager from Firm A, whose firm specializes in the construction installation and decoration business, highlighted that as the installation and decoration business uses a large amount of materials and components in the site, the workforce was there challenged to sort out the materials and components according to classifications, locations, quantities and types, and to ensure that these materials and components were easily monitored. However, as far as the project manager from Firm A is concerned,

> Even though 5-S training is conducted in our company, what we had practiced however only correspond to the first two 'S' of 5-S on the site, namely to conduct housekeeping and classify the materials in order. Apart from that, in terms of employees' disciplines of practising 5-S, it has not so far been developed as it perhaps requires more extensive training as well as managers' support in the future.

Although 5-S is not systematically practised, driven by the construction-related bureau, which actively promotes initiatives such as "*beautifying the construction site*", there is a consensus among most interviewees that efforts were only made to demand the workers to have at least half an hour to clean and tidy the site before they ramp up their work.

9.3.3.7 Use of Reliable Technology (P8)

Status Quo of New Technology Adoption

The interview findings indicate that the substantial houses and infrastructures constructed every year benefited the Chinese construction industry in that these allow the construction firms to apply the technical know-how acquired and to advance existing building technologies. When asked whether any types of new construction technologies were adopted that produced a high impact on the projects or added value to the projects, only five interviewees pointed out that no new technology has been sought in their current projects. However, this does not necessarily mean that there is no association with the adoption of new technology at the firm level (head office). One interviewee explained that:

> Each year, the firm will host annual meeting to promote the timely dissemination and exploitation of new technologies that have been adopted in other projects. This is a learning experience for my project team who have no experience with it. This keeps us updated.

The common explanation is that the pace of technological change in the construction industry is slow. For example, most interviewees agreed that the construction methods for a normal residential or commercial high-rise construction were no longer sophisticated and these should be re-categorized as conventional methods compared to the modern methods of construction such as offsite prefabrication. Naturally, for projects that employ conventional methods, there are typically few technical difficulties to overcome. This perception was recognized by Ling et al. (2008) as one of the major factors that would affect implementation of innovation in firms.

To maintain technological leadership in construction, a majority of the firms interviewed understood the importance of identifying and exploiting emerging technologies if any. It is reassuring to note that the large construction firms had already been active in heeding the construction-related bureau's call of selecting the most suitable technologies for their projects. The handbook on "*Ten Emerging Construction Technologies in China's construction industry*" published by China's

Construction Bureau showcases about 108 specific technologies and methods in ten areas that have been implemented in China's construction industry. Some interviewees highlighted that their respective firms have been chosen to contribute their knowledge in the handbook. Because of this, one Project Manager from Firm N explained that:

> ... some of these so-called 10 emerging construction technologies were not new to our companies, but these are indeed useful in certain areas. We have practiced some in various projects and have gained rich experiences.

Table 9.6 illustrates examples provided by five responding firms, which had made considerable efforts in using new technologies and encouraging innovation to a large extent. All of them have the view that the new technologies have the potential to boost productivity, enhance client satisfaction, and quality improvement. Some interviewees also highlighted that the improved level of mechanization and automation was able to relieve the workers from onerous and repetitive work. As one manager from Firm H explained:

> I always prefer to use the best machinery in our overseas projects despite the price is 10% – 20% higher than the average. Take the screw piling method employed in our Angola project for instance, one unit of the piling machine cost us 2millions RMB. After thoroughly tested in China's Hainan Island, where it has the similar soil conditions compared to Angola, we decided to use it. The results proved that the implementation of such new technology is a success and worthy, as it has solved the problem that we encountered in the previous projects that no piling method are suitable to Angola's soil conditions.

It appeared that driven by external pressures such as rising competitiveness, tighter deadlines and shortage of skilled workers, more construction firms are likely to look beyond traditional construction methods towards emerging technologies. In such circumstances, Toyota Way Principle 8 is useful in guiding the firms towards new technology adoption.

9.3.4 People-Related Practices

Similar to the Process model, this section presents the analysis of how the People and Partner model could be adopted in China's construction industry.

9.3.4.1 Leaders and Leadership (P9)

Long-Term Employment

Of the 27 building professionals interviewed, a majority (17) stated that they had been working for their respective firms since they started their careers. Two vice-presidents shared a very similar career path in that both were "home-grown leaders". They started their first job as a foreman on site, and were gradually promoted to be engineer-in-charge, then project manager, and eventually were placed into a managerial position in the head office and recently promoted as the vice president. They understood the firm's value and culture thoroughly. In addition to direct day-to-day affairs, formulate and implement strategies, manage work-

Table 9.6 Adopted new technologies and their associated outcomes

Firm	New technologies adopted	The benefits	Adhere to the Toyota Way Principle 8
G	• Procured the "tower-belt" machine from the USA and used it for concrete placement for dam construction	• Improved project productivity	• Yes. The adoption of "tower-belt" machine aims to add value to the process as its conveyor system is able to provide long reach, wide coverage and continuous placing of concrete in the accurate position
H	• An innovated method of screw piling implemented in Angola's residential project as a pilot test	• Cost saving: this piling method proved to be cost effective from a long-term perspective • Improved client satisfaction: this was much appreciated by the client (Angola's government project)	• Yes. Given the unique nature of Angola's soil profile, this method generates least disturbance to the soil profile during installation and also creates least soil that needs to be removed
I	• Special lightweight wall panels were designed and adopted to meet the tight timeline for the construction of Universal Studios project in Singapore. These wall panels (known as ACL panels) have secured patent in China	• Improved project productivity • Reduced the installation time from 8 months to 1.25 month	• Yes
O	• Procured a testing equipment and conducted non-destructive testing to check the steel structure requirements	• Improved quality: to ensure quality of the steel structure	• Yes
P	• Placed in inflated tubes along the rebar to reduce the load prior to the slab concrete job	• To meet the design requirements approved by the client	• Yes. This is the first trial on this construction method by Firm P • A number of sites which had used such method were visited for reference. Also, a pilot experiment was conducted in the laboratory before it was officially applied

related problems, and others, their current responsibilities also include laying the groundwork for the firm's culture, and giving formal and informal talks and speeches to employees at different levels.

Technical Knowledge

Most interviewees agreed that the project manager was the key person that made significant contributions to project performance. The Chinese building professionals, as described by a majority of interviewees, are very knowledgeable in terms of the technical skills possessed. This was largely derived from their working experience in various projects that profoundly shaped their technical know-how. At the project level, they were able to help the engineers to deliver technical analyses and provide resolutions in a crisis situation. One project manager explained that:

> Few problems are expected in my project compared to others. This is because our leaders are able to anticipate the risks and contribute to overcoming the significant constraints before these actually turn into real problems. This not only requires a skill set but also a visionary mind, dedicated heart as well as good relationship with clients and other parties.

People-Centric Leadership

The highest calling a project leader has is to guide, motivate and support each person in the project to enable him or her to contribute to the project success (Tener 1993). To satisfy workers' needs and boost their morals, almost all the respondents pointed out that the "people-centric" philosophy was adopted as a guiding principle in the firm. However, unlike Toyota leaders who are often credited with supporting employees while they are doing the work, the interview findings suggest that leaders in China's context also place high importance on valuing people and in caring for them through various means. The examples given for people-centric activities were more focused on the well-being of the workers rather than in supporting their work at the workplace. For example, people-centric leadership surfaces in the way that the project team provided clean and comfortable accommodation (e.g. air-conditioned dormitories in summer for the workers), standard set of meals at site, entertaining performances, playground and others. These can alleviate the pressure on the workers. The caring environment was observed by Chen and Partington's (2004) study, which showed that the Chinese project managers perceived the team members as family members and were willing to take care of and to support each other.

9.3.4.2 People Management (P10)

The interviews sought to understand how Chinese construction firms implement the Toyota Way-styled people management principle that relates to people selection, training, teamwork and motivation. The interview results are summarized as follows:

People Selection

All the interviewees acknowledged that selection was probably one of the most, if not the most important step to bring the right people (including the sub-contracted workers) into the firm. This echoes the survey findings where "*select the best person for given job*" (P10.1) was prioritized and highly practised by the responding

Table 9.7 Types of training programmes provided by the responding firms

Types of training	A	B	C	D	E	F	G	H	I	J	K	L	M	N	O	P	Total
For frontline workers																	
Basic trade skills	X	X	X	X	√	√	X	X	√	X	√	√	X	X	X	√	6
Migrant worker school (facility)	√	X	√	X	√	√	X	X	X	X	√	√	X	X	√	√	8
For employees																	
Firm orientation	√	√	X	√	√	√	√	X	√	X	√	√	√	√	X	X	11
Mentorship programme	√	√	X	√	√	X	√	X	√	X	√	√	√	√	√	X	11
Staff meetings	X	√	X	√	√	√	√	X	X	√	√	X	√	√	√	√	11
Work preparatory meetings	√	√	√	√	√	√	√	√	√	√	√	√	√	√	√	√	16
Exchange programme with other firms	X	√	X	X	X	X	X	X	X	X	X	X	X	X	X	X	1
Self-development	√	√	X	X	√	√	√	√	√	√	√	X	√	√	X	√	12
Cross-department training (multi-skills)	X	√	X	X	X	X	X	X	X	X	√	X	X	X	X	X	2
Training topics																	
Quality-related	X	X	√	√	√	√	√	X	√	X	√	X	X	√	√	√	10
Health and safety-related	X	X	√	√	√	√	√	X	√	√	√	√	√	√	√	√	13
Technical aspects	X	X	√	X	√	√	X	X	X	√	√	X	√	√	X	√	8
Project management-related	X	√	X	√	√	√	√	X	X	X	√	X	X	X	X	X	6

"√" denotes that the firms are dedicated to provide this type of training
"X" denotes that the firms are not dedicated to provide this type of training

firms. In addition, the interview findings highlight the increasing challenges that the industry is currently facing in terms of attracting and retaining qualified employees at all levels.

Training and Continuous Learning

Training in the context of lean production contains various forms and is seen to have an important role. It includes pre-job training, on-the-job training, multi-skills training, etc. Table 9.7 lists the types of training programmes that target three different groups (i.e. frontline workers, site personnel, and management) within the firms interviewed.

1. Frontline workers training

With respect to frontline workers, most responding firms held the view that, as far as possible, training on basic trade skills for the frontline workers is the responsibility of the labour-only subcontractor. From Table 9.7, it can be seen that a majority of these responding firms (10 out of 16) viewed that there was little provision for training frontline workers. In contrast, Firm P was one of the few, which provided vocational training for skills development of frontline workers. One project manager stated that the training covered the basic competencies of masonry, bar-bending, carpentry, etc., and these were implemented with the help of the local construction associations and top management. It is worth noting that the experienced foremen or site engineers are also encouraged to explain instructions and methods pertaining to their own trade skills to the workers in greater detail.

Moreover, about half of the responding firms said that at the site level, basic training relates to construction quality, occupational health and safety (OHS), etc., and that this was available at the migrant worker schools, and where the training was usually conducted during rainy days or at night. However, in most cases, the migrant worker schools are either inactive or have been changed for other uses, such as for a warehouse or a site meeting room. In summary, the results suggest that site management only paid lip-service to this type of training, and as one interviewee from Firm O noted,

> The attendance records are quite disappointing given the effort that is made to providing training for workers. It seems that the training does not hold interests for the workers as they are probably too tired to attend the training courses.

2. Site personnel training

Almost all the responding firms indicated that proper work training is provided to newly recruited employees as well as to the existing employees. They might not have specified what the types of training were, but from their spontaneous answers, these could be categorized into three types: pre-job training, on-the-job training and off-the-job training.

Pre-job training Firstly, all hires begin with the firm orientation which includes an introduction to the firm culture, firm values, personal rules, etc. Most firms do not put much effort in detailing the work instructions during pre-job training but only give basic introductions. This seems to suggest that the new hires are somehow accepted by their employers so far as their technical competence is concerned.

On-the-job training From Table 9.8, it can also be seen that various forms of training were given in the work place. Most efforts were made in the following three areas, namely: (a) mentorship, (b) staff meetings and (c) work preparatory meetings.

(a) Mentorship

Several respondents recalled that at the time when they were recruited, they were assigned a mentor, who was a senior in terms of technical knowledge and skills-set. The mentor continuously offered guidance and advice for them. One site engineer from Firm L recalled that:

> My mentor was very technical knowledgeable. Every time I went to him with civil engineering questions, he always answered it using his experience, and that was exactly what I was trying to ask.

The mentorship programme usually lasts for 1–2 years or a period which is equivalent to an entire project life. In most cases, once the mentorship completes, an assessment is followed to evaluate trainees' technical knowledge, among other skills attained during the mentorship. It was then up to the firm to decide whether the mentor scheme should continue, or considered successfully

Table 9.8 Responses to the practice of *genchi genbutsu*

Firm	The first-line leaders	Leaders from all levels	Remarks
A	√	X	It is more appropriate for leaders to deal with higher-level matters rather than practise *genchi genbutsu* every time
B	√	√	*Genchi genbutsu* is required but the problem is that the leaderships usually engaged in, e.g. project bidding, among other matters so that they have no time for the current project
C	√	√	PM is expected to be visionary and keep an eye on what could possibly occur on site. It is more helpful in preventing potential risks and constraints before these become real problems
D	√	√	Other than *genchi genbutsu*, all the new and recurring problems were profiled and presented to all the parties in the weekly meetings
E	√	√	*Genchi genbutsu* is practiced. However, in order to achieve a state of harmony within the firm, the SOEs leaders favour that the problems could be solved in the way that "*reduce major issues to minor ones, and minor ones to nothing*"
F	√	X	Leaders at the firm level should be more focused on strategic issues. However, they are frequently asked to be present on the project site to display their commitment as it was specifically requested for by the client
G	√	√	For complicated problems, it requires the leader to *genchi genbutsu* and set up a technical committee
H	√	√	NA
I	√	√	*Genchi genbutsu* should be applied to all the management staff at site
J	√	√	Only the problem relating to quality issues, project leaders are asked to practice *genchi genbutsu*
K	√	√	The practice of *genchi genbutsu* is written in their project management handbook
L	√	√	*Genchi genbutsu* should be applied to all the management staff on site
M	√	√	NA
N	√	√	The project team requests the project leaders as well as team members to be on call at all times
O	√	√	Not much problems occur on site
P	√	√	NA

"X" denotes *genchi genbutsu* is not practised
"√" denotes this practice is present

completed. If the assessment results turned out to be merit orioles, the mentor is awarded bonus for his or her efforts.

(b) Staff meetings

Staff meetings were also repeatedly mentioned as a valuable training opportunity. However, the meeting frequency was relatively low, given that the site personnel were busy. Interviews with the building professionals revealed that it was held, on average, four times a year. One project manager from Firm M pointed out that:

> My firm does not have formal training programmes. Instead, what we do is to run a quarterly staff meeting that requires all the site employees to attend. I see this as a good training opportunity; not only can we meet colleagues working in other projects, but also leaders from head office will equip us to better understand the new policies that may affect the project and company.

(c) Work preparatory meetings
Similar to staff meetings, this type of meeting was mentioned earlier as a standardization process to convey the construction methods, quality objectives, and others to trade supervisors and the foremen. It was pointed out such meetings were treated as a valuable learning opportunity because most important issues were addressed and discussed here.

These results seem to indicate that the current training system relied on a more informal approach (e.g. meetings), rather than formal training and presentation. Other types of on-the-job training include exchange programmes, cross-department training, and so on, but these were only mentioned by a few firms. For example, one site engineer from Firm B revealed that his firm had training exchange programmes with several leading domestic design institutes (DI) and project management consultants (e.g. Arup). There were ample opportunities for engineers and project managers to be exposed to different disciplines and to be trained for up to 3–6 months to enhance their understanding of different aspects of construction management. Furthermore, some engineers would be transferred to overseas projects of their firms to learn more about international project management. When compared to Toyota, where tremendous efforts are made to standardize and detail on-the-job training for one particular job, the on-the-job training system within Chinese construction firms still appears to be at the early stage.

Off-the-job training This form of training was also adopted by most of the responding firms. As pointed out by most interviewees, firms usually post calendars of training activities through the firm's own online portal, most of which relate to self development programmes. A number of interviewees recalled that, each year, technical employees are encouraged to renew their qualifications in their own disciplines (e.g. registered builder, registered engineer, registered quantity surveyors and others). In the course of preparation to retake the test, the necessary expenses for attending lectures, seminars, etc., would be covered by the firm. One interviewee from Firm H pointed out that:

> We understand the importance of training. It however depends on individuals. They are in charge of their own future, and if they feel they need to learn in certain areas, we will be very supportive.

3. Management training

Compared to the site personnel, there appears to be very little provision for management training and if any, these only focused on limited areas. The firms interviewed that belonged to the China State Construction Engineering Corporation (CSCEC) claimed to have the privilege to tap on "in-house" training resources provided by the CSCEC management school, a subsidiary firm of CSCEC. Moreover, a deputy manager from Firm F stated that one of the main themes of management training (at his level or above) was to interpret the national policy that may affect the firm's business. This was echoed by a vice-president from Firm G, who shared that:

> In response to the government's call to venture overseas, training relating to policies, procedures, etc., in doing international construction business is given to the leadership of our firm. For example, during an intensive couple of days workshop on FIDIC, the instructor uses many cases to explain how the successful overseas projects were undertaken by Chinese contractors.

Skill Levels and Labour Shortage

The Chinese construction industry is staffed with less-educated migrant workers, most of whom are unskilled and where the quality of the skills possessed is uneven. As one interviewee from Firm N explained:

> The unskilled workers could be the peasants working in the farm during the peak time. When the peak time ends, they put down the agriculture tools and look for jobs in urban cities. Quickly, they turn out to be temporary construction workers.

This results in a labour strategy, namely the use of mixed crews of skilled and unskilled workers commonly adopted by most interviewees. However, the portion of skilled workers among the entire frontline workforce at the project level is uneven. Surprisingly, the structure of sub-contracted workers is largely determined by two factors: (1) the negotiated contract value between general contractor and labour subcontractor and (2) the client's requirements. If the contract price is low, it is highly likely that the labour-only subcontractor would not send the entire qualified workforce to the site. By contrast, unskilled or less qualified workforce would be combined and that often takes up a large portion when compared to the qualified workforce. This would seriously affect the project quality and client satisfaction. Secondly, the requirement on labour is highly associated with client's requirements. One interviewee from Firm F (SOE) pointed out that:

> If the client has priority in the project completion, we make request to our partnered labour contractor that the majority of workforce structure should be skilled labour, especially those who can work hard in the changing environment. In contrast, if the client is from the public sector, who shows less requirements on project schedule, a different strategy on labour will then be employed which permits the workers to be mixed with a larger portion of unskilled or average workers.

The interview results revealed, overwhelmingly, that all the firms interviewed suffered from a high workforce turnover, which forced the construction firms to bring in temporary workers to the site. The most convenient way was to hire them from the nearby sites by offering them higher salaries. In cases like this, a less meticulous selection process was carried out. This directly affected the project quality. To resolve the problem, several leading SOEs revealed that efforts had been made to establish long-term partnerships with several local governments in small Chinese farming counties, which have good reputation for exporting construction workers. Some have established their training centres in the counties to recruit and train the young workers. Before they are sent to the site, they must go through some basic training. This also ensures a sustainable supply of qualified workforce in the long-term.

Multi-skilling

The interview results revealed that there was an overwhelmingly sceptical view with regard to multi-skills training for frontline workers. Not surprisingly, there are presently no multi-skilled workers being employed on site. It is therefore necessary to understand if there are any reasons for the lack in providing for multi-skills training. The following reasons were given by the interviewees:

1. Independent trade skills: the labour-only subcontractor sector comprised of various single-skilled trades such as "carpenter", "rebar workers", "concrete finisher", etc. Each category of workers only executes their own specialized work, i.e. a worker who is a "carpenter" does not perform the duties of a rebar worker even if rebar workers are in short supply. Similarly, the concrete finishers only come to the site when they are needed and they are not involved with the works of other trades.
2. Satisfied with the current skill set: most interviewees seem to be satisfied with the state quo of current skill set that different trades possess. It implies that multi-skilled workforce is not considered during a project's long-term scheduling process, as they put it: "*They know what they are doing and some are good at it. If we (site staff) are placed to do the same job, we are not as productive as they are*".
3. High mobility: given the high mobility of frontline workers, this would be entirely understandable for construction firms to show less interest in fostering multi-skilled workers. Several interviewees expressed their hesitations to invent in multi-skill training as they highlighted that it will be a huge loss when the trained multi-skilled workers left the firm for other purpose.
4. Increase pay: concerns about pay increase were voiced out that: "*The multi-skilled workers will probably demand higher pay when he/she is trained with multi-skills. We must consider whether it is a cost-effective means and we are not ready to pay them differently within the trades*".

In summary, the industry is characterized by different trades that specialized in their own skills. Nevertheless, it is still necessary to consider issues of multi-skilling during the planning phase of a project by combining activities that involved similar crafts and assigning a complete work unit to a single crew without considering craft boundaries.

Teamwork

The frontline workers, most of the times, work in a team. According to several interviewees, among all the workers who are on the site at one time or another, it is easy to find relationships like father/son, siblings, and friends who came from the same farming villages. Those who are from the same family or having similar background tend to be working together in the same occupational trade. For example, in a group of approximately 70–80 construction workers performing the rebar-fixing job at the same time, 3–4 teams can be identified with each team having a team leader. Usually, the team leader is not assigned a regular job, but keeps an eye on his team members, maintains good communications and handles minor problems.

"External" and "Internal" Motivation

This study also investigates the motivational strategies adopted within LCCFs. From the viewpoint of most interviewees, monetary incentives were extensively used on site. At the project level, the monetary incentives for performance measures—i.e. quality, health and safety, as well as progress, were clearly indicated in the contract. For example, before the project commences, mutual agreement would be obtained on the financial incentives provided if the project wins any prestigious awards, such as "*Luban*" or provincial quality award. With respect to motivational strategy at the individual level, most interviewees agreed that the project team imposed few fines on construction workers because they were afraid that the workers' responses to fines were not always positive. The workers knew that they would always have some other alternative construction sites to go to work for. Several interviewees stated that although the frontline workers did not fully comply with safety rules, and did wrong things such as smoking on site, not wearing protective equipment, etc., they were not fined but only received light verbal warnings. One senior project manager commented that a mixture of motivational methods was implemented in his project, namely monetary incentives and small fines, but the latter would mainly depend on who the parties involved were. He explained that:

> A mixed approach is employed in my project in dealing with different parties. In most cases, we impose fines towards the subcontractors if the recurring quality problems were discovered or they did not respond to the order for rectification in a timely manner. So far, a number of rectification orders have been issued to subcontractors; one subcontractor did not make timely rectification and was fined 1,000 RMB. In contrast, we never fine our workers.

On the other hand, interview results revealed that non-monetary or intrinsic incentives were also adopted. Several interviewees stated that there is a number of recognition programmes adopted as non-monetary incentives. The high-performing foreman or trades would be recognized for their efforts and contributions. He or she would be given the title such as the role model frontline worker, the role model trade member, etc. Moreover, workers would be recognized for their participation in QC activities or their contributions of constructive feedback and ideas that resulted in process improvement. Another commonly adopted non-monetary incentive is direct from the leadership, which conveys the message that the company cares for the ordinary workers. In this case, leaders visit the site, the firms provide a comfortable environment for their workers in their living areas, etc.

9.3.4.3 Partner Relationship (P11)

In this section, the interviews assess how the Toyota Way principle of working with partners can be applied in Chinese construction firms.

Partner Relationship

The following outlines the interview findings pertaining to the relationships between the responding firms and their suppliers, subcontractors and clients. It was found that efforts had been made by the construction firms to establish long-lasting relationships with them. One piece of evidence for this was that all the firms interviewed had established an internal accreditation system for monitoring their suppliers and subcontractors. Once a project is completed, the project manager is in charge of evaluating the performance of their chosen partners (e.g. suppliers, subcontractors, etc.) on a number of criteria, including quality of construction, business relations, schedule, cost control, contract fulfilment, etc. The results of the performance appraisal are to be submitted to the head office, where a decision would be made on whether this partnership would continue, or should be terminated for possible future projects. Moreover, it is common to see that the clients tended to nominate their own preferred suppliers or subcontractors throughout the project. Often, these nominated suppliers or subcontractors might not have been included previously in the contractor's database. In order to satisfy the clients and minimize disagreement, these nominated suppliers and subcontractors were likely to be awarded the subcontract.

Apart from collaboration with suppliers and subcontractors, all the firms interviewed have stated that aggressive targets set by the clients, which the interviewees felt to be somewhat unrealistic to achieve, were the most challenging part of their relationships with their clients. This state of affairs is not surprising, given that when the main contractor does not comply with the client's requirements, substitutes can be easily found. Construction firms were therefore trying their very best not to damage their good relationships with clients by engaging more resources or workers, even though, for example the above scenario may give rise to working brutally long hours.

Despite this, a few interviewees indicated that their head office (parent company) had gone an extra step by partnering with some big clients (e.g. local governments, private developers, etc.). This means that these construction firms would be automatically considered as general contractors if the clients intend to expand their business in China. One senior manager from Firm F confirmed that the top 10 of these so-called big client partnerships contributed to 22 % of the firm's total revenue in 2010, and this figure was expected to grow in the future.

Multi- and Single-Sourcing of Suppliers

All the responding firms have indicated that, for certain types of materials or components, there are at least five or more suppliers which they collaborate with, and their particulars were all well documented in the firms' databases. The main reason for keeping multiple sources of suppliers is to help the construction firms minimize risks and enjoy certain level of flexibility. Moreover, unlike Toyota or other Japanese manufacturing firms that do not change their suppliers (Womack et al. 1990), it was found that the firms' databases of qualified work partners were quite dynamic. This is because every year, new and high-performing work partners would be added to the databases after evaluation, whereas problematic partners would be removed. When asked about the trends of changes in the numbers of suppliers, most respondents replied that, due to the rapid development of the Chinese construction industry, more suppliers and subcontractors are needed as the construction firms benefit from securing increasing volumes of construction work.

Quality Products Provided by Suppliers

The interview results highlighted most firms interviewed showed little concern for the quality of materials provided by some suppliers selected from the firm's qualified supplier database. In most cases, the sources are usually regarded as responsible, as their products (e.g. materials, machinery, etc.) must pass a thorough evaluation to determine whether they are suitable before they can be accepted. One of the key dimensions is product quality. In most cases, however, the products were covered by the National Inspection Exemption scheme, and so were not inspected to ensure these comply with the quality standards. This has become the principal reason why suppliers compete mainly on price. More strictly, in the Shanghai area, it is mandated for suppliers to register with the local construction authority a list of what materials (e.g. concrete) were supplied to which project. This serves as a tracking system to assist construction firms in monitoring the material quality and in seeking the root cause in a timely manner if a problem occurs. All these efforts reflect a sense of built-in quality thinking.

Assistance for Partners

However, the real difference between the Toyota Way of treating its partners and the Chinese construction firms' supplier relationships is that limited technical support is provided in the latter even if there is a problem on the supplier's side. It is understandable that manufacturing and construction are two different domains. Unlike Toyota or other Japanese lean enterprises, which can in the first instance try to help the suppliers deal with their problems, Chinese construction firms simply change any suppliers who experience delivery problems or whose prices are no longer competitive. This is because for any given material or component, the construction firms usually have multiple choices. One project manager from Firm M said that:

> We are private firms and we are very sensitive about the price. Given that the quality of materials from different vendors is indistinguishable, we are not willing to pay anything extra for increased reliability, quality, etc. Low price is the king!

In contrast, as explained by most interviewees, their firms tended to collaborate with subcontractors and were willing to provide the necessary technical assistance if the subcontractors experienced a technical problem on the site, especially given that the delay was caused by one subcontractor in the preceding job, which would in turn affect the "downstream" activities undertaken by a different subcontractor. To achieve continuous work-flow, the firms interviewed felt that collaborating with subcontractors is the key to ensuring that project progress is always on the right track. Of all the firms interviewed, Firm P had worked for a Japanese contractor in China for many years. One project manager from Firm P shared that his team had learned from and benefited from this long-lasting relationship with their Japanese counterpart. Whenever they had a technical problem on the site, their Japanese counterparts have always engaged enough technicians to help resolve the problem.

Supplier Delivery System

There is a general understanding in the literature of how Japanese manufacturers, such as Toyota, enjoy frequent deliveries of small shipments. This is because a number of suppliers, especially the first-tier ones, are located close to Toyota's plants. However, this was not the case in the construction context. Having discussed the pull *kanban* system (in P3), the firms interviewed have generally practised bulk lots with few deliveries for a majority of the purchased materials and components. According to some interviewees, if their projects, especially large infrastructure projects, are located in remote areas, then the materials need to be procured locally in order to reduce the transportation costs. This opens new opportunities for new partners. In other circumstances, geographical proximity does not seem to be an issue, especially in the case of speciality contractors whose replacement would be hard to find. One vice-president of Firm K recalled that:

> There were three speciality contractors in LNG storage tank construction and installation who have been working for us since probably day one. It is always one of them who would be chosen to work for us wherever we go. Yes, they are treated as part of our enterprise.

9.3.5 Problem-Solving Practices

This section evaluates the interviewees' perceptions of how the three underlying principles of the problem-solving model can be adopted in China's construction industry.

9.3.5.1 *Genchi Genbutsu* (P12)

The problem process begins by defining a problem, or what the Toyota Way refers to as the "current condition". To grasp the current condition, leaders are required to implement *genchi genbutsu* to understand the situation. It was found from the interviews that the idea of *genchi genbutsu* had the support of a majority of the firms interviewed. As indicated in Table 9.8, all the interviewees agreed that the first-line management should practise *genchi genbutsu*, however, in terms of whether this management idea is worth being applied across all levels of organization, only 3 out of 27 interviewees voiced out alternative opinions. As one Manager from Firm F explained:

> Good leadership should be more focused on the firm's strategic development. It may not be necessary for them to come and see the problem in a timely manner. Instead, in the case of problems that are discovered at the project level, a more appropriate means could be that managers select the right candidates to help solve it.

Even though *genchi genbutsu* was appreciated and strongly encouraged by most of the firms interviewed, nevertheless when it came to the actual site (see Table 9.8), their attitudes and reactions varied. It is particularly true that while it may not be management's intention to avoid wasting their time to walk around the site, this might be attributed to the fact that they were pre-occupied with many other matters, which did not allow them to do so. For example, one project engineer from Firm B revealed that the project manager was sometimes asked by head office to join a temporary team to work on bidding strategy for the next project. As a result, the project manager was not on the site all the time, and the project engineers were tasked to supervise the project.

A few interviewees stated that only when the problem was related to serious quality or OHS issues, project leaders would not be alarmed nor present themselves on site. Good practice of *genchi genbutsu* included:

- Relevant norms and codes were established in the firm's internal project management handbook, stating that not only the site engineer but also the project manager needs to *genchi genbutsu* first hand when a problem occurs (e.g. Firm D).
- More stringent rules established that require the PM and engineers to be on-call for emergencies all the time, and always be ready for *genchi genbutsu* (e.g. Firm N).

Attitudes Towards Problems and *Genchi Genbutsu*

Apart from the organizational culture or leadership factors, a closer investigation of the interview results showed that the decision of *genchi genbutsu* actually depends on the "difficulty scale" of a problem. The interviewees noted that problems can be broadly characterized as "minor", "major" and "mega". It is defined in relates to the employee's ability to fix the problem. Through an analysis of the interview findings, three emerging scenarios that characterized attitudes towards problems and *genchi genbutsu* were identified and explained as follows.

Scenario 1: this is the most optimistic scenario that would see only minor problems occurring on the site. As noted by several interviewees, site engineers are required to spend 70–90 % of their time on the site to spot problems. They are approachable and knowledgeable enough to come up with countermeasures to most minor problems. Most interviewees agreed that there is no need to wait for instructions from supervisors if the site engineer is able to solve the problem in this regard. Moreover, recurring quality problems such as "*leaks in concrete*", "*cutting or bending the rebar with wrong size*", etc., were frequently mentioned in this scenario. Several interviewees outlined that there were numerous recurring quality problems in their projects, partly because the client demanded the crew to focus on speed rather than project quality. In addition, some interviewees attributed the cause of recurring problems to workers' attitude, namely that they "*do not care about such a problem*" or "*do not realize this is a problem*". One project manager from Firm M added that:

> In most cases, these problems are caused by human factors. The variation will be introduced when workers are carelessly performing the job with a 'do not care' attitude. Despite all the efforts made by the management highlighting that such issues are prone to errors again and again in our weekly meetings, it still keeps popping out. My view is that these problems are of little value to learn and document.

Scenario 2: although most site engineers possess outstanding problem-solving skills, some may not cope with the problems in the first place. When a problem confronted is categorized as "major", with no obvious solutions, the site engineer or foreman needs to get their supervisor—the more experienced project manager or engineer-in-charge—involved for more guidance into the investigation. Frequently, this type of problems cross boundaries, and only someone with a much broader perspective, or what some interviewees called, "sophisticated" skills can quickly and effectively facilitate resolution. As described by a large number of interviewees, when the project manager was involved with the investigations, he typically called a meeting, bringing together participants whom he felt were appropriate to assist. At this stage, the project manager needed to update the head office about the progress of solving the problem (e.g. countermeasures, outcomes, etc.).

Scenario 3: the last scenario was described as high-impact but uncommon event. The situation was not previously encountered or a specific solution from past experience was unknown. Thus, such problems cause the entire project team to struggle to come up with appropriate solutions. The project team can quickly turn to the head office for more technical assistance. One interviewee shared that:

> In the head office, more resource can be tapped such as professional experts and experienced technicians, who will be tasked to investigate the problem.

An example to illustrate this scenario was given by one project manager from Firm D, who shared that a project he had undertaken contained deep foundation pit construction. Since the project team had no relevant experience in similar foundation construction, experts from the head office were brought in as back up and their involvement was written in the bid proposal. At the construction stage, before the foundation construction commenced, the expert panel accompanied by the project team made several site visits to investigate geological conditions of the site, and the surrounding environment, etc. In addition, several rounds of meetings were conducted to discuss how best to control the underground water, the supporting structure for the foundation, among other matters during the construction. In circumstances like this, efforts were accordingly made to document the problem, constraints, as well as the solutions proposed. This in turn provided the employees with valuable learning opportunities.

9.3.5.2 Consensus Decision-Making (P13)

Genchi genbutsu is the first step in seeking out root causes. The following steps employ various means to determine the root causes.

Root Causes

Root cause analysis is essentially a method or series of actions taken to determine why a particular failure or a problem exists, and to establish a means of correcting the causes (Hall 2001). During the fieldwork, only four interviewees precisely mentioned that the "root causes" need to be thoroughly investigated and that these should be done step by step. For example, one project manager from Firm E explained that his team used "5W + 1H" to systematically understand the problem. A project director from the same firm outlined this as follows:

> If the problem is relating to a health and safety issue, everyone must be very alert. In this circumstance, we must find out the root cause. Generally, our employees are reminded to follow three principles that have been outlined by top management in dealing with problems. This includes: (1) do not proceed to the next step until the root causes are determined; (2) do not proceed to the next step until all the relevant parties are satisfied; and (3) do not proceed to the next step until the one who should take responsibility is identified.

The remaining interviews shared that they generally asked a few questions to try to discover the symptoms of problems, but did not go beyond to find out the root causes.

Employee Participation

It was found that the frontline workers were often undervalued as a source of information for resolving actual problems, and of opinions about possible corrections. This is because they are usually excluded from all types of site meetings and do not participate in any decision-making. In most cases, such as the weekly meetings, preparatory meetings, problem investigation meetings, and so on, the foremen—the lowest level in the project hierarchy—were requested to attend. Hence, the foremen played a critical role in reporting progress to their direct supervisor. For firms relying on first hand information of what was happening in the site, foreman should be trusted as a reliable source of information. With respect to employees' participation in the decision-making process, one interviewee from Firm G stated that:

> It depends on employees' capacity. If they are capable of proposing solutions, we certainly welcome them offering their opinions. In contrast, what we are more concerned about is that there are few constructive countermeasures directly from the workers.

Consensus

Other issues discussed here include whether the final decision or solution will achieve consensus from the team. According to most responses, there was basically "no consensus" within the project team, in which the project manager usually dictated decisions. However, one project manager from Firm K welcomed the idea of consensus and highlighted the fact that in his project, top management encouraged participants to brain-storm until a better solution surfaced.

9.3.5.3 *Kaizen* or Continuous Improvement (P14)

Kaizen Mindset

Problems that emerge repeatedly in day-to-day operational activities are important for identifying improvement opportunities (Tucker et al. 2002; Liker 2004). A large number of interviewees claimed that, in their firms, *kaizen* activities were completely non-existent. They also reported that neither employees nor frontline workers have a *kaizen* mindset. This actually corresponds to the survey findings in which a few *kaizen*-related activities received low ratings from the survey respondents (see Chap. 8).

Lack of a Non-blaming Culture

Culture can be described as the collective, shared thinking, and behaviour of the firm, or the team. It has a huge impact on the problem-solving process, as well as on *kaizen* activities, within the firm. In the fieldwork, most interviewees replied that when they encountered a problem, they certainly have to consider the consequence of surfacing it, and not simply the act of surfacing the root cause. In most cases, exposing problems will in turn bring potential economic loss. Moreover, the interviewees stated that when faced with problems, some project managers were more concerned about who should take responsibility for the problems, or whether their leaders from SOEs desire problems to be solved according to their own philosophy, such as "*reduce major issues to minor ones, and minor ones to nothing*". Under such leadership, techniques such as the 5 Whys have no place. If people do not look at problems as opportunities to build a better problem-solving system, then they will just take the shortest path to remove the symptom. Regardless of how much professional know how and skills they possess, such wrong behaviours can deter all progress. One project manager from Firm A highlighted that:

> Construction production is usually one-off in nature, and its uniqueness prevents kaizen activities from being conducted. Once this project finishes, our next project could be a totally different one and in a case like this, how can kaizen apply? Continuous improvement only works well under the manufacturing settings where everything is performed under repetitive flow.

A manager from Firm H also commented that a recurring problem explained how problems were repeatedly exposed in the construction site. Efforts were made, but this does not seem to work well. This relates ultimately to working habits and attitudes of workers, which from his perspective are lacking in PDCA or kaizen thinking.

Reflection

One of the common means adopted by the interviewees for conducting "reflection" on site matters is to diary the project progress. Minor or major problems would be documented in detail. In most cases, the diary serves as a technical "bible" that one can rely on when seeking resolutions. This practice was adopted by almost all the firms interviewed. One young engineer from Firm L pointed out that:

> In my current project, I diary a lot of things, for example project progress, what we have completed each day, solutions to certain problems, lessons learnt, etc. I even use this as a checklist for following up certain unresolved issues. By the end of my last project, I had completed 2 volumes of diary, which benefited me a lot.

Learning from Other Projects

It is important to observe *gemba* objectively and to have many ideas relating to possible improvements. In reality, as many of the interviewees reflected, many housing projects were homogeneous in nature, and so they tended to be less critical of their own projects, as one can easily get used to the current conditions of the site. It is not always easy to have many ideas for improvements based only from one's own internal resources and previous experience. Therefore, benchmarking is highly recommended. By visiting other good project sites, one can easily observe them objectively with severe *gemba* eyes. The points for improvement found at these good project sites can be mentioned to the firm running the site and can also be reflected in one's own project site. Similarly, by inviting external people to visit one's own site, a similar effect can be expected, receiving opinions of other project teams from the third person's point of view, and obtaining some useful advice.

9.4 Summary

This chapter presents the results of the interviews and elaborates on the relevance of the current practices of large Chinese construction firms to the Toyota Way principles. Overall, the interview findings supported the survey results in indicating that some of its principles and derived attributes were not only familiar to interviewees, but also appreciated by them. In some cases, the firms interviewed have effectively adopted several Toyota Way principles, such as commitment to reliable plans (P4), promotion of standardization at firm and project levels (P6), partnership with suppliers and subcontractors (P11), *genchi genbutsu* (P12), and others. These were supported by various pieces of evidence during the interviews. Additionally, attempts were seen to be made by these firms to modify some principles and practices of the Toyota Way to better fit the Chinese construction industry. For example, in the case of standardization, the standard operating procedures (SOPs) were actually perceived as less important in operations, given that site management resources are so limited that it becomes impossible to keep a constant watch on workers' SOPs. In view of such limitations, it is suggested that the scope of the standardization principle should be broadened beyond the frontline site and should include standardized management approaches, such as technical meetings for communicating the standardized procedure, using the standardized project management manual for site management, concerning every aspects of the operations. Similar endeavours were also made for the other principles, taking into account the uniqueness of construction projects. These are the encouraging results. However, compared to Toyota's achievements, the interview results seem to suggest that implementation was still insufficient. In addition, the implementation of Toyota Way practices are always associated with challenges and barriers, i.e. from clients, construction firms, their employees, and partners, as well as from conventional industry practices. As such, interviewees are often open to selective adoption, but not to the whole suite of the Toyota Way principles. For some specific principles—such as pull or single-sourcing—the resistance to change is obvious, and attitudes are that implementation of such principles would not necessarily benefit day-to-day operations, and may result in the opposite outcome. Such concerns imply that the preconditions are not yet fully established, and these still remain as threats to full implementation.

Case Study 10

10.1 Introduction

This chapter presents the results of three case studies conducted in selected Chinese construction firms. Section 10.2 describes how the cases were selected. An evaluation of the three cases is presented in terms of their experiences implementing Toyota Way-style practices.

10.2 Case Study Selection

The target companies investigated in the case study were not randomly chosen, but were selected according to their characteristics, and how these characteristics contribute to the research question (Eisenhardt 1989). The ideal candidate firm for this research should possess the following characteristics: (1) it must be a leading construction firm in China and (2) it must be easily accessible and willing to offer information, if required. To reiterate, the chief objective of the case study is to provide a practical example of how the Toyota Way model could be implemented in the context of Chinese construction firms. This study, however, did not present the Toyota Way principles to the case-study firms. Instead, endeavours were made to evaluate their daily operations, personnel-management systems, problem-solving strategies, and so on, at the project level, using the Toyota Way model. Moreover, the case study findings are expected to validate the applicability of the Toyota Way model within large Chinese construction firms. Three projects were chosen, with the details shown in Table 10.1. Two projects (Project A and B) from BAOYE Construction Group are included, and another one project (Project C) is from the a subsidiary of China Construction Engineering Third Bureau (hereinafter referred to as CCETB).

S. Gao and S.P. Low, *Lean Construction Management*,
DOI 10.1007/978-981-287-014-8_10,

Table 10.1 Overview of case firms and project descriptions

Case no.	Project name	Contractors	Contract value (RMB)	Quality objective	Project duration (days)
1	Hangzhou *Huafeng* project	BAOYE	240 million	"Qiantang river" quality award	1,030
2	Wenzhou *Guomai* project	BAOYE	Not provided	"Qiantang river" quality award	960
3	Wuhan *Wanda* Centre project	CCETB	1.23 billion	"Chu Tian" quality award	1,025

10.3 Case Company A

10.3.1 Background Information on Company A

BAOYE Construction Group is one of the four construction firms from Zhejiang province, China, that was accredited as a "Premier" contractor in the late 1990s. It is a large private Chinese construction firm specializing in government and public buildings, and residential and commercial buildings. The company operates mainly (50–60 % of the turnover) in Zhejiang province and Shanghai, but in recent years, its leadership has attempted to explore the new emerging market in China's western regions, partially due to intense competition in the two places mentioned. According to BAOYE's 2011 annual report, by the end of year 2011, BAOYE's construction business achieved a revenue of approximately RMB 12.18 billion (US $1.9 billion), an increase of approximately 26 % over 2010. Meanwhile, the total contract value for construction-in-progress projects was RMB 40.53 billion (US $6.43 billion) by the end of 2011, representing an increase of approximately 18 % over 2010.

10.3.2 Company A's Guiding Principles

The guiding principles of BAOYE are its vision, mission and core values. Mr. Pang Bao Gen, Chairman of BAOYE (2011) clearly states that: "*Going forward to the future development, BAOYE will continue to adopt entrepreneurship approach instead of a businessman like approach in running its business*". The following actions are derived from and guided by BAOYE'S vision.

1. Not only focus on grasping business opportunities, but to pitch on the corporate mission on short-term gains without satisfying longer term perspective.
2. Preserving sustainable profitability and innovation capability.
3. Making profits from end consumers rather than from downstream dealers or intermediaries by fulfilment of demands and requirements.

Messages from Chairman Pang in response to the firm's vision are encouraging. The first one reveals that BAOYE is thinking differently, not to go for more immediate gains, but to urge all the employees to resist the temptation to sacrifice the firm's mission for short-term gains. The second highlights BAOYE'S constant purpose—the highest calling and purpose is to preserve sustainable profitability. The last one implies the company's changing focus, which is on the end-users. It seems to be a win-win strategy that company wants to gain more profits from end-users, while making sure that they get the maximum return. Moreover, as can be found on BAOYE's website, the company's mission is described as "*from construction to manufacturing, leads the construction industry towards industrialisation in China*" (BAOYE 2011). This reveals the company's determination in adopting industrialized means to construct buildings. It also shows where BAOYE is heading, and what is important to the firm. BAOYE's core value consists of four parts, including:

1. "More pay for more work" under the contractual management model.
2. Human capital is the most valuable asset of the company.
3. The value of BAOYE is made of financial assets, hard working spirit, employee's knowledge and management's capacity.
4. To maintain an effective human resource management system, and to eliminate those having unsatisfactory KPI scores.

It can be seen that maximizing clients' values are not directly listed in BAOYE's guiding principles. It seems that BY is more concerned with the interests of employees, from their knowledge, skills, to salaries, etc. If BAOYE chooses to implement or adopt some underlying principles of the Toyota Way, these guiding principles should be revisited in the light of the Toyota Way principles.

10.3.3 Company A's General Human Resource Strategy

BAOYE's human resource strategy states that "*we do not purposely chase 'overseas degree holders', or hunt 'successful CEOs'; we only chase after those who in BY's view are the most appropriate candidates*". This is in agreement with Toyota's strategy in leaders, where the company is focused on growing their own leaders, rather than bringing in outsiders (Liker 2004). BY does not justify the reasons behind this strategy, but from the interviewees' responses, it is clear that leaders must fist thoroughly understand the company culture at BAOYE. This explains why most middle-level managers (i.e. project managers) in BAOYE have been working for BAOYE for a very long time—long enough to witness this private firm, based in Shaoxing, Zhejiang, expands its market across China, and eventually becoming a listed firm on the Hong Kong's stock market.

10.3.4 Company A's Organizational Structure

An examination of the firm's structure reveals that project managers and their teams played key roles in the firm's success. It was found that BAOYE has presently 107 project teams belonging to 19 regional offices (i.e. Hangzhou, Shanghai, etc.).

For convenience, each project team is named after its project manager (PM), who may commit to several projects at the same time. Internally, case studies Project A and Project B were managed by Manager Ma's team and Manager Chen's team, respectively. To some extent, these PM teams are very independent and have a sense of ownership over their projects. In addition, the agreement between these PMs and the head office allows the PMs to tap on BAOYE's established management system, company brand, tendering, etc. However, they do need to follow the site rules and management standards established by BAOYE, and they are subject to various quality audits as well as health and safety checks. In return, as much as 8–10 % of the total contract value will be allocated as a management fee to reimburse the head office, with the project team sharing the remaining profits.

10.3.5 Overview of Two Case Study Projects

Two building projects were provided by the case-study Company A, which were contracted by its Hangzhou and Wenzhou offices, respectively. The Hangzhou project (Project A, "*Huafeng*") is located in Hangzhou's future central business distinct, with a built-in area of 74,335.5 m^2. The Hangzhou team has been working on this RMB 240 million (US$38 million) project since October 2010, and is expected to deliver the project on budget by 30 March 2013 (a total duration of 1,030 calendar days). The height of the building is 150 m, and the structural form is a framed tube structure, with 29 floors above the ground level and three underground levels used as basements. Similarly, Wenzhou's *Guomai* project (Project B) is an office tower, with a total built-in area of 73,785 m^2. The tower consists of two basements provided for car parking. The upper 23 floors are intended to be utilized as offices for the city's Department of Communication. Both sites are relatively small and their organizational charts are also similar. Most of the project team members at Project A and B were interviewed, including the project managers, the materials managers, engineers, several site engineers and others (see Table 10.2).

10.3.6 Project A

This section presents the findings of Project A, which involves the main components of Toyota Way-style practices.

10.3.6.1 Process-Oriented Practices

One-Piece Flow (P2)

At the time the fieldwork commenced on project A, the concrete work for the second-floor basement ceiling was intended to be carried out in about 2 days. Under the principle of "one-piece" flow thinking, the whole site team was clearly aware of the progress of the project. To ensure that the concreting team could mobilize into the site in a JIT manner, engineers urged the carpentry and rebar tradesmen to speed up their work, without however compromising on quality or safety. The chief

Table 10.2 Information on the interviewees from Project A and B

	Project A		Project B	
No.	Role of the interviewees	Working experience	Role of the interviewees	Working experience
1	Project Manager (Mr. Ma)	15	Project Manager (Mr. Zhou)	6
2	Deputy manager (Mr. Chen)	11	Chief Engineer (Mr Pan)	7
3	Materials Manager (Mr. Shen)	7	Site Engineer A	2
4	Chief Engineer (Mr. Zhang)	9	Site Engineer B	4
5	Site Engineer A	5	Site Engineer C	5
6	Site Engineer B	4	Foreman A	6
7	Site Engineer C	12	Foreman B	7
8	Site Engineer D	7	Carpenter A	10
9	–	–	Rebar worker A	6
10	–	–	*Jianli* supervisor A	25
11	–	–	*Jianli* engineer A	13

engineer interviewed pointed out that, when the date of concrete pouring was approaching, the contractor employees became more diligent in aligning their progress with the concreting team and in keeping them updated about the project's progress. It is worth mentioning that concreting work is often performed by a licensed concreting trade, which is only employed to perform concrete work as and when the site requires. The concrete placing team has very flexible working schedules, and is typically committed to a number of jobs within the same region. One of the value-adding activities observed on site was that the chief engineer held a pre-pour meeting with the concreting trade before pours, with the intention of discussing issues associated with the process of concrete pouring.

Moreover, in a typical "one-piece" flow scenario, materials are expected to be supplied in the right quantities, in the right time, and to the right place. In Project A, although it is hard to see the daily consumption of material being specified in the plans, it was one of the themes of the weekly meeting agenda, discussed as part of the weekly materials resource plan. In most cases, as observed on site, the materials were ordered in quantities a lot greater than was needed on a daily or weekly basis, and so there is no need for engineers or foremen to worry about materials being in short supply and affecting the implementation of the "one-piece" approach.

Pull *Kanban* System (P3)

At Project A, the project team pays much attention to materials management. The materials manager explained,

> This is because given that the nature of our firm is private, the price of materials can easily affect us in a negative way. We have to effectively manage materials, especially price-sensitive ones.

Based on the necessary procedures associated with the incoming materials, the materials manager reflected on the materials procurement processes and commented,

> Actually, the so-called 'pull' concept is practised in a way. Take this incoming batch of rebar for instance: it will not be allowed to stay in the storage area or fabrication yard for long, but will be quickly consumed. Because our strategy is based on calculating the procurement time backwards from the date when the tradesmen ask for it [the material], and we will arrange procurement and delivery accordingly.

Following this thought, the materials manager continued, to indicate the cycle time between the point of delivery of an incoming batch of rebar, and point of use is about 3 days. This is the case, because shortly after a new batch of rebar is delivered to the site, at least 2 days are needed for a sample of the rebar to be delivered to the local authority, where it is examined for quality, specifications and other details. This is a mandatory procedure, since the *jianli* (the supervision firm) will not authorise the use of the arriving materials (i.e. rebar) unless test results show that the new batch is defect-free and complies with all the specifications. Concurrently, another day or 2 days is needed to fabricate the rebar for different design requirements. Apart from this, the following two reasons highlighted why inventory might sometimes be held:

1. In order to receive a discount or a low price when buying a lot of materials, the supplier may require the payment to be paid in full, and may demand that the materials be transported away from supplier's location.
2. All the incoming rebar needs to be fabricated into different pieces. Some of this may not be used immediately, and may thus become on-site inventory.

The materials manager seems not to be bothered by this temporary on-site inventory, as he even commented that when this Project A is completed, the estimated wastage of rebar will properly be limited to less than 10 tons, which is acceptable and perfectly within contingent considerations.

Planning and Control (*Heijunka*) (P4)

1. Plans and the planner

   Like most projects in China, agreed milestones are established in the master plan between client and contractor, prior to the commencement of the project. Project A's deputy project manager pointed out that a monthly plan was developed based on three key inputs, namely (1) the master plan, (2) the monthly materials resource plan and (3) the actual plan which reflects real-life progress. With respect to who contributed most to the planning, it was not surprising to learn that the planning power was very much concentrated in the hands of the project manager (PM Ma), who outlined that,

   > The leadership style determines who should take the planning job. At my site, I am the one making the plans and have the final words on it. The progress actually reflects my intention.

   Under such leadership, it appears that only the PM and perhaps his colleagues in the same office know the schedules, whereas little information pertaining to work schedules and plans was expected to cascade down to the worker level. Workers were only informed of the workloads over very short periods (i.e. 1 or 2 weeks), and also the locked deliverable dates. Hence, most workers in Project A do not have a complete picture of work progress.

2. Weekly meetings and weekly plans
   In contrast to LPS, which encourages foremen or last planners to make "commitments" in what they can actually do in the plans, Project A reveals the schedule information to lower level workers in a top-down way. The weekly meetings become important moments in which the PM discussed the plans. In a weekly meeting (11/04/2012) that the researcher was permitted to attend, it was confirmed that the weekly meeting agenda was being closely followed. The meeting began with the PM summarizing the performance in the last week, and announcing the work contents, along with the achievable targets of the coming week to the participants (i.e. the foremen, site engineers, etc.). He also reinforced several key deliverable dates. What followed were essentially the PM's responses to issues raised by the *jianli* at the workplace. For instance, the most urgent concern expressed by the *jianli* was that progress on the formwork for the second-floor basement appeared to be slow, which might affect succeeding work. Without seeking out the root causes, it was quickly pointed out by the PM that the insufficient number of carpenters on site was the cause of the slow progress in the formwork erection. To resolve this issue, the PM then agreed to take actions to urge the supervisor of the carpenters to increase the size of the workforce by any means.
3. Challenges in achieving even workload
   On levering the workload, in the PM's word, currently, the amount of available labour is not very stable. This hinders the daily work from being assigned in an even manner, which in some ways is negatively affecting the progress of the project. On the positive side, the shortage of labour in a way helps to eliminate the idle time within certain trades, since there were always working spaces available for a smaller number of workers to undertake. However, the cost of alleviating this "short-of-labour" condition was to demand that frontline workers work overtime. They then started their work at 6:30 AM, and finished at 6:30 PM, with a 2-h break in between.

Built-In Quality (P5)

Being aware of the difficulties in using the *andon* system on site (Nakagawa 2006), this case study intends to investigate whether the project team's (including frontline workers') attitudes to quality and their resulting actions are in agreement with the "built-in quality" (BIQ) principle. What matters most is to foster a culture in which workers are encouraged to stop and fix defects before they proceed to the next operation. What contradicts this BIQ thinking is the traditional mindset that relies on quality inspection. As shown in Table 10.3, Project A employs BIQ thinking in some limited areas, whereas inspections appear to play a bigger role in quality control.

Table 10.3 Implementation of BIQ thinking and quality inspections in Project A

Category	Areas	Details	Participants
BIQ	Material management	• Perform quality check on materials • Any defective materials must be rejected before use	• Materials division
BIQ	Operations	• Technical review meeting • Stand-up meeting	• Engineering division
Inspection	Operations	• Engineers' site walk • Workers' self-check • Handover check	• Site engineers • Frontline workers • Project manager, *jianli* and client representative
Inspection	General	• Quarterly inspections	• Head office's quality department

Clearly, the contractor personnel did devote efforts to quality control. Apart from two areas in which BIQ thinking was observed (see Table 10.3), it was discovered that inspections were widely adopted as a key approach when completed operations were waiting to be checked. Unlike Toyota workers who are trained to execute operations with BIQ thinking, in Project A, the participants engaged with BIQ are actually contractor employees, who safeguard the materials quality and help to create awareness of quality among workers. In the project manager's words:

> We need to be realistic, as we cannot expect that all the employees and frontline workers will be able to detect problems in a timely manner and report them to nearby supervisors. The quality inspections are still very much needed.

1. Engineers' site walk
   Several site engineers claimed that most minor problems could be discovered during their site walk. These site walks seemed to be quite spontaneous, as the site engineers were seen with no standardized check-lists or other tools for evaluating workers' performance; they instead proceeded based only on their experience. The chances are high that defects or noncompliance will be entered into the subsequent operations without being rectified. In contrast, *jianli* engineers seemed more professional and more rigorous in their checks, as they followed their own audit system (i.e. a standard check list) to evaluate quality and to reflect on the issues in the weekly quality reports.
2. Workers' self-check and handover check
   No matter what level of quality has been built in, a quality check is still needed, and in many ways is helpful to assure that the quality is in line with requirements. The "*self-check*" concept has been institutionalized, and is also stated in the project plans. This practice requires employees to conduct a self-check prior to higher-level quality assessment or handover to the succeeding trade.

3. Company-wide inspections
   Furthermore, BAOYE's head office performs company-wide quality assessment of all ongoing projects on a quarterly basis with announced dates. This is part of the company's initiative to promote "transparent" competition. A recent quality inspection was conducted on 1/3/2011 in Project A, and the problems spotted by the head office's audit team include (1) part of reinforced threaded connection is not in place, and (2) poor welded techniques on several steel columns. It is worthwhile mentioning that the brief assessment results are published via the company's office automation (OA) system, in which the high-performance project with the least problems found is announced here, as well as the worse projects which will be subjected to a fine of up to RMB 10,000 (US$1,580).

Standardization (P6)

Several standardized practices—developed in the implementation framework of the Toyota Way model—were adopted in Project A. This includes:

1. Management standardizations: at the first level, various "standards" set by the head office were available at the site office, and were referred to by the PM as "management standardization requirements". All the standardized items and procedures can be found in the company's project management guidelines.
2. Technical review meeting: at the second level, "technical review meetings" and "stand-up meetings" were conducted. These meetings serve two purposes here. One is to create awareness of quality, in the hope of building quality into the mindsets of the workers. The other purpose is to reinforce the standardized sequence in the minds of the relevant parties, the activity components, the "Do's" and "Don'ts", etc., that make up certain processes. At the time of the site visits, Project A had conducted 11 key "technical review meetings" to focus on major processes, including bored pile construction, excavation, RC works, formwork, and others.
3. Standardized work at the activity level: In Project A, standardization at the activity level seems "less satisfactory". The challenge is that the contractors' employees do not seem to pay much attention to day to day standardized work. Instead, what they do concern themselves most with is whether the subcontracted work can be accomplished on time. The chief engineer pointed out that, "The project team will not micro-manage matters such as SOPs. Instead, our job is pretty much done once we conduct the review meeting and convey the basic knowledge of construction methods and its associated quality objectives, health, and safety issues, etc., to the relevant parties". In the daily operations, the foremen can be trusted to allocate jobs to their team members more efficiently. To them, "standardization" is not their primary concern. Experienced foremen would let the same team stick to what they were asked to do repeatedly (i.e. floor by floor). For example, those who perform rebar placement for columns will do the same thing repeatedly throughout the building process.

Table 10.4 Site assessment concerns "health and safety" aspects

Items	Total number of observations from Oct 2010 to Mar 2011
(1) Mass of materials piled up in the rebar fabrication area, lacking an attached classification signboard	6
(2) Temporary electrical installations must meet installation requirements	6
(3) Unused formworks, square wood, and other debris are massively placed on nearby roads, leading to road congestion	4
(4) The boundary between site, dormitory, and site office is not clearly specified	1
(5) The operating procedures of machines, and other instructive posters, are absent	3
(6) Site water was not cleaned up in a timely manner	2
(7) Several workers did not wear personal protective equipment (PPE)	5
(8) Masses of rebar are inappropriately placed over the support beams	4

Visual Control (P7)

Overall, visual management in Project A was not so well implemented. This was reflected in several instances which were documented in a number of site assessment reports conducted by the *jianli* concerning "*site health and safety*" aspects, between October 2010 and March 2011 (12 times in total). The results reinforced warnings concerning several weaknesses in visual management exercised at the site. The reoccurring themes are highlighted below (see Table 10.4). These may be contributing factors that make the site less organized and subjecting it to more health and safety risks.

These listed issues were frequently addressed by the *jianli*. Not surprisingly, the whole list can run to some length, and certain items kept recurring. Quite clearly, the majority of items listed in Table 10.5 are relevant to *health and safety* issues. It appears that some basic visual control activities such as sorting (item 8), straightening (item 1), cleaning (items 3 and 6), and other visual signs (items 4 and 5) were either absent or poorly practised. The poor health and safety record can therefore easily become one qualifier for applying visual management and 5-S principles.

New Technology Adopted (P8)

Project A aims at the "*Zhejiang benchmarking project in applying emerging technologies award*", which is intended to showcase the technologies deployed, which have been used in the whole project. It was learned that the so-called *top ten emerging technologies* were not new to the project team, and the application of some of these to certain areas was proposed. In order to qualify for a model project in terms of the use of new and reliable technologies, the Zhejiang Construction Bureau set out the rules requiring that participating construction projects must

Table 10.5 Some examples of new technologies adopted in Project A

Item	Proposed technology	Proposed material	Applied area
1	Application of energy efficiency and new material for exterior wall	Hollow concrete block	Wall
		Extruded polystyrene board used for thermal insulation	Roof and exterior wall
		Rock wool insulation, insulated glazing, etc.	Curtain wall, aluminium window
2	New technology for waterproofing system	Polymer cement waterproof coating applications	Exterior basement wall
		Rubber waterproofing membrane	Roof

showcase some new technologies taken from at least six large categories in "*10 emerging technologies*" Some examples are illustrated in Table 10.5.

Unlike Toyota's philosophy of adopting new technology, the use of new technology in Project A was heavily driven by the possibility of winning an external award. In proposing some of the new technologies, fundamental principles—such as whether the technology would improve the construction process appeared to be ignored. Furthermore, in the two cases mentioned here, the so-called emerging technology actually involves using newly improved materials with better functions, or in more sustainable forms.

10.3.6.2 People-Oriented Practices

Leaders and Leadership (P9)

1. Organizational structure and leadership
   The typical organizational structure of Project A is highlighted in Fig. 10.1. Technically, Project A is managed by one of PM Ma's team, which belongs to BAOYE's Hangzhou office. Unlike other project managers in BAOYE, PM Ma currently manages only one project, hence he has the real power in Project A. Project A's leadership appeared to be more authoritative. Supporting evidence for this includes the fact that the project manager there (PM Ma) was fully in charge of project planning and decision-making, whereas the workforce was doing only what the workers were asked to do.
2. People-oriented
   Unlike irresponsible construction firms, which effectively abuse the legitimate rights and interests of the frontline workers, Project A's PM outlined that: "*protecting our workforce has been integral to BAOYE's Corporate Social Responsibility (CSR) philosophy since 2000*". BAOYE has set up a working group of 30 staff specifically over dealing with issues concerning frontline workers' legal interests. For example, the timely payment of wages was a major concern, especially during the Chinese New Year holidays. The company has also set aside a certain amount of funds for managing workers' disputes, in case any emergency situations occur.

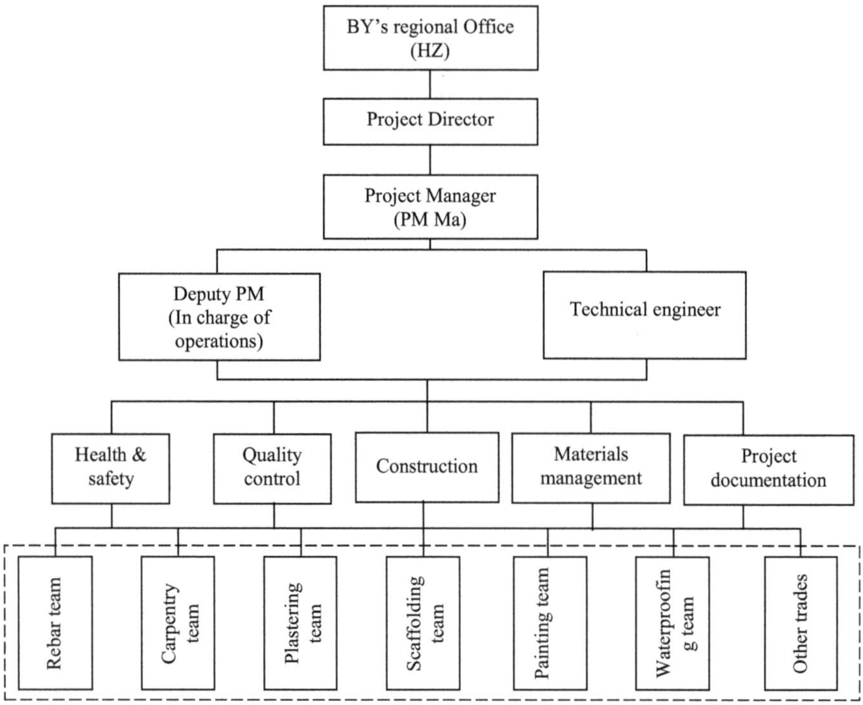

Fig. 10.1 A project chart of BAOYE in Project A

People Management (P10)

1. Workers and teams

   Basically, the different tradesmen are from the labour-only subcontractors (as boxed in by the dotted line in Fig. 10.1). One of the key features of the frontline workers is that their composition keeps changing, sometimes with low-skilled replacements. Moreover, within each trade, there are usually several team leaders who supervise the work, as well as a high-level supervisor who often has multiple teams committed to different projects (i.e. 3–10) for the same trade. The supervisor pays a small amount of management fees to his affiliate agency (a labour-only subcontractor) and that allows them to do business legally. An interview with some of the site engineers revealed that the management of low-skilled labour is ineffective and problematic. One challenge is that Project A was affected by the high turnover in the frontline workforce. According to one technical engineer, it really depends on the workers' moods. If they are not happy, they will probably consider quitting or get a job elsewhere. The engineer added that the trade supervisors are quite reliable in terms of long-term collaboration. Normally, they would still be working with the project team on the next project, while some frontline workers are dismissed at the end of the project.

2. Workers selection
   With respect to the selection of workers, several project team members revealed that supervisors can be trusted and play a key role in hiring workers. In line with Li and Peng's (2006) observation, these supervisors are in charge of recruiting workers and know where and how to seek out cheap workers. The project manager has no preference over the frontline workers, as he knew one cannot expect too much of them. Moreover, when it comes to specialists or machine operators, the project team always carefully verifies their qualification certificates and sends their identification to the project office for records.
3. Training and multi-skills
   It is common to see that the average contractor personnel understands more or less most of the relevant knowledge that the Project A may require. Based on the organization chart of the project, as outlined in Fig. 10.1, the team working on site is small, and so it is easier for the engineers or others to be exposed readily to new knowledge, and to pick this up quickly. When the PM was asked his opinion of multi-skilled workers, he commented that,

   > Although one particular worker left us to work for another site, he will still be doing a similar job. This is because he has no opportunity to be trained for multi-skills.

Relationships with Partners (P11)

1. Multi-sourcing
   Reviewing the material records and other relevant documents confirms that a multi-sourcing supplier strategy is presently adopted in Project A. Take rebar, for example. In the case of this material, every single purchase record was filed. It was found that there are a number of different types of rebar required, which are specified in the structural work plan. These range from HPB235 φ8, through HRB335 (φ10–φ22), to HRB400 φ20. No one single type of rebar was supplied through a single source, but instead multiple sources were used, involving up to six steel manufacturers in total. This is because the steel companies may not have the capacity or business interests in manufacturing all kinds of rebar. Nevertheless, *Hangzhou Huaqing* steel trade company was the only vendor, acting as an intermediary to supply whatever types of rebar are needed from the steel companies for Project A. Meantime, this local vendor can also be found in BAOYE's internally qualified vendor list, for supplying materials exclusively in the *Hangzhou* area. This again validates the interview findings that Chinese construction firms may not necessarily have direct partnerships with large manufacturers (e.g. steel companies), but through its partners/vendors, they use known products from the large manufacturers.
2. Long-term relationships
   A number of suppliers have worked with Project A for a very long time. The project team believes that their prices are reasonable most of the time. Unlike small construction firms, Project A understands the impact of the changing environment (i.e. price spread) could cause some SME suppliers to lose money, as the contract has already been signed. In circumstances like this,

Project A would be willing to agree to a slight increase in price, especially if the negotiation had been carried out in the hope of achieving a long-term relationship.

3. Lean thinking is lacking at the supplier side
   Part of Toyota's success is credited to its suppliers, which also practised the Toyota Way-style practices or lean thinking. However, Project A's materials manager asserts that "lean thinking" is absent from the suppliers they collaborate with. For example, it is convenient for the contractor to offload materials or components with better packaging and organization. The challenging part is that some suppliers are still poor at this. One example relating to material packaging was given by the materials manager interviewed. When steel tubes are delivered by one tube vendor, they are never delivered in carefully organized packages. Due to this poor preliminary classification of tubes, non-value-adding activities need to be performed by the recipients (the contractor), to re-count and reclassify the different tubes according to the right specifications. Although Project A has more than once suggested that the supplier to improve its delivery service by focusing on the details, the supplier has never listened and continues to avoid practicing in a leaner way, despite the small amount of efforts required. There are many cases like this, where it is challenging to extend management skills to downstream partners or to consider them a part of the so-called extended enterprise.

10.3.6.3 Problem-Solving Practices

Genchi Genbutsu (P12)

As indicated earlier, the project manager among other site-managerial employees, was quite active in *genchi genbutsu*, and hence had a good understanding of the site. The PM is quite close to his site team, although he sometimes appeared to be quite authoritative in delegating certain things such as planning and control. He has good analytical and listening skills. As he stated, he would investigate things personally if time allows, or he would ask the site engineers to report their first-hand observations to him. That became the basis of his and his management team's decision-making, and so they did not rely purely on experience. This is in line with the assessment results of the quality audit carried out by the headquarters, which the "leadership" category, PM Ma, along with his management team was rated as being "*visible on site, and willing to get their hands dirty*". It is worth mentioning that among 102 projects audited in the first quarter of 2011, 29 project managers were rated as "not fully in place", meaning that they are frequently out of reach when problems are encountered.

Decision-Making (P13)

Due to the limited time available to the researcher in Project A, no forms of decision-making were actually observed. Yet according to the responses of the PM and site engineers on this matter, it was revealed that most key decisions, such as those affecting the design of the site layout, project plans, and others, were made through consensus among core members of the project team. Meanwhile, the weekly meeting and other forms of meetings are the best venues in which decision-making can be observed. According to Project A's PM, a QC topic is still being sought for after even though the QC team have already reached out possible alternatives.

***Kaizen* (P14)**

In Toyota's view, *kaizen* is critical to everyday life for those working in Toyota. In Project A, it seems that this was least concerned about, since *kaizen* is not reflected in how they do things at the site level. However, an interview with PM Ma and the chief engineer confirmed that it is more feasible and with greater opportunities out there at the company level for *kaizen* to be conducted. For example, their regional office would help PM Ma and his team to access some other good project sites in the same region for them to learn about other practice. There are plenty of learning points, including a better workplace safety plan, a demonstration of the application of a new material or method for construction, and others. On the other hand, if Project A has shown something special that would enhance project performance, the regional office would also help the visiting teams to engage Project A for learning and sharing.

10.3.7 Project B

During the site visit, it was learnt that Project B was currently under schedule pressure. The project was then at the basement construction stage. There were two basement floors with a total area of 20,000 m^2. Basement level two was completed, and workers were proceeding to carry out the work at level one. In addition, three specialized subcontractors were involved in the project: namely the installation, cable and steel structure subcontractors. Data collection on this site concentrated on daily operations in the basement work.

10.3.7.1 Process-Oriented Practices

One-Piece Flow (P2)

The basement area was divided into five pours, according to a number of "pour strips". In each segment, several activities were performed in sequence: formwork erection, rebar embedment and installation, concrete placement and curing, and formwork removal. It was observed that the work was conducted under the "flow-line" philosophy. One of the young engineers interviewed spoke of the "flow-line" approach as "*not very difficult, but a common approach for organizing construction*". An example of the flow-line approach could be seen when carpenters completed the formwork erection in Segment 2, and then carried on the same work in Segment 3. When the formwork in Segment 2 was partially erected, free working space became available for the rebar embedment team, who was then able to squeeze in to begin their work, given that their work followed the formwork erection. By doing so, the workforce could be better utilized, and idle time minimized. It should be pointed out here that once the workers' turnover increased, it might badly affect the flow of work. In Project B, the *jianli* supervisor has on more than one occasion warned the engineer that the carpenters' progress affected the entire process, since their turnover appeared to be quite high.

Waste elimination is the chief aim of the one-piece flow principle. In Project B, a number of waste or non-value-adding activities were spotted, including:

1. Waiting: Waiting for materials in Project B appeared to be a normal occurrence. For instance, it was seen that one type of rebar which should have been fabricated was in short supply, and that resulted in the rebar team becoming idled. To quickly reduce the waiting time, an urgent note was sent to the rebar yard at short notice. One rebar worker spoke of the difficulties in getting the right amount of materials: "*It is quite frequent for us to receive urgent orders to fabricate rebar, and to deviate from the original plan*".
2. Inventory: Inventory was frequently found all over the site, which was also claimed by several interviewees to be the safe level of inventory. There seems to be no solution for reducing inventory to the extent that a manufacturer could achieve.
3. Overproduction: In Project B, overproduction might be enhanced by the fact that the ordering of product (e.g. rebar) was faster than its consumption by downstream practitioners in the supply chain. Figure 10.2 illustrates that, as the site was constrained in size, materials such as rebar had to be piled up on the beams. This was inappropriate storage, which also potentially added risk to the supporting structure.
4. Unnecessary transport and motion: It was clear from Fig. 10.2 that some of these rebar still need to be fabricated before use. This would require time and effort on the part of the workers to transport these materials to the right place for fabrication, since the rebar yard was at the side of the site.
5. Defect: Defects or "noncompliance" could easily be found on site (according to the *jianli* engineers). This was evident that the weekly report issued by the *jianli* always mentioned numerous defects and items that do not comply with the standards, and which the contractor was required to rectify. Admittedly, although it was claimed that quality management was implemented, it was implied that the "do right at the source" culture was not fully fostered in Project B. A majority of contractor personnel interviewed seemed aware of the consequence of rework, and would indeed support the idea of reducing rework by all means, but these statements would seem to came across empty promises.

Pull *Kanban* System (P3)

The researcher came to understand how the material was managed in Project B, in terms of whether it follows the "pull" principle or otherwise. There was no evidence to indicate any pull-associated activities or devices similar to the *kanban* system being used on site. Instead, the very conventional inventory-control means was mentioned by one of the interviewees:

> The amount of materials that we procured mainly depends on the material price as well as on our weekly and monthly checks of material inventory. Basically, using the online

Fig. 10.2 Inappropriate storage of rebar at Project B

> reference price, as well as our experience or hunches, we determine whether it's a good time to replenish the material. One thing is certain – if the price appears to be reasonable and the site condition allows for it, we will certainly procure in bulk.

Such a sentiment reflected that the concept of "zero inventory" was hardly feasible in Project B. Two possible reasons cited were in agreement with the interview findings, namely (1) price of materials and (2) people's high tolerance for inventory. The chief engineer conceded that they have found no better solution but to stockpile some materials on the supporting structure, even at the risk of violating the safety rules. Moreover, since the project manager might have committed himself to several projects, thus even if Project B was completed with extra rebar left unused, these would soon be transferred to other projects for use, or simply sold. This attitude explains why the personnel were slow in taking up the pull concept. However, it was pointed out that once the project approached its end, a tighter approach to material control would be adopted to minimize unnecessary wastes.

Planning and Control (*Heijunka*) (P1)

1. Last Planner

   According to the chief engineer in charge of project planning, the inputs to making weekly work assignments were not collected from the foremen or last planner of each trade. The foremen were not called to participate in these meetings, as they knew that their individual interests would not be taken into account when decisions were being made in the weekly plans. Conversely, in line with the interview findings, the chief engineer drew up weekly working plans for each trade and for the subcontractors, based on his own observations of the site process; this was considered to be the prevalent practice. Later, the weekly tasks would be shared with the foremen or the team leaders of the trades. To assist the chief engineer to put the weekly plan into practice, the foremen would allocate their existing labour resources based on the given schedule. Eventually, all this information would be conveyed to the frontline workers.

They might not be informed of the entire weekly plan all at one time, but it would gradually be communicated to them through their daily stand-up meeting. The chief engineer pointed out that the foremen and their trade members preferred to perform the given job within a finite amount of time, rather than to ask for their "commitments" in making their own plans. Apart from their poor skills in planning, this was also because they feared pinning down their commitments in written plans. In one engineer's words, when the delay occurred or the workers failed to accomplish the assignments within the given time frame, they would not be willing to investigate the root causes, but instead quickly sought excuses to justify their poor performance, typically by blaming the plan from the project team for not taking into account their capabilities. This attitude was a big challenge in the implementation of LPS in the Chinese context. One foreman pointed out that,

> Empowering plans to be made in a bottom-up way seems like a trap to me, since a fine or penalty will be imposed if the plan fails to be accomplished on time as promised.

2. Weekly meeting
   The general contractor (BAOYE) met the subcontractors, foremen, trades supervisors and others once every 2 weeks at the site meeting, where the issues that needed to be addressed included (1) issues that require the client or *jianli* to coordinate; (2) recurring problems discovered by the *jianli* and (3) resource schedule for the coming week. In particular, when BAOYE met a client, they would not voluntarily expose their weakness or problems, but would request coordination on certain challenging issues. In (3), the discussion of resource schedule was of great importance. For example, in the event that the tower crane was in high demand, participants would discuss a detailed time table for the tower crane in the meeting to better utilize these limited resources.

Built-In Quality (P5)

1. General quality management
   Reviewing the documentation for quality management, as well as observing actual site operations, revealed that the basic requirements and instructions of quality management practice were followed. In a recent quality assessment conducted by the head office, the evaluation of Project B is not satisfactory in a number of areas. These included: (1) Quality: Ineffective development of rebar in tension in the joint of the stair slab and beam. In several places, vertical rebar is found to be offset. (2) Health & safety: In basement scaffolding erection, an insufficient number of horizontal bars were used. (3) Machinery: One of the tower crane's trolley wheels is damaged, causing constant friction and collision of trolley and lifting arm.

   It is also noted that the leadership was marked as "not in place". In contrast to Project A, more quality problems and health and safety risks were identified by the audit team, thus reflecting that the commitment of the leadership to quality was absent.

2. Built-in quality (BIQ) thinking
   Instilling BIQ thinking into employees' mindset requires a change in their thinking patterns. When asked whether BIQ was implemented into the operations, it was learnt that Project B also seems to rely heavily on quality inspections. An analogy referring to BIQ thinking was given by one site engineer in Project B:

   > The daily 'tool box' meeting could somehow serve as a kind of BIQ thinking, in which we repeatedly address the quality issues with the workforce, in the hope of creating a basic awareness for them to understand and act rightly and conscientiously.

   Project B has a large quality management team, including one chief engineer, five site engineers, one quality engineer, and several quality engineers from the *jianli*. During operations, when the foremen and site engineers were walking around the site to act as sources of technical information, it was observed that they were not stopped by the workers for assistance. This would seem to be a missing link, where the culture of encouraging workers to raise questions and exposing problems has yet to be established.
3. Quality Circles
   Recently, QC in China has become mandatory for projects that intended to pursue quality awards. In order to qualify for such local quality awards, it is not uncommon to see a situation like Project B, where the formation of the project's QC team was already proposed in the project plan. The QC plan was written into the agenda from the beginning, which was indeed a good starting point. In checking the QC-related meeting notes and preliminary results, it was found that, since the time the project commenced in late 2010, the planned QC team had never meet to discussed quality problems. This suggests that there appears to be a desire to delay QC activities to the very last minute, unless an appropriate reason for investigation was established. It is understandable that Project B was then still in the early stage of construction, and searching for a reason to investigate became their primary concern. Moreover, QC is understood here as the publication of improvements for newly discovered issues that have not been encountered in other projects. That explains why the Project B team has taken so long to search for a topic—they do not want to repeat things that other projects had done before.

Standardization (P6)

1. Standard operation procedures (SOPs)
   Standard operating procedures here did not meet the level of details, one would see in a Toyota assembly plant, in which every single step in the entire process follows a standard procedure. In Project B, construction plans for major processes, along with detailed technical drawings and others were checked. The staff only made sure that the elements of standardization—the work sequences—were illustrated. These were good examples of standardization, but could only be accessed and consulted in the project documentation room.

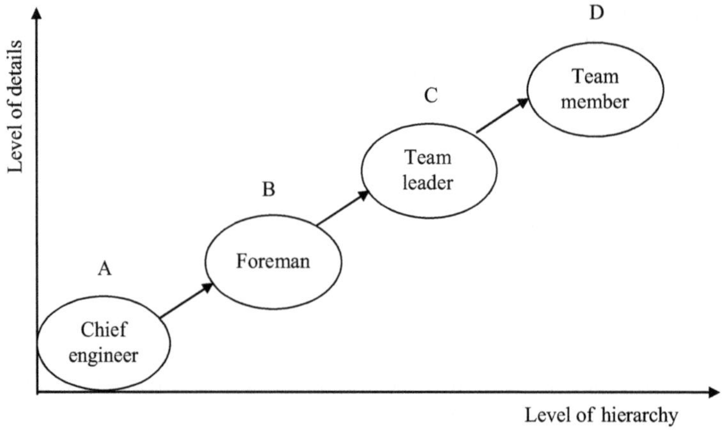

Fig. 10.3 The change of details in technical review meeting

2. Technical review meeting
   At Toyota, information on SOPs is available in the form of standard worksheet and is visible at each worksite. Anyone would be able to check whether or not standardized work procedure is being followed simply by looking at such sheets. Construction is different, but these technical review meetings do serve a similar purpose, in which instructions are visually and verbally presented to the participants prior to the commencement of work. Figure 10.3 illustrates how the level of details of technical clarifications has progressed in Project B. It shows a simple relationship between the level of detail discussed in technical review meeting (Y-axis) and the level of participants' position in the review meeting (X-axis). Simply put, as technical meetings migrate towards lower positions, more details need to be explained. Generally, this starts with the regional office's engineering department conveying the knowledge to the corresponding engineer in Project B. The online project management system ensures there is no need for the regional office engineer to be present on site, because information and knowledge can be shared via the online project management system.

   Once this step is completed, the project chief engineer, in a pre-work meeting or technical review meeting, then highlights key issues and instructions to the foremen, ensuring that they understand how to perform the work. Later, the foremen need to explain the same thing to their team leaders in a more detailed and practical way. This is because team leaders are less educated and less experienced than the foreman. Therefore, the foreman is required to explain more. Lastly, the team leaders will pass this know-how to their team members, using layman's terms. At this stage, the meeting is usually conducted on site in an informal way. If team members encounter problems or hesitate to proceed with unclear issues, team leaders call for a quick site meeting at the gemba (the actual place), and repeat the key instructions. The problem is in the process from A to B (see Fig. 10.3): it was discovered that sometimes no review meetings were conducted at all in Project B. This is to allow the crews to rush their work by cutting short or skipping the meeting. Instead, the project engineer printed out what the regional office has forwarded to them, and simply passed these print-outs to the foremen. As one site engineer explained:

> It is common that such review meetings are skipped. When foremen encounter problems, they will come down and seek your help. That is very common. Project engineers are busy with day-to-day matters and have no time to engage in this. Obviously what is lacking here is the commitment.

Visual Management (P7)

On several site visits to Project B, simple visual control in the form of large visual boards and small signals, dealing mainly with health and safety issues, were observed. However, from the engineers' point of view, these visual efforts are just rote actions, and do not contain much information to reveal the variations that actually occur during operations. The visual control efforts were required by criteria of the "*excellent construction site*" award, initiated by the provincial construction bureau in Zhejiang, China. Because of this, quality engineers and site engineers needed to collect completed project photographs in order to reflect their efforts in keeping the site tidy, organized, etc. Additionally, they were also requested to take photographs to track the quality of part of their project, or of certain parts of the structure such as the levelness of a wall. The site PM claimed that such efforts are a key part of visual management.

New Technology Adopted (P8)

It is more appropriate to examine company B's strategy in acquiring new technology at the firm level. Given that BAOYE has over 100 projects, it is more likely that the use of new technology derived from meeting the varying design requirements of different clients would be seen there. On the other hand, BAOYE is very active in promoting the use of prefabrication technology, and this too was reflected in BAOYE's mission statement. To this end, BAOYE has been committed to research on industrialized housing since the time it became one of the national pilot companies for industrialized housing, nominated by the Ministry of Construction in 1997. So far, BAOYE has invested 200 million RMB to set up an R&D centre with nine individual laboratories focusing on structural safety, interior testing, durability, curtain wall, earthquake resistance, etc. Referring to the overarching significance of Principle 8, the purpose of doing so is not only driven by the company mission, but also bearing the meaningful purpose of providing industrialized housing projects featuring "*comfortable, low carbon, and recycled*" products to society. The group has gained a number of patents derived from its R&D efforts and positioned BAOYE in the forefront of housing industrialization technology.

10.3.7.2 People-Oriented Practices

Leaders and Leadership (P9)

The team under PM Chen's leadership has been working together for quite a long period. The leadership at site was rather confusing especially for outsiders. The project manager was Mr Chen, who was often away from the site. However, when he does show up, it was mainly to deal with issues relating to clients or visiting civil servants. The deputy PM, Mr Zhou, is a relative of the PM. Before Mr Zhou took over this project, he had limited knowledge of construction. The chief engineer (Mr Pan), on the other hand, acted the role of both project manager and technical advisor simultaneously, and he basically needs to take care of every matters on the site. The chief engineer pointed out that he checked the site operation twice a day. To achieve better time management in his position, he delegated some of these day-to-day responsibilities in order to have the time he needed to plan and reflect on daily matters. Some responsibilities, however, must be retained by the chief engineer—for example answering or clarifying technical questions from the subcontractors, foremen and others.

People Management (P10)

1. Recruitment and retention

   Generally, the construction labour force is unstable and subjected to high turnover. In an interview with one carpenter, who recently joined Project B after the Chinese New Year (2011), he explained that he and his friends came to help the supervisor of the carpentry crew, or whom he referred to as the "boss". This supervisor earned a reputation as a good foreman by valuing his team, offering generous and timely pay, and being close to his team members. The newcomers promised to work until the jobs assigned by the boss are done. Their boss usually runs 3–10 projects. This made it possible for him to facilitate labour mobility.
2. Training

   Overall, very little training was given to enhance the capabilities of the contractor personnel, not alone the frontline workers. According to the interviewees, the head office organizes training for the project engineers twice a year, in order to let them catch up on developments in engineering and to update their skills in the areas of, e.g., construction technology in high-rise buildings, and deep excavation pits. More commonly, the head office directly posts training information on a regular basis through the company online project management system. In terms of the on-the-job training, the newly recruited site engineers for example would be assigned a mentor, following the so-called in-house mentorship scheme. According to one engineer, normally they do not have on-the-job training, but the site itself is a good classroom as his direct supervisor encouraged him to ask questions. In receiving the answers, he would learn the skills to tackle similar problems.
3. Motivation

   It is not easy to educate frontline workers, especially in terms of their behaviour and ways of operating. Warnings are among the most common punitive measures adopted by the managers. Fines are the least commonly seen measure, as they are viewed as ineffective. In terms of the financial incentives, the project team has consulted with clients on the issue of the extent to which these incentives should be deployed. The bottom line was that financial incentives would not be given just for following the basic site rules and standards. Further, the project team thought it would be more effective to organize competitions. For example, a "safety awareness competition" was conducted in Project B, aiming to raise awareness of the safety culture at the workers level. The competition contained 100 safety-related questions, and a token award is given to the employees who answered the most questions correctly. It was agreed that good pay is the most effective approach to motivation. An interview with several carpenters confirmed this: one carpenters came to work for Project B because of the good pay of 170 RMB/day. Their counterparts working at nearby sites doing the same job were paid 20 RMB less. In addition, this carpenter has promised to work until the end of this project with the team, when he would receive his total wages as a lump sum.

Relationship with Partners (P11)

BAOYE has its own list of qualified partners. At the outset of the project, the project team can seek the advice of the head office in selecting regional qualified partners. Once the project has been completed, the project team can in turn recommend the high-performing partners to the company's procurement department for consideration, with the purpose of enlarging the company's partner list. Project B has adopted the evaluation criteria created by BAOYE's head office, in order to assess firms for first time collaboration. There are nine areas as shown in Table 10.6 that require suppliers or subcontractors to provide relevant and concrete information. These include three key criteria: business status quo, service and process. Moreover, according to the project manager, all the partners listed undergo a so-called thorough assessment on completion of their work. The next task is to work on partnerships with some of them. So far, for weak or poor performing suppliers, the approach taken was to simply remove them from the list. Trying to help them improve their work is not seen as a practical future plan in Project B.

10.3.7.3 Problem-Solving Practices

***Genchi Genbutsu* (P12)**

The management team at the site level is reasonably visible in the workplace except for Mr. Chen (the PM). On a number of visits, the various managerial personnel could often be seen on site, checking on workers' performance. As revealed by Project B's chief engineers, site engineers are required to spend about 90 % of their time working on the site. Their frequent presence on the site provides easy access for workers if things go wrong, and also for just-in-time clarification and solutions.

Decision-Making (P13)

Overall, many quality problems were detected in Project B in the quality assessment exercise, conducted in March 2011. For example, rebar alignment problems

Table 10.6 Three key criteria and their underlying evaluation items

Criteria	Evaluation items	What to measure
Business status quo	Qualification	Business scope and related qualifications (i.e. "premier", "first", "second", etc.)
	Previous projects	Rewards
	Reputation	Client royalty and satisfaction
Service	Price	Compared to the market price
	Schedule	Delivery ability (i.e. is JIT adopted?)
	Quality of product or service	Customer satisfaction, defect rates, etc.
Process	Production capacity	Production capacity to meet the changing requirements
	Quality management	TQM, ISO certified, etc.
	Health and safety	Plans of health and safety training, records
	Environment management	ISO14000 certification

were spotted. However, this issue did not receive enough attention from Project B's site team, since as weeks after the assessment, similar problems were found: after the basement second-floor concrete pours, the vertical steel rebar in one of the columns, was out of alignment by up to 50 mm, owing to the pressure that resulted from the concrete pours. The chief engineer quickly came to the site to understand the causes. It was quickly diagnosed as a quality problem—the impact resulting from the concrete pours, as well as the poor workmanship of the rebar placement. As the *jianli* engineers were not aware of the problem, the chief engineer—without discussing countermeasures with his site engineers or attempting to achieve consensus—ordered the workers to cut off the misaligned rebar, and to let the carpenters seal the formwork and cover up the defects. The chief engineer did not give reasons for this decision. Later, one site engineer explained that his "cut-the-corner" strategy was aimed at not exposing the problem to the *jianli* and getting him in trouble.

Kaizen (P14)

From the case highlighted above, it can be seen that continuous improvement was also lacking in Project B. Since Project B commenced in October 2010, no QC team members have sat down to discuss improvement. Instead, the list of items that needed to be rectified or improved actually came from the *jianli*'s notes, which as an external source urged the project team to improve on the defects that have been spotted.

10.4 Case Company B

10.4.1 Background Information of Company B

Company B is based in Wuhan, Hubei province and operates in the public and private sectors across China. Company B has a revenue of RMB20 billion (US$3.17 billion) in 2011, and had a workforce of 3,300.

10.4.2 Company B's Guiding Principles

To obtain information on the company's philosophy of conducting business, two interviews were conducted; one with a deputy manager of company B, and the other one with the project manager. The core values of Company B are stated as follows: (1) to provide the best service to clients, (2) to maximize value for shareholders, (3) to enhance the well-being of employees and (4) to improve the living environment for people. It is worth mentioning that company B is a third-tier subsidiary of a leading large-scale Chinese construction firm, which is organizationally complex, with multiple departments and systems. Given these relationships, the core philosophies of the company are actually taken from its mother company. How the guiding principles impact project operations, people management, and other functions at both the firm and project level will be discussed in the later sections.

10.4.3 Overview of Project C

Project C was selected to review the extent to which Company B's daily operations are in line with the Toyota Way principles. Project C is located in Wuhan, Hubei province. Given that this project is poised to become an important venue to commemorate the 100th anniversary of the *Xinhai* Revolution,[1] the stakeholders involved were fully aware of its political significance. The contractor is company B, and the client is a leading real estate developer in China. This was an RMB1.23 billion (US$195 million) project to construct two hotels, three office buildings, two malls and five high-end residential buildings on a site area of 106,200 m^2. Because of the enormous size of the project, it was decided to develop the project in two phases. Taking into account the required deliverable dates of each building, as well as the status of ongoing site demolition works, Project C was further divided into five distinct sites (see Table 10.7).

As highlighted in Table 10.7, Phase 1 contains South plot site A, which has a total gross floor area of 130,000 m^2. This involves the construction of a 22-storey five-star hotel and a 42-storey 5A-class office building with two basements. Phase 2 includes the South plot site B, and the whole north plot, which contains another

[1] The revolution is named after the *Xinhai* Year of the sexagenary cycle of the Chinese calendar.

Table 10.7 Two-phase construction of Project C

Phase and location	Building no.	Buildings (units)	Floor no.	Total gross floor areas
Phase 1 (south plot site A)	–	Five-star hotel × 1	22	130,000 m^2
	–	5A-class office building × 1	42	
Phase 2 (south plot site B)	12 and 13	SOHO office building × 2	30	110,000 m^2
	–	Mall × 1	3	
Phase 2 (north plot site C and D)	1	Mall × 1	3	190,000 m^2
	2 and 3	High-end residential building × 2	43	
	6 and 7	High-end residential building × 2	42	
Phase 2 (north plot site E)	5	High-end residential building × 1	43	180,000 m^2
	9 and 11	SOHO Office building × 2	22	
	10	Office building × 1	10	
	8	Shopping mall × 1	3	

Note: Building no. 4 was omitted because the number 4 was inauspicious in Chinese culture and was thus not used by the project team

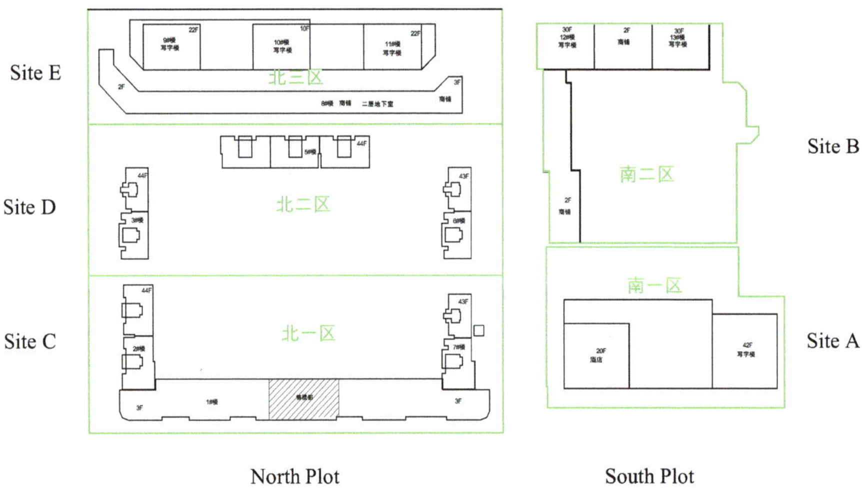

Fig. 10.4 Project C's site plan

three individual sites. The sequence of Project C started with the construction of site A on the South plot, and then proceeded to South plot site B, followed by the construction of the buildings on North plot Site C, D and E, in that order (see Fig. 10.4). Each site consists of five stages of work: foundation and piling, earthworks and internal support systems, basement construction, main structure, and exterior and interior finishes. Although these can run concurrently, each site contained its own subprojects and required distinctive leadership, capacities, and processes.

10.4.3.1 Challenges Encountered in Project C

The difficulties and challenges associated with Project C were noted. Apart from some technical difficulties, such as health and safety issues involved in excavating the phase I foundation area (which covering an area of 13,600 m^2), and the massive concrete pours, several external challenges also needed to be tackled. These include: (1) poor preliminary preparation, (2) the pressure of tight schedule and (3) design changes.

Firstly, one of the biggest challenges in the early stage was some of the incomplete elements that emerged when the project was about to commence. The preliminary preparations appeared to be poor and inadequate. For example, drawings were incomplete, site demolition works were incomplete, and so on. All these "incomplete jobs" contributed to a delayed start to the actual construction of Phase 1. It also caused much inconvenience for project team in planning the work efficiently. Secondly, the total period for construction was set to be 1,025 days. Given that the project commenced on 11 November 2009, the duration would stretch to include three Chinese New Year periods and three winter construction periods. This is because in China, the labour situation, as well as the harsh climatic conditions in winter, would slow the progress of work during these periods. Thirdly, there were a few design changes (see Table 10.8) endorsed by the client, which created more work. Logically, the deadline should have been moved forward, but since the added work was associated with an up-scale hotel and office building, it was natural that the client wanted the project to be completed earlier, so that revenues and tenants could be secured earlier. Consequently, the deadlines were also affected.

Overall, Project C had a delayed start, and was undertaken in circumstances in which the deadlines of sub-projects were drastically affected. There was also the high likelihood of potential penalties being imposed if the project was not completed on schedule.

10.4.3.2 Process-Oriented Practices

One-Piece Flow (P2)

1. The uninterrupted work-flow

   The construction site of Project C is huge, and the work was well in progress on several sites, where people were working in sequence on their given tasks. Guided by well-thought out designed work plans, almost no idle time was expected, and everyone had work to do. An interview with several site managers confirmed that among all the factors, (1) maintaining stable manpower, (2) having a complete set of detailed designs and (3) commitments made in detailed planning were the key factors highlighted in achieving "one-piece" flow in Project C. Firstly, in the case of the selection of workers, the project was in favour of not only those who had prior experiences with similar projects or with

Table 10.8 Design changes and their associate impacts on deadlines

Items	Work scopes	Plans	Start	Ground-breaking	Main structure completion	Added work
1	5-star hotel (Phase I)	As-planned	2009.10.10	2010.4.30	2010.9.15	2 floors added to 20 floors
		Adjusted	2009.11.11	2010.5.20	2010.8.30	
		Difference	Delay 32 days	Delay 12 days	47 days earlier	
2	5A office building (Phase I)	As-planned	2009.10.10	2010.5.20	2011.3.21	8 floors added to 34 floors
		Adjusted	2009.12.29	2010.5.20	2011.1.10	
		Difference	Delay 80 days	No change	150 days earlier	

the same client, but also those high-performing work teams with high levels of morale, commitment and productivity. The project manager believed that retaining high-performance work teams is one solution to make one-piece work-flow successful. Project C also had an agreement with several other teams to keep them as "spare teams" in case of labour shortages being encountered, or of the project going off the scheduled track. Secondly, following a discussion dealing with the problem of the unnecessary waiting time that might be caused by incomplete designs, the project team came up with a proactive initiative to turn up the heat on the design institute: they sent a full-time engineer to coordinate and exchange the information that is needed. Thirdly, the project manager acknowledged that tremendous efforts were made in planning and control. The planning team was very committed and gave much attention to details. Take, for example, the 5A office building on site A: Prior to the concrete being poured for its foundation slab, the concrete plan was already carefully drawn up, with all the details considered. The concrete pour was expected to be undertaken in a one-piece flow manner, to ensure that the work would be done right in the first instance. The plan started by setting out the quantities of concrete needed for each section,[2] as highlighted in Table 10.9. In order to ensure that concrete can be pumped continuously, two nearby concrete batching plants were used. The number of vehicles needed to transport the ready-mixed concrete in a timely fashion to feed the pumps was also considered in the context of the following factors: the average output of a pump (30 m^3/h), the maximum volume of each truck carrying ready-mixed concrete (8 m^3), the speed of the trucks, distance between the concrete batching plants and the site (20 km), and the allowable waiting time (2 h). The end result was that each pump was associated with 12 trucks.

Furthermore, transportation routes were discussed and finalized as illustrated in Fig. 10.5. In Fig. 10.5, the green line represents the route of the trucks, which

[2] The site was divided into six sections based on the number of "pour strips".

Table 10.9 Quantities of concrete needed for six sections of 5A building's foundation slab

Sections	S1	S2	S3	S4	S5	S6
Quantities (m^3)	3,580	1,640	1,200	780	1,050	8,230
Hours (h)	60	27	40	26	35	92
No. of pumps	2	2	1	1	1	3
No. of trucks	24	24	12	12	12	36

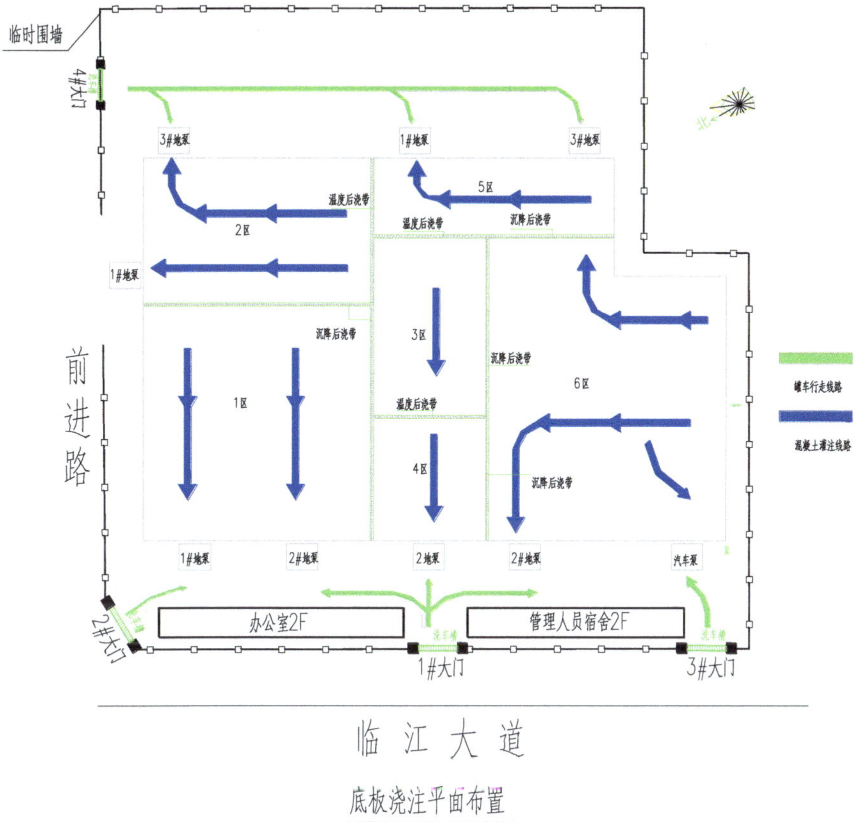

Fig. 10.5 Logistics plan for concrete pouring of the 5A office building's foundation slab

transport the ready-mixed concrete, and the blue line highlights the sequence of concrete pours in each section.

2. Outcomes

Given all the challenges highlighted, the Phase I project was completed within budget and ahead of schedule, and received commendable comments from the clients. It is worth mentioning that the main structure of the 5A office building was completed 152 days earlier than originally planned.

Pull *Kanban* System (P3)

1. Phase I: push to support full scale construction

   Project C adopted mixed pull and push strategies for material procurement in different phases. In Phase I, given that the construction volume was relatively small, and under great pressure to deliver the sub-project on time, the push strategy was adopted. This is because it was the first time that Company B is working with the client. Strategically speaking, if Phase I can be successfully completed in time, the company would win more confidence from the client. The project team decided to prepare the material and equipment resources as much as necessary in order to support Phase I construction to be carried out at full speed. Furthermore, the conditions of the site also allowed for much inventory to be stockpiled. In other words, there is no need for materials procurement to take into account the site requirements; this can simply be based on bills of quantities.
2. Phase II: pulled by site needs

   When the project progressed to Phase II, in which many more working spaces were released on different sites, many more workers were brought in, more constraints were encountered. Subsequently, the project team conceded that the material strategy for Phase I was no longer appropriate and needed to change. Firstly, in the case of rebar procurement, Project C set up a "rebar team"—an independent team under the materials department—to deal with all matters associated with rebar, including rebar planning, procurement, quality checks, storage, fabrication and others. This "rebar team" consisted of five full-time employees with their job scope and responsibilities clearly identified. This also resulted in frequent communications with supervisors in different sites, suppliers and steel vendors.

For example, in one of the sub-projects of Phase II (i.e. residential building no. 2), the rebar team ordered rebar based on the cycle time of 5-floor intervals, rather than for each single floor use. The reason why five floors were taken together as a unit was that the project plans mandated the removal of formwork every five floors for reuse purposes. To the rebar team, this then meant that the use of rebar should accommodate the five floors worth of production. As the deadline window was locked, it was therefore necessary for the crew to construct one floor in about 3 days, or five floors in 15–16 days. The procedures for rebar purchase were explained by the PM as follows. It started from setting out one floor of the no. 2 residential building. Once the setting-out information was available, the rebar team could produce a material resource plan for five floors of use. The plan was then submitted to the technical department for a thorough review, including validation of whether the plan was technically and economically viable. Before finally placing the order, the site manager's review and approval was also required.

This process explains the reason why a moderate level of rebar inventory was observed during site visits. The PM outlined that it was possible to shorten the delivery cycle into 3 days in order to accommodate just one floor, but that this was not economically feasible. Such frequent delivery would result in unnecessary transportation costs; it would add risks for both sides, since the rebar prices changed on a daily basis.

Table 10.10 Project C's hierarchical plans

Level	Name of plans		Planners	Reviewers
1	Milestone plan		Client and general contractor (GC)	–
2	Master plan		Project manager (GC)	*Jianli*/client representative
3	Phase plan		Planning personnel (GC)	GC/*Jianli*/ client representative
	Sub-project plan		Planning personnel (GC) and subcontractors	
4	Operational work plan	Quarterly plan	Planning personnel (GC) and subcontractors	GC/*Jianli*/ client representative
		Monthly plan		
		Weekly plan		
		Daily plan		

Planning and Control (*Heijunka*) (P4)

1. Four levels of plan

   Table 10.10 illustrates the key elements of the plans, along with the relevant departments and employees involved in the planning process. As with the LPS's hierarchical structure, the project plan used in Project C has four sub-elements.
2. Project Planning Taskforce

   At the outset of Project C, the project team set up a "*Project Planning Taskforce*" (PPT), whose main tasks and responsibilities consisted of the following:
   (a) Keeping the master plan and phase plans up to date.
   (b) Releasing daily plans, as well as daily resources plans.
   (c) Coordinating with various site teams on their progress, given that space conflicts were common. A detailed timetable was requested to highlight working sequences and allocated duration.
   (d) Checking and analysing the status of activities on a daily basis, and announcing daily progress at the daily meetings.
   (e) Adjusting the daily plan in a timely manner for the following day, including the adjusted resources plan.

   It was claimed by one PPT member that, in order to achieve a high level of accuracy and reliability in the daily plans, it was necessary for top management to show their commitments. This explained why the management staff of Project C stayed in the office until very late in the day in order to review the daily plans.
3. Daily plan and control

   Most projects break down the master plan into weekly plans, and this approach may work well for smaller projects whose risks are known. However, taking into account all the challenges mentioned above, extra efforts were needed. The taskforce introduced a simple spreadsheet, entitled "*daily plan control*", to break the daily workload further into detailed sub-tasks for the relevant parties. The project would arrive at the situation captured in Table 10.11 (for example)

Table 10.11 An example of daily plan adopted in Project C

No.	Work area	As-planned completion date	Status quo by 21st April	Schedule analysis	Construction volume (remaining)	22nd April plan	Labour resource requirement (required, actual use)	Suggestions
1	Hotel's first floor basement (structure)	10th May 2010	(1) Formwork panels of second-floor basement wall is completely removed (2) Scaffold support beams has been erected (3) Setting out the first floor basement	Progress goes according to plan	3,300 m^2 (3,300 m^2)	Need to carry out waterproofing installation for about 400 m^2	Carpenters (200, 200) Rebar workers (120, 120) Waterproofing workers (20–30, 24) Concrete workers (30, 20)	(1) Waterproofing team can commence their work at several points (2) The concrete team should also squeeze in and finish their job fast
2	Earthwork @ zone 2-2	12th April 2010	Earthwork excavation up to 500 m^3 by 21st April	Weather affects the earthwork	64,800 m^3 (3,200 m^3)	Carry on earthwork excavation up to 1,000 m^3	10–15 dump trucks used to remove the soil	Speeds up the progress of excavation
3	Concrete cushion, waterproofing @ zone 2-3	22nd April 2011	(1) Foundation platform formwork @E, G, and J axis (2) Rebar cage fabrication at foundation platform	Progress goes according to plan	1,211 m^2 (800 m^2)	Finish the foundation platform formwork @ K axis	Carpenters (20, 20) Earthwork workers (10, 6) Concrete workers (10, 10)	Coordinate with waterproofing crew
4	Second-floor basement structural work @ zone 2-1	25th April 2011	(1) Around 1/3 of the beam rebar placement work is completed (2) Around ½ of formwork for wall and column is erected	Progress goes according to plan	Rebar 190 t (100 t) Formwork 4, 000 m^2 (1,000 m^2)	(1) Finish wall formwork and column formwork (2) Complete replacement of rebar within walls and 2/3 of beams	Rebar workers (60, 60) Carpenters (90, 95)	/

5	Foundation platform and slabs @ zone 3	19th April 2011	Clean up the pit	Rebar team mobilizes into the site	Rebar work 840 t (740 t)	(1) Complete replacement of rebar at pit (2) Finish earthwork excavation (3) Concrete cushion are made	Rebar workers (180–240, 100) Earthwork workers (20, 16) Brick layers (35, 35)	/
6	Quality status on 21st April: (1) It was found at two places where a row of beam hoop reinforcement is not installed: beams at podium zone 2-1: K/2-3 axis. (2) Check with *jianli* that elevation of slab at basement first floor, and formwork of podium zone 2-1 is within scope of specifications.							
7	Site health and safety records: (1) Serious water leakage problem between supporting piles @ zone 3. (2) Pit wall slope crack unhandled near gate entrance #1 (@ zone 3) area. (3) The following processes do require extensive monitoring: (a) the installation of 3# tower crane, (b) stability of excavation slope and material platform and (c) shear wall formwork removal @ zone 1).							
8	Main constraints: acceleration is needed in installing 1# and 3# cranes. Potential obstacles would affect the plan: (1) Insufficient numbers of rebar workers at the pit area. (2) The transportation of materials to first floor basement appears to be slow.							

项目	分类	工程项目	计划完成时间			调整时间			完成情况	责任单位 责任人	备注
			开始时间	周期	完成时间	1ST	2nd	3rd			
酒店	土建扫尾	6-20层地面找平基层清理、清洗	2010/12/15		2010/12/20				完成	三局蔡元	
酒店	土建扫尾	6-20层地面找平	2010/12/15		2011/1/10					三局蔡元	紧急，12
酒店	土建扫尾	B2-21层楼梯间地面找平	2010/12/15		2011/1/15					三局蔡元	
酒店	土建扫尾	8层样板层变更钢件整改抹灰	2010/12/15		2010/12/20				完成	三局蔡元	紧急
酒店	土建扫尾	18-20层钢件设计变更钢件抹灰	2010/12/15		2010/12/20				完成	三局蔡元	紧急
酒店	土建扫尾	10-20层消毒间钢件设计变更钢件抹灰			2010/12/20				完成	三局蔡元	
酒店	土建扫尾	5-10层消毒间钢件设计变更钢件抹灰	2010/12/15		2010/12/20	2010/12/23			完成	三局蔡元	紧急
酒店	土建扫尾	5-	2010/12/15		2010/12/20				完成	三局蔡元	紧急
酒店	土建扫尾	5-21层新增管井砌筑，消毒间补图纸。	2010/12/15		2010/12/25				完成	三局蔡元	紧急
酒店	土建扫尾	电气竖井层间修补与增挡水坎	2010/12/15		2010/12/30					三局蔡元	
酒店	土建扫尾	5-	2010/12/15		2010/12/20				完成	三局蔡元	重要
酒店	土建扫尾	给水管道安装	2010/12/15		2010/12/30					三局刘介	
酒店	土建扫尾	管井后开洞口修补、清理	2010/12/15		2010/12/30					三局蔡元	
酒店	土建扫尾	屋面机房门防火门2个洞口复核	2010/12/15		2010/12/20	2010/12/21			完成	梁魏，周雷	门洞尺寸
酒店	土建扫尾	消防电梯1F夹层门洞高度不够，凿除	2010/12/15		2010/12/20				完成	三局蔡元	
酒店	土建扫尾	塔楼屋面电梯机房地面水泥砂浆压光、	2010/12/15		2010/12/20				完成	三局蔡元	电梯公司
酒店	土建扫尾	屋面电梯机房钢楼梯踏步及扶手施工、	2010/12/15		2010/12/20	2010/12/25			完成	三局蔡元	机房照明

Fig. 10.6 Example of part of cross completed item in plans

by diligently checking against the completion status of the daily work assigned, and the availability of machinery, labour resources, material inventory, etc. Thorough factor analysis and the removal of constraints rewarded the site team of Project C a better understanding of the schedule.

4. "Accomplished item" control sheet

   In addition to the daily plan, there was another daily meeting later in the day to review the extent to which the tasks that were given to related parties had been accomplished. The number of pages for the control sheets were long, but the procedure was straightforward: simply select the present date (i.e. 12 December 2010), and a list of tasks with starting date 12 December 2010 was quickly shown to the participants. Figure 10.6 would be more or less in line with PPC (one component of LPS). Similarly, the so-called accomplished item control method was introduced, and was thought of as a very dynamic tool for monitoring "commitments" in completing assigned work.

   As shown in Fig. 10.6, the anticipated dates of completion for each sub-task were also indicated, as well as the level of priority (urgent, important, less important, etc.). To differentiate between completion statuses, those that have been "accomplished" successfully were marked in grey. If a slip occurs, the reasons for the delay must be stated in the margin, in the "remark" column. A first warning would be issued to those responsible. Accordingly, a new completion date must be justified by the relevant party in a timely fashion. At this stage, no fine would be imposed, as the buffers allocated in the original plan to protect against uncertainties may suffice here. All the relevant parties (i.e. subcontractors, trades, etc.) are allowed to adjust their completion dates up to three times. If the assigned tasks remain uncompleted when all three allowable adjustments have been used up, the project team would impose a penalty in proportion to the costs and inconvenience caused to others. Overall, this simple "*accomplished item*" control method embraces the principles of visual management (P7) to present daily work, as well as the principle of pull (P3) to remind the participants of the sequence of related tasks.

5. Are the outcomes LPS practices

   Endeavours were made to provide more details in the plans, as shown earlier. To reiterate, each of the four hierarchical plans were closely monitored, and timely adjustments were quickly made. The PPT members analysed the variations derived from the original plans, removing constraints and adjusting the plans accordingly. All these efforts ensured that the master plan would stay on track. Principally, this scheduling method is in line with the principles underpinning the last planner system (LPS). Both focused on the reliability of agreed daily and weekly plans. As a result, the outcomes were rewarding. The Phase I projects, including two buildings, were successfully handed over to the client, both ahead of the original plan. Furthermore, one engineer interviewed, who was also aware of this practice, explained that to ensure the daily plan works well, workers are required to work in two shifts. At peak hours in Project C, when most workers were needed, there were 3,000 workers at the same time, along with hundreds of machines and equipment. The workers gave up all their public holidays, and turned long holidays into "construction golden weeks".

Built-In Quality (P5)

1. General quality management
   In an interview with a project manager from Project C, it was disclosed that there is no shortcut when it comes to quality issues. According to the interviewee, the concept of "good quality does not result from monitoring and inspection, but from building it in" was employed. However, numerous defects were still spotted every day. Unlike the conventional way in which the quality-control team is set up within the project team, here the established quality team at project C was supervised by someone directly from head office (Company B). In other words, the quality supervisor is independent from the project team, and his duties include conducting monthly meetings to discuss quality problems arising from the site. The status of quality issues would be directly reported to the corresponding department at the company level. This was a strong message to Project C that the head office was very keen on project quality, even though the client seems to be more interested in speed over quality. It is worth pointing out that the quality objective set by the client was merely targeting for each unit of construction to pass the first inspection with 100 % acceptance rate, whereas Company B aimed for a high-level quality award as part of its strategic intent.
2. Built-in quality
   The request to implement built-in quality has appeared in the method statement in several areas, including floor-concrete construction, steel construction, waterproof construction, joint construction and interior finishing. The reason for this is that the cost of rework in these areas would be high, if defects were to be found. Moreover, the project manager believed that the most effective way of instituting built-in quality is to select a high-performing work team with the right attitude to take ownership of project quality.

Standardization (P6)

A number of standardizations for major construction processes could be found on site, including "formwork", "concreting", "windows and doors" and others. The written standards for these were based on lessons learnt from past similar projects,

Table 10.12 Standardized durations for construction of buildings no. 2 and no. 7

Items	Work scopes	Plans	Start	Breaking ground	Main structure completion	Project production cycle per floor
1	Building no. 2 (Phase II)	As-planned	2010.7.30	2011.1.10	2011.10.20	3.6 day/floor
		Adjusted	2010.6.1	2010.9.26	2011.3.10	
		Difference	32 days early	47 days early	165 days early	
2	Building no. 7 (Phase II)	As-planned	2010.7.30	2011.1.10	2011.10.20	3.7 day/floor
		Adjusted	2010.6.1	2010.9.30	2011.3.20	
		Difference	59 days early	43 days early	155 days early	

and from taking this present project's unique characteristics into account. For example, the construction plans for rebar construction clearly described the fabrication of rebar, rebar joints and rebar placement in standard ways. More specifically, the rebar placement for different structural parts was designed to have their own standard operating procedures (SOPs). In addition, other standardized items included quality of work, inspection procedures and others.

1. Repetitive process

   Project C included several high-rise buildings, whose superstructure with standard floor design could be seen as a repetitive process. As highlighted in Table 10.12, each floor was planned to be completed on a 3.5-day cycle.

   Initially, the floor cycle took about 4 days, but after overcoming the learning process, the cycle of building one floor was reduced to 3 days, and the project proceeded smoothly as the workers become more efficient. To successfully apply the same production rhythm on the remaining floors, this should be carried out in accordance with the standards set in completing the first floor (also called the benchmarking floor). In a very restricted way, crew composition, daily construction activities and volume, crew size, and the required tools and equipment are carefully designed.

2. New standardization development

   One project engineer interviewed spoke of the difficulties in getting workers to suggest new ideas to bring current standardization efforts to a new level. The biggest challenge cited was the quality of the workers. These frontline workers lacked creativity, and simply do what they were told to do. However, there would always be positive contributions made by some senior workers, to whom the project engineer did not hesitate to give credit. For example, compared with the conventional methods of supporting formwork used for high-rise elevator shafts—which used traditional metal-frame scaffolds—a new idea came from one experienced frontline worker, who proposed an improved scaffold system using a triangle-steel-structured platform which could be hoisted up. This resulted in a safer work environment. The project team appreciated the potential of this improved method, and offered assistance to the worker to put this new

Fig. 10.7 An electronic board displays the next handover day of façade finishing

elevator shaft formwork on paper. A patent was later filed. This improvement quickly drew the attention of the local construction authority, and its application was encouraged in the remaining high-rise buildings in Project C, and also in projects undertaken in Wuhan.

Visual Management (P7)

This project had all the CI boards displayed to the public, as required on site. This includes project information, project team structure, etc. In addition, at the entrance to the site, a large LED screen was placed, displaying a countdown to the completion date of certain major processes. Figure 10.7 is one example of this, highlighting to the project team and the workers that there were 92 days to 30 June 2011—the deadline for finishing the external façade.

10.4.3.3 People-Oriented Practices

Leaders and Leadership (P9)

1. Leadership quality and development
   Given that Project C was viewed as a key project, the project team was carefully assembled to garner success. The project manager is a leader with a wealth of experience in quality management, scheduling, etc. from his previous work on similar mega-projects. Based on the PM's experience and understanding of the company culture, the PM pointed out that Company B is committed to nurturing two "skills paths" for its future leaders: The first path deals with "width", meaning that leaders are expected to gain experiences from different departments, and to have good exposure to various functions. The second concerns the "depth" of the leaders' professional knowledge: the company values their existing skills sets with specializations in certain areas. It was then hoped that a complete skills sets could help to solve problems and contribute to the project's success. Simply put, prior to placing someone in a leadership position, Company B paid much attention to the candidate's prior experiences, skills sets, attitudes, and multi-tasking abilities.

2. People-oriented leadership

   Effective labour management on site was the key to team stability and harmony, which eventually led to project success. This was especially true when the project progressed to a period, when there are more than 2,000 frontline workers working at the same time—a situation which gave unexpected challenges to management. Reportedly, during the spring festival in 2011, many groceries were bought and stored in a chilled storage place rented by the management of Project C exclusively for the workers. Meanwhile, the project team purchased fresh vegetables and other goods from the market for the workers. These illustrated the concern management had demonstrated for the workers.

Teamwork and Exceptional Employees (P10)

1. People selection

   Company B has an established human resource recruitment policy. According to Company B, four aspects of potential candidates were highly valued. Their moral virtue was valued, so were their work experience, work performance and peer views. In addition, the primary focus was placed on working experience. As explained by the project manager,

   > Experience is the key and the source of aspiration. It also provides leaders the platform to develop, exercise, and improve their (critical) thinking, (long-term) vision, decision making, and problem-solving skills.

2. Mentorship for recruits

   Company B's fast growth and development in recent years has seen urgent demands for both skilled managerial and technical personnel. Company B has employed more than 1,000 employees since 2009, although the general manager claimed that staff recruitment, along with the acquisition of technology, seemed to lag behind, and cannot appropriately accommodate the company's rapid development. At the project level, it was reported that about 50 fresh employees joined the Project C's site team since 2010. The project team encouraged a high-quality mentoring scheme. The site managers interviewed from the five main sites confirmed that it has become a mandated norm that new employees must work along with identified mentors for a period of time. For this reason, workshops were conducted to discuss how to select the most suitably experienced "mentors" for the new recruits. Not only has the training materials describing "*what knowledge mentors need to teach their students*" been standardized, the desirable outcome of the mentor scheme has also been defined. By the end of 2010, Project C, for the first time, ran an "Excellence in Mentorship" competition to recognize the most high-performing mentor–student pairs.

   It was anticipated that when the present project (Project C) was completed, a few potential employees would be recommended to the head office for better future career opportunities and development. Those young recruits would be placed in key positions in various departments, or would continue to work on iconic projects. It is also important to highlight that Company B evaluated employees' performance as one key element, but also took account of their moral and disciplinary behaviour.

3. Teamwork
   When labour demand reached its peak, another matter that increased was the number of working spaces, along with the associated health and safety risks. Traditionally, in smaller projects, the subcontractor has one supervisor to manage the teams and who was responsible for the entire team's results. The site managers interviewed asserted that this model was obviously inadequate in Project C, given that there were substantial numbers of frontline employees working at the same time. Instead, Project C introduced a new approach that demanded empowerment and teamwork. As noted, the site managers expanded their roles to a lower level, to include smaller teams (i.e. 10–12 team members), with one supervisor serving as the team leader. These team leaders quickly became the focal points, and took responsibility for setting goals, making decisions, providing information, removing barriers, and simple planning. It was observed that team members gathered around the team leader at the beginning of each working day, and the team leader would then allocate tasks, specify the work contents in details, analyse the points of potential risks and hazards, etc. The stand-up meeting ended with a check of everyone's *personal protective equipment* (PPE). When the work was done at the end of the day, the team leader also organized the team members to clean areas as necessary. After the team leaders have checked their work, they then pointed comments on the current status of the work.
4. Training
   At the project level, the PM conceded that time was scarce in this project. As a result, not all relevant solutions to the labour shortage, such as training, could be considered. Rather, what they adopted was to simply dismiss workers with poor attitudes and skills sets. In order to replace them, the project manager would then asked the labour-only subcontractors for new replacements. However, in some ways, training has to be provided. With a small investment of time and goodwill, the contractor could overcome out a crisis, for example when it was difficult to find someone with skills that can customize reinforcement cages. The Project C team invited several professionals from a nearby factory that specialized in customizing rebar, and asked them to provide hands-on training for their on-site ironworkers. To make it more formal, officers from the local construction bureau were also invited to assess and issue certificates for these specialists, once the training was completed.
5. Multi-skills
   Multi-skills training is another core element of the training system. Over the past several years, there has been increasing recognition by top management of the need to equip new recruits with multi-skills or cross-departmental training. The project manager pointed out that in Company B, the rotation of young employees through a wide range of positions has been implemented for many years, in order to enhance their experience. For instance, those with technical backgrounds (civil, M&E, etc.) were more likely to be rotated to work in a management-

related department, while those with their primary skills sets in management training would be placed in commercial departments, such as bidding. The top management noted the importance of, and the need for, training in various skills that extend beyond the topics they already trained in. This would allow employees to develop a full range of knowledge and skills now expected by the modern construction industry. It was also good for employees to find out what their true passion and interests were.

6. Outcomes of motivation
   The result of Company B's people management approaches turned out to be rewarding. At the company level, 90 % of the new recruits became the backbone of the business. Among them, a hundred young employees became project management assistants serving in various projects. Moreover, with respect to the workers' contributions to improving certain parts of the construction process, the project team not only gave financial incentives but also recommended one crew to the head office for competing in the "best crew award" of 2011.
7. The undesirable effects of lean
   In the period of schedule acceleration, the workers devised slogans to motivate themselves. These included, "*men work like machines, women work like men*", or "*no leave, no sickness, and no home*", etc. All these implied that the workers were operating under enormous pressure. Furthermore, the on-site first aid station revealed that 80 % of employees had consulted with the doctor during the project. In the Chinese view, these facts reflected the workers' resilience and strong ability to endure the hardship. On the other hand, it should be realized that their sicknesses and hospitalization were caused by fatigue and hard work, which was precisely what Liker (2004) explained as another form of waste—*Muri* (overburden).

Relationships with Partners (P11)

1. Supplier and subcontractor management
   The project team highlighted that there was rigorous assessment of suppliers, vendors, subcontractors, and other partners on a yearly basis at the company level. The purpose was to evaluate their performance and to maintain a long-term relationship. However, if a supplier did not show sustained delivery performance to a satisfactory level, Company B would confirm with the supplier to remove its name from the qualified list. According to the PM, evaluating suppliers' performance in Company B appeared to be quite hierarchical, given that Company B is a third-tier subsidiary of a complex organization. The evaluation starts from the project level, where the project team would recommend good partner candidates to Company B for review. Later, the assessment results would need to be forwarded to Company B. Within Company B, there is a list of all the qualified partners, categorized based on region. In other words, when company B ventured into other regions outside Wuhan or Hubei province, it would firstly consult this list to select local suppliers from the corresponding regions. As indicated earlier, this was the first time Company B worked with the client. It was not therefore unusual for the client to introduce a few capable

long-term subcontractors whom they have long-term partnership with to Project C, whose names were not in Company B's list. When Phase I was completed, the project team took the initiative to recommend these subcontractors to a higher level for consideration and review. In the following year, some of these subcontractors were successfully included in the qualified pool, and became new members.

2. Centralize the procurement function

   Company B has introduced framework agreements with "approved suppliers" for major material supplies. There was actually no necessity for full tender exercises for these suppliers since the client was also aware of the reputation of these suppliers and their approved products. The project manager outlined that they continued to centralize its procurement function to aim to lower the costs of materials, as the PM believed that individual contractual arrangements offered less value for money, compared to those negotiated through centralized procurement. The construction of Phase I revealed that this approach helped the project to save substantial money, up to 1 million RMB, as well as to achieve better quality. The project manager estimated that the centralized procurement for steel would have led to savings of about 10 %, in comparison with traditional means. According to the commercial department's field work, the price of steel using centralized procurement was 300–400 RMB/ton lower than the average market price (approximately 5,500 RMB). In addition, the steel's appearance and quality were also good. Moreover, by using centralized procurement, the supplier partner was able to provide customized products to accommodate Project C's requirements. For example, the project needed 77 tons of 8-m length steel, as well as 23 tons of 9-m length steel, which the supplier agreed to provide as customized products. As a result, this set-up easily helped the project saved 17 % and 12 % of the costs of 8-m and 9-m steel, respectively.

10.4.3.4 Problem-Solving Practices

Genchi Genbutsu (P12)

Owing to the enormous pressure of the project scheduling described earlier, quality defects were commonly seen on the site of Project C. In order to attain the largest decrease in schedule time with fewer quality problems, the quality team kept their responsibilities and objectives in mind, and hence *genchi genbutsu* was implemented at the site all the times. Meetings were held about once every 20 days in order to refresh the team on recent non-compliances, defects, quality issues, etc. The key effort here was to highlight the quality problems that had recently occurred, using PowerPoint slides, to all participants—including the subcontractors and managerial staff. This allowed the participants to raise questions and discuss root causes and countermeasures.

Decision-Making (P13)

Due to the pressures of time, a considerable amount of efforts was put in developing a detailed schedule (e.g. a daily plan) to ensure that the client's requirements could be met. The successful on-time completion of the main structure of Phase I, its

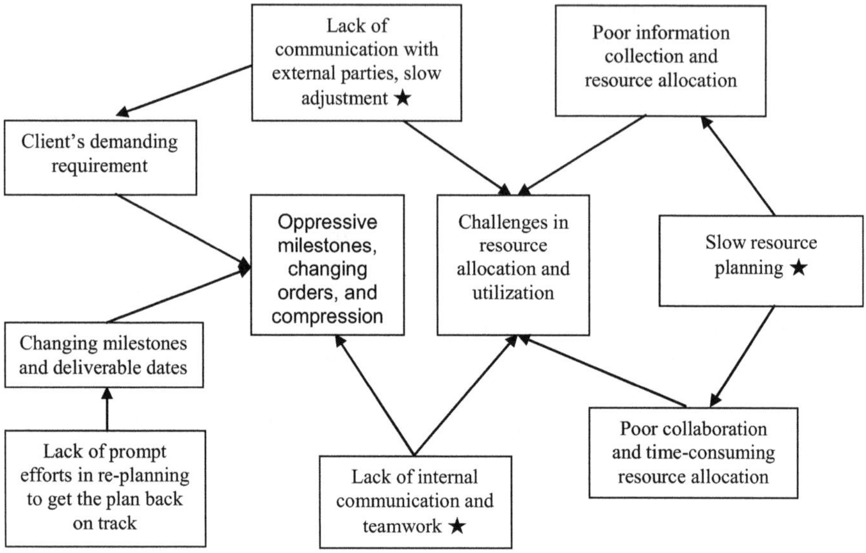

Fig. 10.8 Analysis of causes

subprojects (buildings 2 and 7) was set as the main goal of QC activity. The QC members appeared to be quite astute in accomplishing these by utilizing Company B's strong technical backup, and aligning this with the team and member efforts. The QC meeting records were carefully reviewed, and it was found that during one of the brainstorming sessions, the acknowledgement of the challenges (see Fig. 10.8) was decided as the key agenda. This started from brainstorming the possible difficulties in accomplishing the given deadlines: It was eventually agreed by the participants that the two key difficulties were (1) oppressive milestone, changes in orders, and compression of the schedule, and (2) difficulties in resource allocation and utilization. Following these discussions, a "multiply-whys" method was employed to further seek the root causes that might have contributed to these two challenges. As a result, another brainstorming session was carried out, and eight causes were quickly identified. At the same time, the level of importance was also discussed. In Fig. 10.8, the symbol ⋆ represents the key root causes.

The following meeting discussed possible countermeasures (see Table 10.13) to the problems highlighted in Fig. 10.8.

The outcome of this was rewarding: the main structure of the 5A office tower and hotel was completed in 152 days, and 52 days ahead of the original schedule.

Kaizen (P14)

Project C was constantly praised by the top management of Company B as "*a benchmarked project which did an excellent job of meeting unique challenges*". Its outstanding performance in various areas, including site organization, attention to progress, faster scheduling, standardized management approaches, etc. was highly

Table 10.13 Causes and proposed countermeasures

Item	Causes	Countermeasures	Objectives	Actions
1	Lack of communication with relevant parties, slow adjustment	Enhance communication and introduce more dynamic management approaches	Closely monitor project progress	• Enhance communications between various parties, i.e. collecting information pertaining to progress • Project planning taskforce was established to be responsible for this • Hierarchical plans • Daily meetings to check status • "Accomplished item" control sheet
2	Slow resource planning	More efforts are needed in pre-planning	Utilize resources	• Storage and appropriate allocation • Early involvement of suppliers • Select capable teams from various interviewed subcontractors, those who have long-term relationships with contractor or client would be given special consideration • More rewards and motivation

Note: Some of the actions derived from item 1 have been illustrated in earlier sections

recognized by the various stakeholders. Project leaders from other projects both inside and outside of Company B came for site tours to learn about the "best practices", and to reflect on the relevance of what they learned to their own job sites. A number of project managers visiting the site showed interests in many learning points and details that Project C focused on, while others inquired about issues pertaining to people management, cost control, and other management approaches. It was highlighted that recently, such sites tours have increasingly become training programmes held in Company B for project managers to exchange information and to discuss "best practices", with the hope of applying the lessons learnt in their own projects for continuous improvement.

In parallel to what Toyota has implemented to encourage frontline workers to propose new ideas and feedback that might result in continuous improvement, Project C has also experimented with such initiatives to respond to Company B's call for feedback. An initiative known as "golden ideas" was conducted in Company B since 2000. As a result, although the number of the so-called golden ideas was not comparable to what Toyota was able to achieve, an array of creative ideas and feedback have been voiced by different people in Project C. These ideas included:

- To introduce a deep-well system in Phase II, using an appropriate amount of groundwater to solve the problem of insufficient water during Phase II construction.
- Washing vehicles using water obtained from dewatering of foundations.
- In the summer of 2010, Wuhan city imposed limits on the traffic, especially on traffic involved with construction. This was the time when Project C was in the piling phase, in which large volumes of excavated soil needed to be removed. One golden idea proposed, which hinged on "sustainable thinking", was to utilize the concrete mixer truck to load soil. This piece of feedback, which gave multiple uses to the truck, not only helped the project saved money, but also satisfied clients and local government.

Reportedly, at the level of the firm, Company B has received more than 200 ideas and feedback in various areas, including ways to improve processes, to reduce costs, etc. In addition, of all the ideas and feedback collected in a year, Company B rewarded those with the most creative ideas that could most help projects.

10.5 Summary

In this chapter, three projects of two companies, A and B, were selected for evaluation of their site practices, people management, and problem-solving behaviours, as compared with Toyota Way principles. These project-specific case studies revealed that different firms with different characteristics, nature, resources and capacities, would adopt the principles of the Toyota Way to different extent, in order to suit their projects' interests and needs. Overall, in the two companies and three projects visited, many elements of the Toyota Way principles were seen in place, but none of the case study projects implemented the full suite of the Toyota Way principles in a holistic way. Although manufacturing terms were not frequently heard or discussed on site, the evidence confirmed that the terminology used shares common ground with the Toyota Way principles. In the case of Projects A and B, the efforts in terms of site management relevant to the Toyota Way principles appeared to focus on productivity and quality-improvement activities. On the other hand, Project C (undertaken by company B) seems to have picked up some Toyota Way knowledge, especially in the area of material procurement using pull thinking, hierarchical planning, with built-in commitments and others. They were driven by their efforts to meet the client's requirements. These areas were identified as the key factors that might affect the schedule—which happens to be the priority of the client.

These cases also provided examples of how the Toyota Way implementation model may be used in real-life projects. It was seen that the principles of the Toyota Way model could be used as appropriate guidelines, but the interpretation of these principles into action depends on the awareness and understanding of lean, or other similar management approaches. The findings actually confirmed that the crucial role of people in the deployment of lean or Toyota Way practices should be emphasized. It should also be noted that the commitment of management, as well as their awareness and understanding, is the most important prerequisite, without which it is not possible to successfully implement this model in practice.

Lastly, it can be concluded that these three case study projects have arrived to different levels of maturity in their implementation of the Toyota Way. It seems to suggest that firms need not necessarily follow the practices presented in the Toyota Way model too strictly, but should take the project uniqueness, client requirement, partners, employees and others into consideration and thus develop appropriate means based on the Toyota Way thinking.

Discussion and Validation 11

11.1 Introduction

This chapter uses SWOT analysis to discuss the results presented in Chaps. 8–10. The discussion in this chapter reinforces what the findings from the earlier chapters suggest, and how the research results relate to the literature review. The implications and strategies for LCCFs are also included. This chapter ends with a validation of the strategies proposed.

11.2 Overview

This research aims to establish the implementation framework of the Toyota Way model to guide LCCFs in their lean transformation. To achieve this aim, the research has employed mixed methods to investigate "to what extent" and "how" Toyota Way-styled practices (from four layers) can be practised within LCCFs. Generally speaking, all evidence points to the fact that knowledge of implementing Toyota Way-styled practices in the Chinese construction industry is insufficient. Nonetheless, the quantitative studies have shown some encouraging results: a number of Toyota Way principles have nevertheless been adopted, implying that in a few cases the LCCFs have demonstrated their basic knowledge and employed tools that fit the description of the Toyota Way. Yet the face-to-face interview findings portrayed a rather different situation. Since their aim was to investigate how the Toyota Way-styled practices could be (better) implemented through identifying possible constraints or challenges, the interview findings pointed to a gap between the status quo of current practices and the authentic Toyota Way-styled practices. Additionally, under each Toyota Way principle, constraints or challenges were revealed (see Chap. 9). Based on the interview findings, as well as on the case projects examined, it seems that no single LCCF has fully demonstrated its ability, capacity or readiness to implement Toyota Way-styled practices.

DOI 10.1007/978-981-287-014-8_11,

11.3 Discussions and Strategies

In order to discuss the findings arising from both the survey and interviews in a holistic way, SWOT analyses were conducted in this section to summarize the results.

SWOT has become an increasingly popular analytical tool adopted by researchers in the construction industry. For instance, at the firm level, Lu et al. (2009), Zhao and Shen (2008), and Ling et al. (2009) used SWOT methods to examine Chinese international construction companies, foreign construction companies in China, and Vietnamese ACE firms, respectively. Moreover, there are also reports that used SWOT to investigate individual construction firms. For example, one of the largest Chinese construction firms, China Communications Construction Company's SWOT analysis was published in Datamonitor's[1] (2011) database. At the project level, Milosevic (2010) undertook a SWOT analysis from both the investor's and the contractor's viewpoint in the planning, contracting and construction of a project.

This chapter pinpoints the main strengths (S), weaknesses (W), opportunities (O) and threats (T) for LCCFs in terms of the implementation of four different themes of the Toyota Way model. Simply put, strengths are those factors which LCCFs are already performing quite well, which can also help them to facilitate or improve their lean implementation or lean transformation. Despite these strengths, it can be easily recognized that some notable weaknesses also exist in LCCFs, concerning the application of the Toyota Way model. It is important to determine such weakness, because LCCFs need to correct them in order to stay in the right direction for implementing the Toyota Way. In addition, interview findings reflected the opportunities and challenges (or threats) for firms adopting the Toyota Way model in the context of construction. All the findings mentioned earlier in Chaps. 8–10 can usefully contribute to the SWOT analysis. This SWOT information is useful in matching the resources and capabilities to their (future) lean transformation. As such, it is instrumental for strategy formulation. By understanding these four aspects of the situation, a firm can better leverage its strengths, correct its weaknesses, capitalize on opportunities and deter potential threats (Barker and Smith 1997). In terms of implementing the Toyota Way, more attention should be given to the areas described as firms' weaknesses and constraints. This is followed by developing specific strategies and considerations which can then:

1. be used as a set of guidelines for implementing Toyota Way-styled practices, and
2. assist top management in developing long-term Toyota Way implementation plans.

[1] Datamonitor is a leading business information company specializing in industry analysis.

Table 11.1 SWOT analysis of LCCFs in terms of Toyota Way Philosophy model

Toyota Way Philosophy model	Strengths (S)	Weaknesses (W)	Opportunities (O)	Threats (T)
Long-term philosophy	• Established long-term relationship with partners (S1)	• Lack of long-term thinking in employee development, partners relationship, as well as problem-solving (W1)	• The role of government (O1)	• Fierce competition (T1) • Pursuit of short-term benefits (T2)
Constant purpose	• Articulated firm's value, missions, etc. (S2)	• Lack of commitment on improvement initiatives (W2)	• N.A.	• N.A.
Customer-focus	• Always set client requirements as priority (S3)	• Poor focus on internal customer (W3)	• N.A.	• Clients can exert their power intentionally (T3)
Be self-reliance and responsible	• Strong specialization in what they have been doing (S4)	• N.A.	• The role of government (O1)	• N.A.

11.3.1 Toyota Way Philosophy Model

11.3.1.1 SWOT Discussions

The results of the SWOT analysis of LCCFs implementing the Toyota Way philosophy model are summarized in Table 11.1.

Strengths

Statistically, the LCCFs performed quite well in terms of most of the attributes identified in the first principle of the Toyota Way model (see Chap. 8), and hence were placed in the top right quadrant (high–high) of the "implementation-importance matrix" (see Fig. 8.1). The strengths include:

- Long-term relationships appeared to have been established with different types of partner (e.g. suppliers and subcontractors) (S1). This enables their companies' technical departments to quickly select a group of appropriate candidates from the portal where partners have already registered. For projects in very remote locations, a number of partners were also found to be available.
- All the firms interviewed were established firms, whose mission and values had already been articulated (S2).
- Firms showed the ability to rapidly respond to clients' needs, which is always a priority for LCCFs (S3). In order to satisfy clients, LCCFs would typically identify various methods for reducing costs, improving quality, enhancing the safety of the working environment, and others. For instance, to cut costs from the

client's perspective, value engineering would be used, i.e. using alternative materials or components to replace the higher-priced materials of similar quality.
- With respect to self-reliance, most firms interviewed have developed into industry leaders and have demonstrated their specialization in what they have been doing for many years (S4).

Weaknesses

- Nonetheless, judging from the attributes rated in other Toyota Way layers—Process, People/Partners, and Problem-solving, for example—it becomes clear that some attributes which should embrace "long-term" philosophy were not appropriately implemented. This suggests that although some LCCFs claimed to practice "long-term" thinking, this approach may just be present in their written statements (such as mission statements) and have not yet been translated into strategies or actions. For example, employee training and development does not embrace long-term thinking (W1): in most cases, limited or no training was provided.
- Although a long-term relationship was claimed to be established with existing working partners, short-sighted behaviour was still found, including multi-subcontracting, distrust, etc. (W1).
- Problem-solving practices did not feature in the commitment to long-term thinking: the attitude of employees towards problems has nothing to do with continuous improvement, i.e. there is no documentation of the progress made or of problems solved (W1).
- Efforts for improvement initiatives, such as cost deduction, TQM, inventory management, etc., have been claimed to have been introduced into the firms, but in fact they only existed for a short period, as a result of the lack of long-term commitment (W2).
- Poor focus on internal customers was identified as another weakness in customer-focus (W3).

Opportunities

As noted, a significant number of large construction firms are state-owned enterprises (SOEs), which are closely associated with the central government. Such firms' strategies and plans are overseen by the central government. If Toyota Way principles are appreciated by top management, the chances are high that the implementation will benefit from government support (O1), i.e. that specified allocations of funds can be received from government sources. Moreover, the government has put forward an agenda for improving the management skills of firms, and has highlighted that the business and project management of Chinese construction firms must be standardized, normalized, and fine-tuned at all stages of the management process. It is implied that the government may need a holistic system to act as a reference point for policy-making.

Threats

Firstly, the major threats that would affect the LCCFs in developing a long-term philosophy lie in the fierce competition (T1) among Chinese construction firms. Secondly, some players (such as subcontractors and suppliers) blindly pursued short-term benefits (T2) and were not at all good at laying the groundwork for long-term gain. In order to survive in such a competitive industry, reaping a short-term advantage naturally has become primary goals. For example, one potential threat is the use of substandard material instead of first-rate materials, in order to cut costs. In order to reap even very marginal profits, some partners from multilayered subcontracting would use substandard supplies simply for the short-term benefits. This may eventually lead to low quality projects. Moreover, highly demanding clients (T3) are another serious threat. The real threat is that the binding force of contracts in China is so weak that clients can exert their power intentionally. It is not uncommon to see clients changing the contract terms, even when everything has already been "agreed on" earlier. Thus, because of unreasonable requests from clients, companies may subscribe to short-term solutions even at the risk of compromising their long-term goals.

11.3.1.2 Strategies

In Liker's (2004) view, the first principle of the Toyota Way model is the most difficult one to apply, as it will generally involve an enormous change in culture and mindset. Although some of the attributes derived from this principle were rated highly by most LCCFs, it does not follow that they possess a long-term philosophy. It is important to reflect on what strategies should be taken to overcome the weaknesses diagnosed, especially regarding (1) the lack of internal customer-focus and (2) the options for cultivating a long-term philosophy.

Improving Customer-Focus

Customer-focus is one of the central pillars of lean thinking (Womack et al. 1990). Currently, customer-focus is perceived and used simply as a unilateral term, for which the LCCFs only emphasized the external aspects. The LCCFs should, however, be aware that customer-focus involves not only meeting the requirements of external clients and customers, but also those of internal customers and client—that is, colleagues within the firm or project should also be considered as customers, and that employees rely upon the internal services of others to complete their tasks effectively. It is therefore required that all employees pay equal attention to their internal customers. Similarly, employees should also be aware of who their next "internal customer" is, and what their expectations will be. Good communication in this context would result in a better process with less conflicts, and that will eventually add value to the final customers.

Long-Term Philosophy

Secondly, it should be noted that the cultivation of long-term thinking does not occur overnight. A starting point for developing long-term thinking should be based on the current situation, followed by the future vision (Liker 2004). In other words,

Table 11.2 Strategies for achieving both short-term and long-term objectives

	Short-term objectives	Long-term objectives
Business/ projects	• The bottom line is to not act irresponsibly in the pursuit of short-term profits	• Aim for generating value for the clients, employees, and society at large
Process	• Adopt the relevant Toyota Way process-oriented initiatives to compete for projects within the client requirements	• Continuously improve the project process by employing Toyota Way process-oriented initiatives
People and partners	• Be aware of employees as internal clients (people) • Work with capable partners and remove incapable ones (partners)	• Acknowledge people are the most important asset of a firm; respect them, develop and grow with them (people) • Maintain long-term relationships with partners (partners)
Problem-solving	• Solve the problems on-site and learn lessons from them	• Become a learning organization

time should be given for LCCFs to move towards long-term thinking by taking the steps needed to reach each of the smaller short-term goals. To achieve both short-term and long-term objectives, several strategies are proposed for them to focus upon (see Table 11.2):

11.3.2 Toyota Way Process Model

11.3.2.1 SWOT Discussions

In general, the attributes identified in the Toyota Way process model were viewed as being less holistically adopted by LCCFs (see Chap. 8). Apart from built-in quality (P5), according to the matrix of implementation importance (see Fig. 8.1), the largest number of the process-oriented principles was in the "low implementation, low importance" quadrant. In addition, the interview results also suggested that there is a huge gap between the Toyota Way-styled practices and what the LCCFs actually implemented. However, these are not necessarily categorized as weaknesses because within some process-oriented principles—and even though they were rated poorly in terms of implementation as a broader theme (e.g. P7, P6, and P3)—a few individual attributes turned out to be different. Taking this into consideration, Table 11.3 summarizes the SWOT components of the LCCFs in their adoption of the Toyota Way process model.

Strengths

The strengths of LCCFs in this aspect rest in the attributes that were relatively highly rated in the survey, as well as being supported by interviews findings. As shown in Table 11.3, several factors can be viewed as strengths in terms of the implementation of the Toyota Way Process model. A closer examination reveals that the identified strengths pertain mainly to Chinese building professionals' technical knowledge and strong ability in managing projects. Thanks to the

Table 11.3 SWOT analysis of the LCCFs under Process model

Toyota Way Process model	Strengths (S)	Weaknesses (W)	Opportunities (O)	Threats (T)
P2: One-piece flow	• Project managers (PM) possess strong technical know-how in creating uninterrupted workflow (S1)	• Workers lack understanding of *muda* (non-value-adding activities) (W1)	• N.A.	• Shortage of skilled workers (employees)
P3: Pull *kanban* system	• Good cost control of building materials (S2)	• Project team has no idea of "pull" strategy (W2)	• N.A.	• Uncertainty in material prices (industry) • Just-in-case timidity (firm culture) • Client's request to prepare piles of inventory on-site (client)
P4: *Heijunka* (level out the workload)	• PMs possess good planning skills (S3) • Workers are able to work hard under stressful condition (i.e. overtime) (S4)	• Foremen possess poor planning ability (W3) • Workers have no sense of ownership of scheduling (W4)	• N.A.	• Clients are sometimes demanding in terms of project delivery (client) • High labour turnover (employee)
P5: Built-in quality	• Management has high priority for quality (S5)	• Workers lack "do it right" attitude (W5)	• Frequent government's quality audits country-wide • Role of *jianli* • Collaboration with foreign firms	• Firm culture allows "re-do" or "rework" (firm culture) • Multi-subcontracting (industry)
P6: Standardized work	• PM guidelines are available to provide basis for standardization (S6) • Detailed documentation for main construction process (S7)	• Poor emphasis on standardization (W6)	• Prefabrication is promoted by the government	• Clients are not supportive about using prefabrication (client)
P7: Visual management	• Mandatory visual signs are put up (S8)	• Workers' high tolerance of an untidy site (W7)	• Government's efforts in "beautify construction project" initiative • Collaboration with foreign firms	• Messy construction site is firm culture (firm culture)

(continued)

Table 11.3 (continued)

Toyota Way Process model	Strengths (S)	Weaknesses (W)	Opportunities (O)	Threats (T)
P8: Use of reliable technology	• Possess advanced and innovative technology in certain areas (S9)	• N.A.	• Government's call to adopt technologically advanced construction methods	• A labour-intensive industry (industry) • Construction methods are still featured as conventional (firm culture)

booming construction industry in China, where the building professionals have ample opportunities to participate in different types of projects, they have not only developed their technical competence, but their site management skills have also been enhanced. These in turn added to their strengths, which include:

- Good knowledge and understanding of construction-site layout design (S1). Although they may never have heard of the Japanese term "*muda*", the way they work is actually in line with the principle of eliminating "*muda*"; they did understand the way that non-value-adding activities can adversely affect projects. In projects, much effort was made to optimize the site layout, for example, a number of factors have been considered in order to meet the needs of site logistics (i.e. minimizing on-site traffic congestion), material flow and storage (i.e. avoiding double handling of materials), labour movement (i.e. minimizing walking distance) and others. Such valuable knowledge has developed out of their study of contract documents, diligent site investigation, and timely communication with the local authorities. This seems to suggest that Chinese building professionals are capable of planning site layout with the aim of achieving the maximum efficiency. They acknowledged that removing non-value-adding activities was essential duty in their jobs, as these activities were not to be tolerated.
- Good cost control of materials (S2). Since materials make up a substantial part of the cost, project leaders keep a close tab on the market prices of materials. Experiences have taught them to be price-sensitive for building materials and have also equipped them with the ability to negotiate with their suppliers. Although large amounts of bulk materials are visible on-site, they explained that when the project completes, the actual wastage of building materials will be under control. They also have good knowledge of quality, price, storage and transportation of materials.
- Good planning skills (S3). The four hierarchical plans were generally adopted at the project level. In contrast to the LPS principles, the schedules in the Chinese context are commonly developed by the project manager or engineers, rather than by empowering the last planner (i.e. the foreman) with planning

responsibilities. This is because the foremen's experience and understanding of the project has trained them to be confident in their planning skills. They believe that delays will rarely occur as long as the subcontracted teams are willing to follow instructions and are capable of facing the challenges in the schedules they produce. Apart from their good planning skills, the workforce in China is typically hard-working, and willing to obey their supervisors in order to remedy a slipped schedule back on the right track as quickly as possible (S4).

- Commitment to quality (S5). Managers have high quality expectations from employees, subcontractors and others. It has been mentioned that the quality requirement stated in the contract is the minimum standard, and that leaders actually look for a higher standard. As far as motivation is concerned, the manager's commitment to quality largely arises from the desire to win the "*Luban*" award, which is the highest quality award in China. This "*Luban*" award could benefit them in various ways in terms of enhanced career development.
- Strong execution ability. This is reflected in a number of activities that were well-executed, including standardization and visual management. For example, project-management guidelines were established for project use (S6), and main construction method statements were also available and are reviewed with the foremen, supervisors and site engineers prior to the commencement of specific processes (S7). Project managers had played a key role in implementing such procedures in the project. With respect to visual control, mandatory signages, notice boards, etc. were displaced prominently (S8).

In terms of technology, some of the firms interviewed have participated in very complex projects, both in the domestic market and overseas. They are capable of taking on major projects due to the strong technological skills they possess (S9).

Weaknesses

It is clear that the most reported weaknesses in this aspect were those posed by people, especially by frontline workers. The consensus among the interviews pointed to two major weaknesses:

1. Lack of awareness of lean and Toyota Way terminology among employees.
2. Lack of skills or capacity to practise lean or Toyota Way process-oriented initiatives.

1. Lack of awareness

   It is widely acknowledged that ultimately lean or Toyota Way process-oriented initiatives should be implemented by the frontline workers (Liker 2004). However, in the context of the Chinese construction industry, the problem is with the rather low awareness or prior knowledge of lean terminology (i.e. *muda*, 5-S, visual tools, and pull/*kanban*, to name a few) (see Table 11.3) among employees and frontline workers at the site. Without a basic understanding of the underlying mechanisms of the lean terminology, it is not possible for the workforce to take the initiative to improve the processes by employing the lean tools derived from the Toyota Way process initiatives. This seems to be a common problem that can be found elsewhere (see Picchi and Granja 2004; Salem et al. 2006). Hence, the introduction of an awareness program on-site is strongly recommended for a start. Workshops and training should be given to deliver the key principles, to explain applications, and to provide guidance on learning.

2. Lack of skills or capability to practise lean initiatives
   Apart from the low levels of awareness, another major weakness is the lack of skills or the capability to perform these process improvement initiatives at the project level.
   - Lack of skills in the elimination of *muda* (W1): Although the Chinese workers enjoy a reputation for being hard-working, they are nonetheless not capable of identifying non-value-adding activities for higher efficiency. To identify *muda* requires a basic understanding of processes, familiarity with lean terms, and a commitment to continuous improvement. One interviewee stressed that workers were perceived as either not hard-working or lazy. Hence, it is the project leaders who need to work hard on the layout planning, work planning, and others, in order to minimize the chances for workers to engage in non-value-adding activities. For example, a greater workload could be given, or the time frame to complete the job could be frozen. This approach of assigning work would result in less idle time, as workers are then aware that they need to speed up their work or face punishment for not delivering the job on time.
   - Lack of knowledge of "pull" strategy (W2): The project team has zero knowledge of the "pull" strategy in terms of material procurement. Instead, the push approach is widely adopted on site.
   - Lack of planning skills (W3): The workforce performed by simply completing the given tasks along with a (locked) time constraint. Several project managers thought of engaging their foremen in the planning process, but this ended up with poor efforts, because most foremen or supervisors are not qualified to analyse where they presently are to identify the constraints. In addition, the foremen are not ready to engage in the planning process, as they are not comfortable yet with taking on responsibilities instead of merely taking orders (W4).
   - The lack of a "do-it-right attitude" (W5): this weakness comes mainly from the frontline workers. The problem is that, due to the financial constraints and cost considerations, the workers supplied to the construction project are a mix of skilled, semi-skilled and poorly skilled labourers. Their attitudes towards quality also vary, and this depends on their experience, the training they have received, the level of commitment of their supervisor, and other factors. These workers are not part of the firm's internal team. Hence the chances are high that they may be working in a different work culture, which may not put a priority on quality. Because of this, site engineers are required most of the time on-site to correct their attitudes and, such operations lack the "do-it-right" attitudes.
   - Poor emphasis on standardization (W6): site management supervises the site at the micro level—focusing on the end results, without caring for the details of the procedures adopted in their operations. In this context, the workforce's

understanding of and skills in standardization develop slowly. On the one hand, the workers work purely based on their experience and with limited consultation with the written standardized procedures due to their poor literacy standard. In their views, practice makes perfect, thinking that their skills will develop through doing the work repeatedly.
- High tolerance of site untidiness (W7). A high level of tolerance for untidy construction sites is one of the weaknesses for implementing the P7—5-S program in particular. The workforce is not motivated to improve its working environment, as it takes away their spare time in order to tidy, clean, sort, and so on. When the workers are exhausted after their work, they are not willing to perform even simple cleaning tasks.

Opportunities

Although weaknesses and threats hindered the successful implementation of the Toyota Way process model, various opportunities exist to support and improve implementation. Opportunities lie in three aspects, namely (1) government support in various areas, (2) role of *jianli* in "built-in quality" and (3) collaboration with foreign firms.

Firstly, government support is reflected in four ways:

- The government bureau has initiated quality audits, and most visits occurred with prior notification. The audit is not to opportunistically issue punishments or fines on the quality infringement but to enhance their awareness and quality skills.
- The construction bureau has made basic visual displays of health and safety, risk identification, and basic cleaning tasks mandatory on-site. This could be an opportunity for them to extend the existing norms and policy into a higher level in a restrictive way, and to include other 5-S principles if possible.
- Prefabrication is actively promoted by the government: it represents an opportunity for the principle of standardization to be implemented. This is in line with Arif and Egbu's (2010) observation that manufactured construction would appear to be an attractive and strategic direction China should adopt. For example, a video clip from YouTube in June 2010 showcased how a high rise hotel was built by a Chinese construction firm in 6 days (Broad Group 2012). Moreover, it is comforting to see that several responding firms have also strategically invested heavily in prefabrication; one respondent from a private firm stated that its focus on prefabrication is reflected in the firm's mission which is set out as "from construction to manufacturing".
- The government pushes for use of technology. The availability of the so-called *top 10 emerging new technologies* is a good opportunity and a good starting point for the adoption of technology in construction. Again, P8 (adoption of reliable technology) of the Toyota Way provides a series of guidelines for better technology adoption.

Secondly, the role of *jianli*—known as the supervision firm—was very unique in China, which became compulsory in the 1990s to monitor if contractor's quality programmes are put in place. *Jianli* has since then gained much experience in assisting construction firms to fulfil quality management in practice. This has laid a much needed foundation for the adoption of lean construction because it shares some similarities with TQM (i.e. customer-focus and continuous improvement). Moreover, it opens an opportunity for *jianli* to upgrade their skills and to offer a role as a lean champion or a change agent in assisting the construction firms in promoting the lean construction concept in the industry.

Thirdly, the increased involvement of international construction firms in China represents an opportunity for several principles to be implemented. As the interview results revealed, three responding firms have had experiences in working with their Japanese counterparts, and they were deeply impressed by the authentic Japanese management style which pays much attention to details and commitment to quality, schedule and tidiness of construction sites. These international construction firms have brought about not only competition but also management know-how from the developed countries.

Threats

Threats can come from a number of sources, both internal and external, that hinder the implementation of the Toyota Way process model. As shown in Table 11.3, these threats can be categorized into (1) employees, (2) firm culture, (3) clients and (4) industry practices.

Firstly, there are two threats from the employees: namely, the shortage of skilled workers and the high level of labour turnover. The former may prevent uninterrupted workflow from being achieved, because skilled workers tend to work in an effective manner and with an in-depth understanding of their work. Almost all the firms interviewed expressed their desire to recruit a greater number of skilled workers, but it has become a lot more difficult to find skilled workers than it was in the past. Without a qualified workforce—regarded as key assets in embarking on and implementing lean—the implementation process will be slowed down. This has the potential to become a major threat to affect the workload from leveraged effectively. When the labour turnover appears to be high, it is more likely to affect the way workloads are designed and allocated. In such circumstances, Chinese building professionals have no better solution than to request their workers to work overtime, with 1 or 2 days off per month, in order to ensure that the work can be completed on schedule. This also resulted in reluctance of management to train their worker, as the firm would potentially suffer a loss if their workers simply leave the site.

Secondly, the threats pertaining to the firm culture are as follows:

- Just-in-case mindset. This might be the reason why the pull/*kanban* system has yet to be adopted in material procurement. Uncertainties about material prices caused project teams to be cautious, thinking, "what if the inventory runs out on-site when the price is still at a high level". Therefore, the Chinese way is to adopt a just-in-case approach: it involves preparing on-site a safety stock—which usually contains a week or two's worth of materials. Meanwhile, the purchase of material is based on the planner's forecasting skills and adjustments:

whenever they feel that the price is reasonably low, they will procure the price-sensitive materials in bulk.

- Firm culture allows the attitude of "re-do" or "rework". This potential threat is the opposite of "built-in quality" (P6). This explains the fact that a team of site engineers would be tasked to patrol the site and spot as many quality, health, and safety problems as they could.
- Firm culture regards having numerous construction sites as a good business sign. The firm culture of most LCCFs encourages bulk materials to be stored on-site, and tolerates tools, materials, and other items being strewn all over the place, once everyone is busy at their work. The firm culture considers that these are also good signs that the project is progressing.
- Firm culture views construction methods as conventional. This could be the reason for slow technology development in China's construction industry. In the respondents' views, the conventional methods are the most appropriate means to undertake the current projects. Such views prevented the construction firms from adopting new technologies such as prefabrication.

Thirdly, clients play a critical role and pose another major threat that hinders the implementation of several process-oriented principles, including:

- Clients wish to see stockpiles of inventories on-site (P3). This performance can hinder the contractors from pursuing the pull system on-site.
- Clients are sometimes demanding in terms of project delivery. For example, clients who drastically reduce the project delivery time is a common phenomenon in the Chinese construction industry. Even large construction firms are helpless but to accept it, because they are afraid of jeopardizing their relationships with the clients. In order to deliver the project on time under such time pressures, the "locked milestone" strategy is widely adopted for specific portions of projects. This gives subcontractors and specialized trades locked deliverable dates in which to complete the work. If the work is not completed, large penalties are imposed. In cases like this, some interviewees pointed out that it is not possible to "pull" the work from the workers; the majority agreed that their commitments at this point are unreliable. Hence, a top-down approach is preferable, where the plan is generated by the project team in the absence of any concerns from frontline workers and foremen.
- Private clients are not supportive in using standardized components (i.e. prefabrication). The contractors complained about the client's lack of interest in prefabrication technology. One interviewee highlighted that "*in terms of standardized components, we have no say in it unless it is supported by clients. They might not be a fan of prefabricated materials/components, although we understand it can boost productivity here. But we are not likely to change, as the client will not take the risk for us to use this so-called unconventional approach to constructing the project.*"

Lastly, there are external threats stemming from industry practices:

- Uncertainties. Most of the uncertainties are related to price changes in materials. This is the major factor that prevents LCCFs from implementing the pull system, opting instead for buying the materials in bulk at a price they feel to sufficiently be low.

- Multi-subcontracting. Alarmingly, many contractors in China seek profits by illegally leasing their licenses or subcontracting their jobs to unqualified firms (Lan and Jackson 2002). This illegal practice compromises quality, as each subcontractor within the different layers of a multi-subcontracting system will always reap benefits; obviously, the consequence is low quality performance.
- A labour-intensive industry: This nature of the Chinese construction industry is still seen as a major threat for the development of technology that requires heavy financial investments. However, from a short-term perspective, the abundant labour resources and cheap labour have undermined technology adoption.

11.3.2.2 Strategies

In the Chinese construction context, many LCCFs seemed to adopt the Toyota Way-styled process initiatives in a piecemeal fashion. That is, they adopted some of its aspects and methods while ignoring or rejecting others. Changing this status quo takes time, and therefore training is inevitable for both labour-only subcontractors and the main contractors. As pointed out earlier, for a start, programmes should be introduced to increase the awareness of lean or Toyota Way principles on the site, at least to improve their understanding of what these lean tools are, and how these tools can be used in the daily work. In response to the weaknesses and threats discussed earlier, strategies are formulated below for each principle of the Toyota Way Process model.

One-Piece Flow (P2)

2.1 Enhance project leaders' understanding of site issues; more *genchi genbutsu* practice is encouraged to identify the constraints for better improvement of workflow.

2.2 An awareness programme on the theme of "eliminating *muda*" should be introduced on-site.

2.3 Involve subcontractors/suppliers in the planning process, so as to improve the availability of manpower and material resources.

Pull *Kanban* System (P3)

Resistance will always be encountered when implementing pull *kanban* systems when major material prices fluctuate. The common reaction to such fluctuations is the adoption of the purchase-to-stock strategy. The approaches to eliminating external factors (such as changing prices) include:

3.1 For those material that are subject to fluctuations in market prices, purchasing partnerships with trusted working partners should be established for material purchase to reap mutual benefits.

3.2 However, for those materials that cannot be stored on-site, and which are vulnerable to burglary, or which have to meet with customs requirements, a pull system can be adopted.

3.3 Concurrently, it is important to track material usage as well as the needs from the foremen. For the necessary inventory to be on-site, good material management is required.

Heijunka (Level Out the Workload) (P4)

The large gaps in the planning ability of project managers and foremen prevented collaborative planning, such as LPS, to be implemented in the Chinese context. Rather, there is a one-way flow from project manager's office to conveying schedule information to the frontline workers. Nevertheless, in order to better achieve the aim of levelling out the workload, as well as gaining more reliable working plans, the strategies below are proposed.

4.1 Following the existing structures of project planning, which are widely adopted within LCCFs, attention can be paid to the following aspects: (1) updating weekly plans and look-ahead (monthly) plans in a timely manner if unexpected events occur; (2) communicate these updates and changes with the frontline workers in a timely manner through their supervisors, in order for them to understand where they are and what the follow-up actions are.

4.2 From a long-term perspective, trust should be established between foremen, supervisors and contractor's employees. The foremen and supervisors need to be trained to understand where they are and where the project is heading. They should also be encouraged to participate in the planning process, at least to provide key information and commitments in the weekly plans. This should be carried out in an open, understanding and trusting atmosphere.

4.3 Given that PPC is not very much practised by Chinese building professionals, alternatively, tracking the reliability of the weekly performance is of importance; root causes should be revealed and lessons should be learnt—which can be used as the basis for the next week of planning and for better implementation.

4.4 Adopt appropriate motivational strategies to maintaining the core workforce at least until the project is accomplished. This is because a high turnover in the workforce can severely affect the workload to be levelled. To understand the capacity of the workforce, efforts should be made to alleviate the burden of the workers, i.e. overtime, which can fuel discontent.

Built-In Quality (P5)

Overall, the difficult part of quality control has been relatively well executed (i.e. rejection of defects, stopping work when problems are found, etc.). In response to the weaknesses and threats diagnosed earlier, a few areas remained to be improved, including:

5.1 Improve the firm's culture relating to quality, i.e. promote "zero tolerance" for substandard quality. To achieve that, employees should be empowered to stop the process when an abnormality occurs or when problems are detected.

5.2 Apart from the difficult part of quality control, employees should also be encouraged to give feedback on quality issues, engage in quality improvement dialogues, provide teamwork in problem-solving, etc.

5.3 QC should be done more frequently. The management must not ignore the importance of recognizing and rewarding QC achievements.

Standardized Work (P6)

It is encouraging to see various forms of standardization being practised in projects. However, in order to better implement standardized works, adequate groundwork should be completed in advance, including:

6.1 Standardized operating procedure (SOP). In particular, the construction work statement should not remain in the project filing room. The essence of the construction methods should be made easily accessible for convenience, and frequently shared with workers in order for them to truly understand the procedures.

6.2 It should be acknowledged that standardization cannot be maintained forever, as Imai (1997) puts it: it goes hand in hand with continuous improvement. Hence, any improvement in the current processes (i.e. in the construction work statement) should be encouraged. Recognition for innovative ideas should be made to replace the conventional approach, even for small improvement.

6.3 In the event that the construction industry plays an important role in realizing sustainable development and the transition to a green industry, LCCFs are advised to keep track on the development of prefabrication components and materials and the potential these represent to the construction industry in China. For appropriate projects in which standardized prefabrication components can be applied, the use of prefabrication should be promoted.

Visual Management (P7)

Visual management is one of the most poorly implemented principles of the Toyota Way process model among LCCFs. The opportunity is that some of the basic ideas for visual management (i.e. putting up visual labels, signage, etc.) parallel the government's campaign of "beautifying the construction site". To better improve the implementation of visual management, three strategies are proposed:

7.1 Highlighting the visual management components in the government's "beautifying the construction site" campaign to all the employees. These should be set as priorities to be implemented.

7.2 Prior to the implementation of "5-S", this concept should be introduced through training to all employees. Given that 5-S contains five core elements, the process can start by implementing 5-S using the basic principle, i.e. site cleaning, sort redundant materials out, etc.

7.3 Efforts should be made to change the clients'—as well as the contractors'—mindset of the site environment. The goal of working towards a tidy, organized, clean construction site should be set.

Use of Reliable Technology (P8)

Chinese construction firms lacked the initiative to use new materials, new technology, and new construction methods, as they continued to operate in the conventional way. With the government's encouragement to adopt emerging technologies, the firms' experience and profile in applying new technologies will soon be linked to the evaluation of performance, which will encourage more Chinese construction firms to adopt new technologies. In this context, it seems more meaningful to introduce Toyota Way Principle 8 to guide LCCFs in adopting the new technology. The strategies could include:

8.1 The adoption of emerging technologies, as promoted by the Ministry of Construction, should be based on the characteristics of the project, the actual situation of project location, and other considerations.

8.2 The handbook of "*Ten Emerging Construction Technologies in China's construction industry*" can be used as a reference, but the actual adoption may not necessarily be limited to the handbook; other forms of innovation are also encouraged, as long as this leads to the improvement of construction tools, machinery, etc.

8.3 If resistance to adoption is encountered, long-term interests should be taken into consideration, as it may be helpful to the employees, their work, and the firm at large.

11.3.3 Toyota Way People and Partner Model

11.3.3.1 SWOT Discussions

The literature highlights that the people-oriented principles of lean are surprisingly absent, or insufficient to support the take-off of lean initiatives among the so-called lean organizations (Liker 2004). In a similar vein, the implication of lean in relation to HRM in the construction context has been the subject of further scrutiny by Green (2002) and Green and May (2005). Table 11.4 outlines all the factors affecting LCCFs in the SWOT analysis for implementing the Toyota Way People and Partner model.

Strengths

Toyota Way-styled leadership (P9) was regarded by most survey respondents and interviewees as an essential factor for a firm's performance; these were thus relatively highly implemented. Strengths include the fact that (project) leaders from LCCFs possess in-depth technical skills (S1), they know about things and physical processes, and they also possess problem-solving skills (S2). These results, however, seem to be inconsistent with the assertion of Lu et al. (2008), who identified the lack of general management skills as a common weakness of Chinese construction firms. In contrast, China's booming construction market has benefited a large number of project managers, by developing and enhancing their management skills, especially in the technical aspects (Gao et al. 2012). Because of their experiences gained through various types of projects, they have now become valuable assets for companies and projects. Among those who have been interviewed, some have worked for their firms for a very long time, and they thoroughly understand the company culture, values and mission. This appears to be very similar to what Liker (2004) described about the experience of Toyota's

Table 11.4 SWOT analysis of the LCCFs under the People and Partner Model

Toyota Way People and Partner model	Strengths (S)	Weaknesses (W)	Opportunities (O)	Threats(T)
P9: Leaders and leadership	• In-depth technical skills (S1) • Good problem-solver (S2) • Strong willingness to support the employees doing their work (S3)	• Lack of motivational skills (W1) • Lack of teaching skills (W2)	• Collaboration with the foreign firms (O1) • Massive construction projects (O2)	• N.A.
P10: Employees management	• Good teamwork among project team members (S4)	• Lack of trainings at the workforce level (W3)	• Government support (O3)	• High level of turnover (T1) • Resistance to change (T2)
P11: Partners relationships	• Good relationship with various working partners (S5)	• Lack of collaboration with partners (W4) • Lack of sharing in terms of information, best practice, etc. (W5)	• Large base of suppliers (O4)	• Competitive bidding still prevails (T3) • "Guanxi" or relationship (T4)

leaders. Moreover, the Chinese building professionals also showed their great willingness and commitment to support their employees to perform better (S3). This echoes Han et al.'s (2010) finding that putting people first is the most important leadership characteristic shown in China: assisting subordinates in problem-solving with their assigned duties. Overall, their down-to-earth personality and family styled leadership (Chen and Partington 2004) as well as in-depth understanding of construction projects (from the interview findings) have allowed many project managers to become successful and able leaders. In terms of the strengths of people management, it was revealed earlier that individuals (such as site engineers, QS, and others) within project teams are also technically competent and dedicated to their work. They usually work as a project team, and meet regularly (S4) on-site to discuss urgent or unresolved project issues in order to satisfy the client's requirements. It should be noted that other than this nature of teamwork—the weekly meeting (Bryde and Schulmeister 2012)—teamwork such as QC teams or *kaizen* teams was generally limited at the project level.

Another strength of the LCCFs' relationship with their working partners is that all the firms interviewed have a database of qualified suppliers, subcontractors, and vendors across China, and the number of their working partners is increasing. The strength lies in that, for the same item or material, the LCCFs have multiple suppliers to keep price competitive, and this ensures alternative channels of supply in case one fails (S5). This reflects the Chinese way of maintaining relationships with working partners, which is to take risk allocation into consideration, and in "not putting all your eggs in one basket". Additionally, with respect to the frontline workforce supply, several firms interviewed took the initiatives to establish relationships with some Chinese counties that are well known for labour outsourcing. Endeavours were made to set up training centres in such places to provide basic skills training before the workers are sent to work in the projects (Gao et al. 2012).

Weaknesses

In terms of the weaknesses of Toyota Way-styled leadership in the Chinese context, apart from technical skills, it seems that the importance of some soft skills was undervalued—for example, there is a lack of motivational skills (W1) and of teaching ability (W2).

- Lack of motivational skills (W1): project leaders seemed to struggle to formulate effective motivational strategies for their workforce. On the one hand, being afraid of losing their workforce, leaders were reluctant to issue "fines" or "penalties", but only gave verbal warnings. The bottom line was to maintain a sufficient workforce as a priority whenever possible.
- Lack of teaching ability (W2): Although the so-called mentor system is available in most LCCFs, it is generally designed for new recruits and only lasts a short time period. This contrasts with Toyota's requirement of its leaders: not only should they be constantly on the shop floor to provide necessary guidance for employees (Liker 2004), but the leaders also need to teach the employees to perform their work better under the "learning by doing" thinking. It seems that the Chinese building professional are the "fire-fighters" on-site, who always turned up to help resolve certain problems. Rather, the (project) leaders should take their time to *genchi genbutsu* on the site and provide hands-on teaching to their employees for them to develop their own problem-solving skills.

In people management (P12), although the training system was said to be available, it was only paid lip service most of the time (W3). The problem is that there is far less training provided at the project level. For some reasons, the firms seem to expect people, especially frontline workers, to know how to perform and behave on the job. They tend to assume that everyone knows and understand the importance of being on time, taking the initiative, being friendly, producing high quality work, and being readily competent for the work. In contrast to Toyota's intense efforts to develop its human resources, the LCCFs have much room to improve in the following areas:

- Enhance on-the-job training.
- Upgrade frontline workers' skills based on evaluations.
- Training should be cultivated as part of the firm's culture, rather than exists only as long as when the top management conducts an audit.

The last major weakness comes from the low level of collaboration between the contractors, the suppliers and the subcontractors (W4). It was revealed that project effectiveness (P11.4) and technical capacity (P11.5) are the two areas least considered for possible improvement. There are several reasons for this weakness. Firstly, the multi-tiered subcontracting system in China confuses the contractors; they do not know who they should be collaborating with. Moreover, their definition of long-term relationships narrowly focuses on whether their names are included in their working partners' portal. Even if they are one of the listed partners, it only means the relationship has been established. It also means that they are more likely to receive some portions of the work in future projects. This, however, does not mean that the contractor would help the working partners grow, and that they are now part of a big family or the "extended" enterprise of the general contractor. This shallow interpretation of a "long-term" relationship prevents the collaboration between the two parties from developing into possible improvements of their projects. Furthermore, low awareness among suppliers and subcontractors of lean practices and principles is yet another problem. More effort is thus needed to promote lean principles in the same language, and to ensure that all practitioners understand—especially those regarded as belonging to the extended firms of the general contractor.

Opportunities

Multiple opportunities present themselves as areas for improving the implementation of the principles of the Toyota Way People and Partners model. For leadership development, the increased involvement of international construction firms in China represents an excellent opportunity (O1) for Toyota Way-styled leadership practice to be implemented. As the interview results revealed, three responding firms have had experiences working with their Japanese counterparts, and they were deeply impressed by the authentic Japanese management style that pays so much attention to details and has such great commitment to quality, schedules and other project features. Moreover, as indicated earlier, the numerous ongoing projects (O2) also provided valuable opportunities for leaders to pick up skills, broaden their knowledge and enhance their understanding of construction projects.

For human resource development, the opportunities once again lie in the hands of the government (O3). Efforts relating to protecting and improving the situation of construction workers have been made by the government in the past decades through various means. These include establishing migrant schools for training the workforce, pay increases, and other benefits. With all these encouraging efforts, the implementation processes should become smoother. Furthermore, for the partners' relationship, there is the opportunity for a large base of suppliers and subcontractors available in the market. Because of the booming construction industry in China, subcontractors and suppliers are mushrooming and their numbers are growing (O4).

Threats

Several major threats need to be considered. Firstly, a threat may come from resistance for employees (T1) especially from the introduction of multi-skilled programmes. It is understandable that multi-skilled training would result in both more workload and greater levels of stress. Moreover, a high level of workforce turnover (T2) is regarded as a major threat: it contributes to firms being reluctant to invest in training workers that can readily leave the company. On the one hand, the LCCFs, especially the SOEs, are a perfect workplace for fresh graduates to begin their career; yet, such employees typically use them as stepping-stones to seek greener pastures. On the other hand, the turnover at the site level is also high, as employees are tempted away by higher wages—even by small salary increases. There are also threats to building relationships with partners, including (1) the competitive bidding practice that still prevails in China (T3), and which affects the trust between the two parties. If LCCFs desire to enter into a true partnership with their suppliers and subcontractors, then they must try to be as nurturing as possible, and focus less on the cost; (2) "guanxi" or relationship (T4). This threat can result in unqualified subcontractors being brought into a project. The potential problem is that if their company culture or philosophy of conducting business is not aligned with that of their main contractor's, things may quickly go wrong.

11.3.3.2 Strategies

Some relevant strategies are proposed in this section for both LCCFs and their external partner firms, to better implement the Toyota Way People and Partner model.

Leaders and Leadership (P9)

Firstly, there is a soft skill gap among Chinese building professionals. It should be acknowledged that soft skills are what accompany hard skills, and these help the firm to use its technical expertise to full advantage. This was identified as the biggest weakness, and the corresponding strategies for dealing with it include:

9.1 Begin by building the company's culture in a way that its leaders are encouraged to *genchi genbutsu* on the projects, to understand the projects and their employees' work, as well as their own capacity. In this way, the relationships between them will be established with the use of soft skills.

9.2 The management should develop a long-term plan for leadership development, with a focus on developing their teaching skills. This solves the weakness of Chinese building leaders' tendency to become problem-solving "fire-fighters", while overlooking the importance of cultivating their employees' own problem-solving skills.

People Management (P10)

The principles relating to people management (P10) were only moderately practiced within LCCFs. These principles include basic HRM activities, such as people selection, training, teamwork and motivation. Some forms of these are present in a majority of the firms surveyed. However, the interviews revealed a huge gap between the status quo of LCCFs in this aspect and the genuine Toyota Way-styled practice. In order for these activities to be better practised, some strategies and considerations for improving people management within LCCFs are given below:

10.1 Select the right people: In the Chinese construction industry, selecting the frontline workforce is more challenging than selecting "blue-collar" employees, given that the former is notorious for its high mobility. It is more urgent for LCCFs to build up their labour sources. It is worth taking a good suggestion, currently practised by a few leading contractors: to sort the workforce into three layers based on their abilities and skill sets—that is, to maintain the appropriate levels of skilled and semi-skilled workers. Such a flexible labour resource structure can easily accommodate the varying needs of different projects.

10.2 Training: the LCCFs should make sure that adequate resources are available to carry out a variety of on-site programmes for executive and workforce training. Addressing the identified weaknesses in Table 11.4, the breadth, types and variety of training should be considered and enhanced. Apart from quality, health and safety, and construction skills, other themes such as "identifying the non-value-adding activities in work", "possible improvement in current working procedures" and so on are worth incorporating into current training programmes. For workforce training, training facilities should be utilized (e.g. by moving the training school to on-site), and the training plan should be further developed and linked to the performance of the workforce. The workers may actually be excluded from the firms, but strategically the best-performing teams should be maintained within the above-mentioned flexible labour resource structure.

10.3 Teamwork: although teamwork is practised, it is nonetheless advised that within LCCFs, the working environment should allow various forms of teamwork to emerge in the daily work. Possible forms of teamwork may include QC teams, *kaizen* teams, 5-S teams, and multidisciplinary teams to tackle a particular problem. In addition, such forms of teams provide their members with multi-skilled development.

10.4 Motivation: for improving the presently used motivation strategies.

Partner Relationships (P11)

Toyota treats its suppliers as partners and as integral elements of the Toyota Way (Liker 2004). The interviews with LCCFs suggest that this principle would seem to be considered unnecessary to their operations, given that low levels of implementation of this principle were described by a majority of the interviewees. Taking the SWOT factors into consideration, the strategies for the LCCFs to develop meaningful relationships with their partners are presented below.

11.1 Limiting the selection of partners: carefully review the existing portal that integrates the information of suppliers and subcontractors. Efforts should be made to remove working partners with unsatisfactory records and performance. It may be unnecessary, and not to mention impossible, to transform the multi-sourcing partners into single-sourcing partners overnight. Yet from a long-term perspective, it should be reasonable to reduce the number of working partners, and in this way, trust and collocation would be enhanced.

11.2 Eliminating the illegal multi-subcontracting practice: given that the multi-subcontracting system prevails in the Chinese construction industry, contractors must establish ground rules to eliminate such unlawful practices.

Table 11.5 SWOT analysis of the LCCFs under Problem-Solving Model

	Strengths (S)	Weaknesses (W)	Opportunities (O)	Threats (T)
P12: *Genchi genbutsu*	• Commitment to problem-solving and professional responsibility (S1) • Site engineers are encouraged to stay at *gemba* (S2)	• Heavily rely on experience (W1)	• Clients' requests (O1)	• Busy work schedule compromises the *genchi genbutsu* practice (T1)
P13: Decision-making	• N.A.	• Limited use of decision-making tools (W2) • Limited participation from the lower level (W3) • Poor capability and negative attitude to problem-solving (W4)	• N.A.	• Rare use of statistical control (T2) • Seeking someone to blame (T3)
P14: *Kaizen*	• Management had good attitude as well as action on *kaizen* (S3)	• Only serious problems draws attentions (W5) • No documentation of improvement and/corrective action from mistakes (W6)	• Introduction of QC (O2)	• Lack of supporting culture of carrying out *kaizen* (T4)

The partners' portal should be used as a reference for choosing reliable firms for material delivery and subcontracting work.

11.3 Encourage collaboration: with respect to the issues relating to the lack of collaboration raised in the SWOT discussion, LCCFs are advised to create more opportunities for team-working with their partners to improve potential areas and work problems of the project. They should capture the lessons learnt and share best practices with other partners.

11.4 It is also important to establish training to raise awareness of the Toyota Way practice (or lean practice) for partners, and to periodically evaluate their compliance with the Toyota way principles, so as to maintain an ongoing dialogue with the contractor to foster continuous improvement.

11.3.4 Toyota Way Problem-Solving Model

11.3.4.1 SWOT Discussions

This section outlines the SWOT analysis of problem-solving practices used by LCCFs (see Table 11.5).

Strengths

The LCCFs have many strengths in their implementation of the Toyota Way problem-solving model. First of all, the project leaders are committed to problem-solving, and this is reflected in their high professional responsibility (S1). This is consistent with the study of Han et al. (2010), which stated that Chinese leaders are dutiful, diligent, reliable, and take the initiative at work. It is true that without these qualities, leaders will react slowly to the problem. Another strength lies in the accessibility of the project engineers (S2), given that they were requested to spend most of their time on-site, to walk around the site, to spot as many defects and problems as possible, and to monitor workers' performance. Overall, it is encouraging to see that the interviewees' attitudes towards *genchi genbutsu* were positive (see Chap. 9).

With respect to the practice of *kaizen* or continuous improvement, it was found that management had higher awareness and understanding of *kaizen* activities (S3) than their subordinates. The reason for this might be that the preparation for pursuing ISO 9000 certification as a part of the TQM efforts of LCCFs has contributed to continuous improvement—a key component of ISO 9001/2—familiar vocabulary (Gao et al. 2012).

Weaknesses

One weakness of the *genchi genbutsu* implementation included the interviewees' statement that experience was what they have heavily relied on during decision-making processes (W1). There is a chance that narrowly relying on experience without consulting the data collected around these problems may prevent the root causes from surfacing. The weaknesses in decision-making include the following:

- Limited adoption of decision-making tools and techniques (W2), such as the "5 whys" for finding the root cause. Employees are easily satisfied with superficial solutions, and do not attempt to find out the root causes.
- Limited participation of employees in decision-making (W3): this arises because of the more centralized system that is adopted within the firms, whereby major decisions relating to all aspects of the project (cost, scheduling and problem-solving) need to be finalized by senior leadership.
- Employees' poor capability and negative attitudes (W4): it is generally believed that the employees' untapped potential in problem-solving and their "don't care" attitude lead leaders to distrust the employees, and therefore prevent them from joining the decision-making process.

With respect to the *kaizen* implementation, the identified weaknesses included (1) only serious problems drew the attentions of management (W5), and (2) poor documentation of *kaizen* resulting from mistakes (W6). The former is a result of the common culture that not all the problems are treated equally. Also, due to the busy schedule of the project management team, other day-to-day matters need its pressing attention rather *kaizen* problems. The evidence for the latter is that even when a problem has been solved, the Chinese building professional does not value the documentation for the process, unless the problem is particularly severe. Not many of the interviewees appreciated the importance of documentation. This too was reflected in the results of the questionnaire survey. Without thorough documentation of the resolutions to problems, new knowledge and new standards were prevented from being shared and learned.

Opportunities

The client plays a key role and provides an external opportunity (O1) in response to better *genchi genbutsu*. A client's request is usually taken seriously and treated as a priority. Hence, if there is a request from the client asking the project leaders to be present on the *gemba* (site) in order to investigate a problem, the project leaders would not usually reject such a request and would in fact show up. However, caution should be taken here: a client's request for *genchi genbutsu* is an enabler or facilitator, but is not in itself a truly Toyota Way-styled *genchi genbutsu* practice. Therefore, it requires LCCFs to cultivate a company culture that always sees problems as opportunities for potential improvement. In the case of *kaizen* or continuous improvement, the growing recognition of QC in construction projects in China (O2) represents another opportunity. The interviews confirmed that QC was conducted in the projects that were being undertaken. The successful introduction of QC enables the workers to understand that there is an opportunity to improve quality, for example, in a teamwork fashion. The opportunity could lead to a QC improvement motivating the firm or employees to engage in further, similar improvement initiatives.

Threats

At first glance, the threat to leadership in being *genchi genbutsu* might come from the project leaders' busy schedule (T1). A common excuse repeatedly heard during the interview was that: "*I wish I had more time to spend on-site. However I am always too busy handling other issues*". A closer examination reveals that the major threat to the implementation of Toyota Way *genchi genbutsu* within LCCFs is that the traditional firm culture within large Chinese construction firms undervalues the importance of "*going to see the source of problems*", unless it is related to major issues. In contrast to Toyota's culture—which sees every problem as an opportunity—LCCFs were operating in the opposite way, treating the problems as troublesome responsibilities. "Minimize the big problem into a smaller one, and then into no problem" ("大事化小, 小事化无") sums up the firm culture encountered during the fieldwork.

The threat concerning consensus decision-making comes from the use of data. In the Toyota Way, data measurement, collection, analysis, summary and interpretation provides the basis for joint decision-making (Liker 2004). However, a majority of the construction projects in China do not require advanced information technology or data-mining techniques, and hence far less statistical control was employed in the decision-making process (T2).

Finally, unlike large-scale manufacturers with in-house *kaizen* teams continuously striving for gradual improvement, the LCCFs did not present convincing examples of how *kaizen* was actually implemented in their projects—with the exception of the recently introduced QC teams, and of learning by visiting other projects undertaken by the same company. The large differences lie in the fact that Chinese construction firms would not engage in the operations or processes needed to seek out potential improvement opportunities. In fact, this may be the very last thing they are willing to consider, as substantial portions of construction work are subcontracted, and there is tight delivery pressure. What stays at the top of their agenda is the rush for the project to be completed by managing and controlling perceptibly more important matters. In addition, other possible threats were also raised by interviewees, including:

- Limited time windows (T3): as indicated earlier, the pressure to deliver means that there is no time left for improvements to be considered. Moreover, once the project completes, the next project may be a totally different one, in which implementing *kaizen* becomes very challenging, as there is now a new set of standardizations to be developed; in such cases, it is not possible to simply build upon the previous standards.
- Lack of supporting culture of carrying out *kaizen*: unlike Japanese workers who treat problems as opportunities and who are encouraged to expose problems rather than simply bury them, clients in the Chinese construction industry do not appear to be happy to see annoying problems recurring. The same attitude was found among leaders on the contractor's side. Under this organizational culture, workers thus become numb to problems, and start to hold attitudes like "avoid trouble whenever possible", "don't care" and so on.
- Cutting corners: it was generally agreed by most of the interviewees that financial incentives are at present the most effective means for motivating Chinese construction workers to work "faster" and "more efficiently". This does not mean, however, that a continuous improvement effort can be expected in this way. Quite the opposite: some workers might think themselves as being "smart" enough to cut corners or skip a few essential procedures in order to "fabricate" an artificial improvement.

11.3.4.2 Strategies

The Toyota Way PSM was developed by and works well for Toyota's employees (Liker 2004). The top hierarchical level of the Toyota Way is its problem-solving model, which is based on three principles, namely *genchi genbutsu* (P12), decision-making through consensus (P13) and continuous improvement (P14). It is not simply a "how to" concept, but is clearly related to key management issues: the development of *genchi genbutsu* leadership, empowerment, participative decision-making, continuous improvement and other features. This section presents strategies for improving the implementation of the Toyota Way PSM within LCCFs. The strategies are outlined here in response to the SWOT factors discussed earlier.

Genchi Genbutsu (P12)

Genchi genbutsu is one of the guiding principles of the Toyota Way (Liker 2004). Given that Chinese building professionals have shown their commitments and professional ethics in problem-solving, and as there is an increasing demand from clients to request project or company leaders to be present on-site, it is suggested that:

12.1 A *genchi genbutsu* firm culture is established: the attitude of *genchi genbutsu* should be strengthened, even going so far as to write this into the core values or guiding principles of the companies.

12.2 Details are relied on: Decision-making should not only be based on experience, but also needs to incorporate an understanding of the condition.

12.3 *Genchi genbutsu* is set as a priority: time management is very important for leadership, as leaders are always busy with unexpected activities and events. *Genchi genbutsu* should therefore be set as a priority to show the determination of the leadership, and to boost the morale of employees. Excuses such as "I'm too busy at work and hence have no time" should not be tolerated or allowed to compromise the practice of *genchi genbutsu.*

Consensus Decision-Making (P13)

It was revealed earlier that consensus decision-making received the least amount of effort and attention (see the survey results) from Chinese building professionals. However, it is understandable that each individual has his own way of making decisions, and that the process also exhibits certain cultural features. Understanding the Toyota Way-styled decision-making process requires good knowledge of its relationship to Japanese culture (Low and Gao 2011a). Earlier research reported that Chinese culture, being embedded in its tradition of Confucianism, is focused on group harmony, stability, and cooperation, and is also collectivist in nature. The survey findings, however, reported that Chinese employees from construction firms are still far from practising consensus decision-making. In order to narrow the gap, the strategies proposed include:

13.1 Establish a "no blame and no complaints" work environment: in a workplace without blaming, all employees should be encouraged to voice their opinions and their suggestions for decisions. In addition, ideas from the lower hierarchy should also be respected and appreciated. This is the first step to introducing a consensus.

13.2 Encourage the use of decision making tools in a systematic way: decision-making tools, such as the 5 whys, cause and effects, and pareto—all of which are widely used in industrial engineering—are seldom practised within the Chinese construction industry. These decision-making tools should be encouraged and introduced in the form of workshops or classroom training.

Kaizen (P14)

Kaizen is not simply a set of tools for implementation, but is a long-term mindset in which employees commit to make things better (Imai 1997). Taking into consideration the SWOT factors pertaining to *kaizen*, the strategies proposed are as follows:

14.1 Leaders' initiative: *kaizen* activities are more widely known and recognized among management, who should play the role of champions in taking the initiative to facilitate continuous improvement in the lower levels of the firm. Unlike Toyota employees who may form *kaizen* teams in the spirit of volunteering (Imai 1997), the initial strategy for LCCFs should be to introduce recognition rewards for any *kaizen* improvement achieved in the project.

14.2 Moreover, it is very important to understand the importance of the role of culture, which should be shifted to allow seeing problems as opportunities. The common practice within the Chinese construction industry is to treat serious problems with greater emphasis, while for all other problems one of the strategies is "to turn big problems into little ones and little ones into none at all". This attitude requires tweaking. Learning from the Toyota Way, the opportunity can be taken to solving problems in order to make improvement. These are the first few steps to cultivating a *kaizen* culture.

14.3 Efforts are also required to document the improvements made, especially the valid solutions that are generated. In response to this identified weakness, it is suggested that the project team emphasizes the importance of documentation, which can be used for future learning and sharing purposes. Several good practices were already discussed during the case studies which relate to *kaizen* activities.

11.4 Toyota Way Model: Guidelines for Implementation

11.4.1 Introduction

The conceptual framework of the Toyota Way in the context of construction was developed in Chap. 4. Acknowledging its comprehensiveness, the pyramid structure of the Toyota Way model has been retained, and lean construction concepts and practices have been linked to various relevant principles to make these more applicable in the context of construction. This section discusses improvements to the Toyota Way framework in the Chinese construction industry, based on the fieldwork results, the SWOT discussion, and the validation exercise.

11.4.2 A Review of the Toyota Way Model

The conceptual model has proven to be robust for this research, as it was tested in the Chinese construction industry and shown to be valid and applicable to the large construction firms in China. However, this is not to say that no further development or adjustment of the model is needed, given that the findings have highlighted a great number of constraints that could become potential threats to the implementation of Toyota Way-styled practices (see the discussion of SWOT). This would seem to suggest that some Toyota Way-styled attributes or practices may need to be tweaked before these can truly be accepted and better implemented by Chinese building professionals. Taking this into consideration, the implementation of the Toyota Way model would undergo certain minor changes. These changes will be reflected in the implementation guidelines.

11.4.3 Framework Implementation Guidelines

The framework implementation guidelines are provided in a series of tables below, with the aim of narrowing the gap between the Chinese practices that currently prevail and the Toyota Way-styled practice. In doing so, it is expected to facilitate LCCFs to manage their organizations and construction projects more efficiently and effectively from the beginning to the end. The framework implementation guidelines not only list the Toyota Way-styled practices and depict how these should be implemented in a holistic way, but these also offer strategies for implementing them effectively. These strategies are valid in the Chinese construction context. Moreover, the implementation framework can be used by project managers and project directors in particular, given that it is they who plan and manage the delivery of the whole project from the very beginning to the end. This framework can also be used by the top management of the firm, especially as this is essential for effecting organizational culture changes, mindset changes, etc. All such changes require the commitment of top management who need to take the initiative to become champions for facilitating implementation. Overall, this framework can be used as a practical guideline covering a number of areas, including organizational philosophy, process, people and partners, and problem-solving. The purpose of the Toyota Way implementation framework with the toolkit is to assist the construction firms in a few ways:

1. To enable the construction firms to assess themselves, using a series of questions and scales to evaluate where their firms stand relative to Toyota Way implementation.
2. To provide guidelines for operationalizing the activities of the Toyota Way implementation.
3. To monitor and audit best practices for the application of Toyota Way principles.

For actionable attributes that received sufficient support, the strategy is that no adjustment is needed. In contrast, for those attributes that were discovered to be subject to implementation challenges and difficulties, attention was paid, with the caution that the mechanism of the Toyota Way principle must not be changed; actions, however, may be altered. The bottom line is that these attributes cannot be treated negatively or abandoned. The previous sections provided strategies and considerations for these specific attributes for enhancing the reliability of the modified Toyota Way model for the Chinese construction industry. Eventually, the strategies and considerations for these specific attributes, as well as for those that have been received well, can be used as a set of guidelines for implementing the Toyota Way.

11.4.3.1 Philosophy Model

The Toyota Way Philosophy model implementation guidelines are shown in Table 11.6. The table summarizes the key attributes towards the Toyota Way-styled long-term philosophy. For those that the LCCFs did not perform well, the corresponding strategies are already highlighted in the earlier section.

Table 11.6 Toyota Way Philosophy model: implementation guidelines

	Long-term philosophy (P1)
Objectives	Establish a long-term philosophy and apply it to various aspects of the firm
Requirements and actions	• Sustain a constant purpose (company vision, mission and values) • Have a high purpose or mission to generate value towards employees, society and customers • Formulate a plan towards the realization of a company's long-term vision • Short-term losses affect decision-making, but are less important than pursing long-term goals • Have a clear view of core competencies and endeavour to become an expert in this area • Be responsible for products, employees and society • Understanding customer's requirement is priority work • Be able to rapidly respond to meet the changing requirement of customers (e.g. design change) • Treat employees and suppliers as internal customers
Expected outcomes	• Improved understanding of customers and customers values • Improved customer satisfaction • Increased competitiveness • Contributing more value to employees, firms, and society at large

11.4.3.2 Process Model

Liker and Meier (2006) described working on a couple of process-oriented initiatives as an ideal starting point for lean implementation. The tangible benefits reaped in the short term will make this approach undeniably attractive to most practitioners. However, the implementation of process-oriented initiatives should embed the first principle highlighted in the previous section, which is to recall what customers or clients really want, and what the value to be maximized within the process is. Table 11.7 presents the guidelines of the Toyota Way Process model. The objective is to highlight what actions need to be taken to improve the process within each different principle. There are three elements associated with the actions for each principle: the role of employees and workers (manpower), the role of materials and machines and the role of workplace design. A better implementation of the above-mentioned process-oriented initiatives requires, collectively (1) in terms of manpower, that those working in the lower levels of the hierarchy in projects need to be aware of what they can do to contribute to a better process; (2) in terms of materials and machines, that an understanding is gained of how materials and machines should be treated in order to assist people to achieve better processes and (3) in terms of workplace design, that those aspects which affect the design of workplace design need to be considered, and improvements incorporated at the project site level (i.e. of layout) to result in better processes (Liker 2004). The proposed strategies were presented earlier. These are more relevant to Chinese construction firms, given that they are formulated to improve their weaknesses in implementation.

Table 11.7 Toyota Way Process model: implementation guidelines

	Requirement and actions	Expected outcomes
P2: One-piece flow	P2.1 Employees are concerned with waste elimination P2.2 Material flow is adhered to consistently throughout the daily work activities P2.3 Materials, equipment, and other resources are provided in a "just-in-time" manner when needed P2.4 Site layout is organized to enhance material flow, employee movement and so on, in order to minimize waste due to movement P2.5 Strive to cut to zero the amount of time any work is sitting idle or waiting for someone to work on it P2.6 Make flow evident through organizational culture	• Waste is eliminated as much as possible • Achieving uninterrupted workflow • Sufficient numbers of workers can be maintained on site • Materials arrive in JIT manner
P3: Pull "*kanban*" system	P3.1 Materials are ordered as close as possible to exact needs P3.2 Strive for as low as possible levels of material inventory (even stockless) on the construction site P3.3 Use simple signals—cards, empty bins, and so on, to monitor the level of inventory and to order the needed materials or components P3.4 Monitor the quantities of materials, components and equipment that the teams actually take away P3.5 Clear job contents, work time, material requirements, and other information should be prepared before releasing a work task to a crew	• Low level of inventory at project level • Good practice of material management • Increased reliability of work plans • Enhanced ability in shielding the downstream work
P4: Level out the workload	P4.1 The project manager plans the work with inputs from other parties, including subcontractors, clients and suppliers P4.2 Daily work activities are planned to balance material availability, manpower, machine availability and workload between operations P4.3 Foremen (the last planners) make commitments as to what their crews will do each week based on what is ready to be done P4.4 Weekly and daily work assignments are completed in accordance with the weekly and daily schedules P4.5 Levelling the daily work activities without overburdening workers and machinery	• Improved collaboration between project teams and other stakeholders in project planning • More empowerment can be seen on site • Enhancement of foremen's skills in job planning • Less overtime resulting from uneven workload

(continued)

Table 11.7 (continued)

	Requirement and actions	Expected outcomes
P5: Built-in quality	P5.1 Employees are dedicated to providing quality "built-in" to every aspect of operations P5.2 Preventing defective or "no inspection" assignments from entering the next process P5.3 Rejecting defective materials, components, and equipment P5.4 Employees are encouraged to seek support from their supervisors when something goes wrong at work P5.5 Employees are empowered to be responsible for quality P5.6 Employees who work in the same team meet on a regular basis to discuss quality problems and lessons learned P5.7 Feedback about quality is routinely given by employees	• Improved quality: reduction in rework and less reoccurring quality problem • Improved skills in detecting problems • More empowerment in letting workers stop the operation if a problem occurs • More teamwork and communication relating to quality improvement
P6: Standardization	P6.1 Established standard operating procedures (SOPs) (e.g. for work processes) are practised by employees for each major operation and process P6.2 Employees play a key role in creating the SOPs P6.3 Employees are encouraged to improve the existing SOPs based on their own practical experience P6.4 Incorporate employees' creative improvements of the standard into new SOPs P6.5 Use standardized prefabricated components from offsite yards	• Improved understanding of standardization • Improved productivity resulting from the implementation of standardization • Improved ownership of worker operations • More standardized components or materials can be introduced and used
P7: Visual management	P7.1 Adopt visual aids to make wastes, problems and abnormal conditions readily apparent to employees P7.2 The information posted on job status, schedule, quality, safety, and others appears in a place that most workers can see on a daily basis and is kept up-to-date P7.3 Appropriate signage is used to identify layouts, traffic flow, safety concerns and so on P7.4 The construction site is kept clean at all times P7.5 Employees take pride in keeping the construction site organized and clean P7.6 The workplace follows the principles of 5-S	• Increased use of visual tools • Higher awareness of the 5-S programme and fuller participation • More organized and tidier site • Employees are more disciplined to keep the site clean

(continued)

Table 11.7 (continued)

	Requirement and actions	Expected outcomes
P8: Use of reliable technology	P8.1 New technology must support the company's values P8.2 New technology must demonstrate its potential to enhance processes P8.3 New technology must be specific-solution-oriented P8.4 New technology must be thoroughly tested and proven to provide long-term benefits	• An improved procedure in adopting new technology

11.4.3.3 People and Partner Model

It has become clear that actionable attributes and strategies within this theme are generic, and thus are applicable to construction with no adjustments needed. However, there is a difference between the people elements within construction and manufacturing (see Sect. 3.4.2): the quality of frontline workers in the Chinese construction industry is not comparable to that of Japanese blue-collar workers. To narrow this gap, and to achieve a maturity level similar to Toyota in this category, Table 11.8 provides a checklist for LCCFs wishing to implement the Toyota Way People and Partner model in the Chinese construction context.

11.4.3.4 Problem-Solving Model

The guidelines for the Toyota Way Problem-solving model are shown in Table 11.9, highlighting the key attributes and strategies for better implementation.

11.5 Validation

11.5.1 Overview

Validation is a process closely related to triangulation and consists of member validation or member checks. In other words, validation participants determine whether the research's interpretation of the findings and its recommendations accord with their own (Bryman and Bell 2003; Lyons and Doueck 2010). Peer or expert validation is widely used, where findings are shared with others who have expertise in the research phenomenon or the population (Lyons and Doueck 2010). In this study, validation aims to seek opinions on the Toyota Way model within Chinese construction firms, as well as to identify the validity of the strategies proposed for implementing the Toyota Way model better. Due to time constraints, the validation exercise was conducted with Chinese building professionals currently working in Singapore. Since Chinese construction firms are currently well represented in Singapore, this mode of carrying out the validation exercise greatly assisted the researcher in finding sufficient numbers of validation participants. The key requirement for the validation participants was that they must have had some

Table 11.8 Toyota Way People and Partner model: implementation guidelines

	Leaders and leadership (P9)	People management (P10)	Partner relationships (P11)
Objectives	Toyota Way-styled leadership (servant leadership)	Treats people as the most important asset in the firm	Challenge and grow with partners
Requirements and actions	P9.1 Leaders are motivated to inspire people to achieve goals P9.2 Leaders must have in-depth job knowledge P9.3 Leaders possess teaching ability and are able to pass their knowledge on to others P9.4 Leaders must support the employees doing their work P9.5 Leaders will take time to understand problems and root causes before acting P9.6 Leaders strongly encourage employees to develop "continuous improvement" in thinking and action P9.7 Leaders must understand company policy and procedures, and communicate these effectively to their team	P10.1 Select the best person for a given job P10.2 Training is provided to equip employees with the required skills before they are assigned to work P10.3 On-the-job training is provided to further develop employees' exceptional skills P10.4 Employees are cross-trained to perform additional function P10.5 Training materials are standardized P10.6 Employees are encouraged to cooperate with others to complete the whole task P10.7 Daily work activities are organized into team functions P10.8 Internal motivation methods P10.9 External motivation methods	P11.1 Respect partners' capabilities P11.2 Challenge partners by setting collaborative targets P11.3 Take part in partners' production process P11.4 Work with partners to improve project effectiveness P11.5 Work with partners in various areas to develop their technical capabilities P11.6 Share information with partners in a structured manner P11.7 Conduct joint improvement activities with partners to solve problems P11.8 Strive to establish a long-term relationship with reliable partners P11.9 Limit the number of suppliers
Expected outcomes	• Higher commitment to *genchi genbutsu* • A greater willingness to teach their followers • Working with employees in more continuous improvement programmes • Being more respected by employees	• Employees understand their work contents clearly and are skilful in their work • Multi-tasking • Good team working spirit and willingness to work in a team • Highly motivated	• Small base of partners (suppliers, subcontractors) • Improved project effectiveness resulting in more collaboration • Enhanced capabilities of partners • Collaborative problem-solving • Mutual benefits in the long-term

working experience in the Chinese construction industry. This is because eventually the Toyota Way model, along with the strategies recommended, is intended for implementation in the context of the Chinese construction industry. In addition, the

Table 11.9 Toyota Way Problem-Solving model: implementation guidelines

	Genchi genbutsu (P12)	Consensus decision-making (P13)	Continuous improvement (P14)
Objectives	Establish *genchi genbutsu* culture	Value the decision-making process and achieving consensus	Practise continuous improvement and become a learning organization
Requirements and actions	P12.1 Solve problems by going to the place (e.g. on the construction site) where the problems were discovered P12.2 Analyse and thoroughly understand the situation before making decisions P12.3 Making decisions based on the verified data P12.4 Making decisions based on management team's past experiences P12.5 Allow *genchi genbutsu* to become part of the company culture	P13.1 Use appropriate problem-solving methodologies (e.g. the 5 Whys) to determine the root causes of problems P13.2 Conduct experiments to test the potential cause of a problem P13.3 Broadly consider alternative solutions P13.4 Value the process through which the decision was reached P13.5 Build consensus within the team, including employees and outside partners P13.6 Address the root causes of problems via effective communication vehicle	P14.1 Reflect on mistakes (defects, rework, safety issues, etc.) on a regular basis P14.2 Management should treat problems as development opportunities for employees P14.3 *Kaizen* activities are conducted in the workplace P14.4 Management supports *kaizen* activities P14.5 Improvements will be codified into documents and/or policies used by organization P14.6 Each hierarchy of the organization develops measurable objectives and actions to support executive-level goals P14.7 Managers are keen on measuring the objectives and give feedback P14.8 PDCA methodology is used to solve problems
Expected outcomes	• Increasing commitment to *genchi genbutsu* • Improved appreciation of the *gemba* • Rely on reliable data, rather than the past experiences	• Increased use of decision-making tools, even if these are not widely adopted yet • A more consensus decision-making process respecting different voices • Improved persistence and skills in finding out the root cause	• A change of culture to appreciate the importance of revealing problems, rather than hiding them • More proactive continuous improvement programmes • An improved system for documenting and sharing lessons learnt

Table 11.10 Particulars of research validation participants

No.	Position	Organization	Work experience in China (years)	Work experience in Singapore (years)
1	Project Director	Contractor A (Nantong)	12	16
2	Project Manager	Contractor B (Jingye MCC)	5	6
3	Project Manager	*Contractor C (BHCC)	10	13
4	Project Director	Contractor D (China Construction)	8	9
5	Project Manager	Contractor D (China Construction)	7	5
6	Project Manager	*Contractor E (Poh Lian)	5	11

Note: "*" indicates a Singaporean firm, while unmarked firms are subsidiaries of Chinese construction firms in Singapore. The average work experience in China is approximately 8 years, less than the average 10 years in Singapore

validation participants must have prior understanding and experience pertains to lean, so that they are competent to comment on the Toyota Way model and the strategies. According to O'Keefe et al. (1986) as well as Bryman and Bell (2003), the number of validation participants does not matter quite as much as the expertise that the research participants have, and usually their number is no more than ten. Hence, in this study, the group of six participants invited for validation is considered sufficient. Their profiles are presented in Table 11.10.

The validation exercise was conducted between August 2012 and September 2012. Prior to validation, briefings of about an hour were given to these six research participants. The validation process was carried out in the following stages:

1. Fieldwork findings on the state of the level of implementation of Toyota Way-styled practices, as well as on the SWOT factors that might affect the implementation of Toyota Way-styled practices, were presented to the validation participants.
2. The validation participants were then asked to confirm whether they agreed with the strategies proposed in (1): the set involved 45 strategies, falling within the 14 Toyota Way principles. Each strategy requires the research participates to rate their level of agreement on a scale of 1–5, from "strongly disagree" to "strongly agree". The ratings of the six participants are presented in Appendix E.
3. The respondents were then presented with a framework for improvement. The participants reviewed and commented on the framework developed in the study, and assessed:
   (a) the completeness of the strategies in dealing with all the issues that prevent effective implementation of Toyota Way-styled practices, and
   (b) the willingness of management to carry out the implementation model.

11.5.2 Discussion of Validation Results

The outcome of the validation is discussed in this section.

11.5.2.1 Strategies

The results, presented in Appendix E, indicate that the validation participants agreed with the various strategies proposed on the basis of the SWOT analysis. The ratings given to these strategies ranged from "agree" to "strongly agree", or from 3.83 to 5. In the opinions of the respondents, some strategies were already in place, despite their experiences with projects in China (a number of years ago), or in Singapore (currently).

- Philosophy: in the views of the participants, the strategies outlined in this category were useful and generally valid. However, they also mentioned that the long-term philosophy is not an easy task to implement in a short time. Once it is decided upon, it also becomes hard to change in the short-term. The current practices in China are such that they still place too much focus on profit, and try to maximize it by all means, given the fierce competition.
- Process: the ratings for the strategies listed in the process model showed a strong consensus. For example, one strategy under principle of *heijunka*, which states "tracking the reliability of weekly plans and revealing the root causes where possible" was rated "5". While the strategy concerning the adoption of prefabrication was rated "3.5", indicating some hesitation. This is because they understood the pre-condition for the adoption of prefabrication given that they all have similar experience in Singapore, where prefabrication is widely used. They argued that prefabrication may not be an entirely valid strategy for the Chinese construction industry since the possibility of earthquake would prevent them from adopting such a technology.
- People and partners: the strategies highlighted in this category, in the views of the participants, were all considered to be timely, appropriate and effective. They agreed that managing people and people's performance is the most difficult area at the project level. There is a great need to learn from best practices in effectively managing people. Although the strategies related to people and partners sound effective in theory, in real-life projects, it is reasonable to doubt the extent to which these would be implemented. For example, "adequate resources to carry out on-site training programmes" and "removing working partners with unsatisfactory records and limiting the base of parties" all translate to cost increases, and their implementation could face obstacles.
- Problem-solving: all of the validation participants mentioned that whenever they are working for Chinese or Singaporean projects, the problem-solving procedure that they followed has much common with the Toyota Way style. That is, it is simply a cycle involving (1) the timely discovery of problems, (2) the discussion of possible solutions, (3) the resolution of the problems and (4) the lessons to be learnt. In their views, it does sound like a valid strategy to focus on better problem-solving, especially with keeping the continuous-improvement thinking

in mind, along with the culture shift towards greater openness and tolerance to exposing problems.

All that needed to be done with these already-in-place strategies was to ensure that these are properly communicated for their effective implementation.

11.5.2.2 Validating the Framework

When the Toyota Way model and the attributes identified were presented to the respondents, they generally found the implementation framework to be acceptable. Again, they acknowledged that some of the so-called Toyota Way-styled practices have already been implemented. This is where they found the framework useful, as it does not introduce a new technique, but rather identifies familiar practices—but using terms that are more widely known in the manufacturing setting. There are challenges concerned with full implementation and, in their opinions, with the SWOT analysis. In fact, for those weaknesses associated with frontline workers, they also commented that similar generalizations can be applied in the Singapore construction context, as most of the workers are from overseas, and they sometimes have even greater problems than would be the case with the Chinese workers alone in China.

11.6 Summary

This chapter has presented the SWOT analysis of LCCFs in terms of their implementation of Toyota Way practices. It has also presented strategies for better implementation in response to the weaknesses and threats identified. It is necessary that proper strategies be derived when strengths, weaknesses, opportunities and threats are understood, and where improvements are required. This chapter has outlined the generic strategies under each of the four themes of the Toyota Way model: long-term philosophy, process, people and partners, and problem-solving. The strategies for Toyota Way implementation were validated with a group of Chinese building professionals in Singapore. The lessons learnt from the Toyota way model were also shared with them, and their opinions were sought on the resulting model. An attempt was made to overcome the problems and bottlenecks identified during Toyota way implementation within the construction industry. While these guidelines may appear generic, they do point out the areas which Chinese construction firms need to improve. Unlike other studies which may provide more specific strategies for firms at different stages of implementing other management initiatives, this research, anchored on the Toyota Way *kaizen* thinking, strives for perfection. The checklist requires firms' commitment, determination, resources, etc. to accomplish what they can, using their maximum efforts.

Conclusions 12

12.1 Introduction

This final chapter presents an overall summary of the study. It starts with describing how the aims and objectives of the research have been met, along with the major conclusions and potential contributions to both theoretical and managerial practices that arise from the research. The limitations of the study, along with recommendations for construction firms and the Chinese construction industry and future work are described at the end of the chapter.

12.2 Realization of the Aim and Objectives of the Research

This study aimed to investigate the Toyota Way-styled practices within the construction industry, with a particular focus on establishing an implementation framework that would embrace the Toyota Way-styled practices for LCCFs. Figure 12.1 illustrates how the aim and objectives of the research are realized.

12.2.1 Key Attributes Arising from the Toyota Way in the Context of Construction

The first objective of this research is to break the Toyota Way principles down into measurable or quantifiable parameters. This contains two sub-objectives, namely:
1. Reviewing the state of the art in lean and Toyota Way, and
2. Identifying the actionable attributes of the Toyota Way-styled practices.

For a start, a survey of the relevant literature was performed to gain a theoretical and critical understanding of lean production and lean construction (Chap. 3), and Toyota Way-related literature (Chap. 4). Lean production was investigated in terms of its development (Liker 2004; Shah and Ward 2007; Sugimori et al. 1977; Womack et al. 1990), its principles (see Womack and Jones 1996; Koskela

DOI 10.1007/978-981-287-014-8_12, © Springer Science+Business Media Singapore 2014

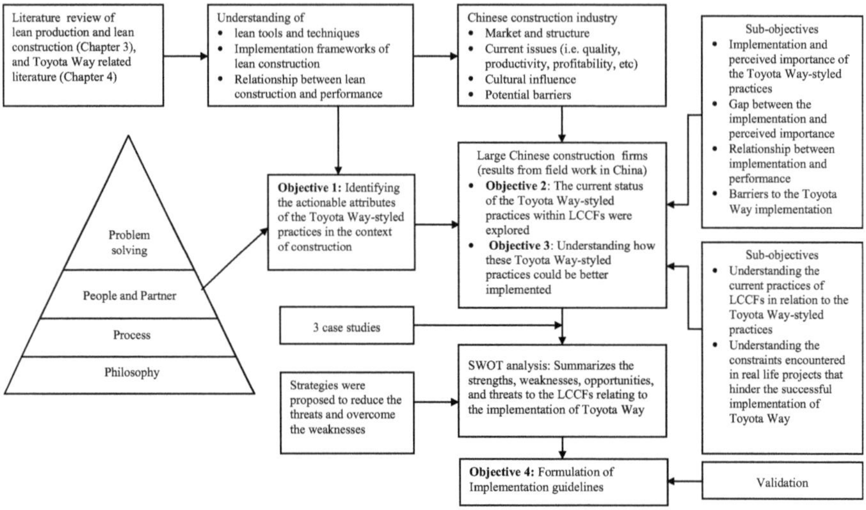

Fig. 12.1 Flow chart showing the realization of how the aim and objectives

1992), and the models and frameworks it employs (Forza 1996; Karlsson and Åhlström 1996; Paez et al. 2005; Sanchez and Perez 2001; Shah and Ward 2007), along with the evolution of management theories (Mullins 2006; Santos et al. 2002a).

Although lean production has received growing recognition and application within manufacturing and other industries, it has been criticized as suffering from shortcomings, including:

1. It has a vague definition (Hines et al. 2004; Jørgensen and Emmitt 2008; Shah and Ward 2007).
2. It exclusively focuses on process-oriented initiatives for short-term benefit (Liker 2004; Emiliani and Stec 2005; Hines et al. 2004; Paez et al. 2004).
3. The implication of lean construction for people in relation to HRM, in construction, is a subject that needs further scrutiny (Green 2002; Green and May 2005; Low and Gao 2011a).
4. It has insufficient details for its implementation guidelines: it was still lacking a generally accepted toolkit that would address all the key Toyota Way principles.

In this regard, neither lean production frameworks nor lean construction models were chosen as the basis of this study, but these can be used instead as references. To contribute to the resolution of the shortcomings highlighted above, and to provide a better understanding of lean in the context of construction, this research acknowledges the comprehensiveness of the Toyota Way model (Liker 2004), and has selected it as the conceptual framework. The Toyota Way model contains four key elements: long-term philosophy, process, people and partners, and problem-solving.

1. The foundational layer is the long-term philosophy, which requires construction firms to be prepared to see the long-term benefits of their actions, rather than being short-sighted.
2. The underlying constructs that have arisen from the two middle layers of the Toyota Way are process (technical) improvement and human resource (social) development. The process improvement represents those attributes (from P2 to P8) that cater for continuous improvement on a project level, while the human resource development construct refers to the aspects of leadership (P9), people management (P10) and the cultivation of a long-term relationship with working partners (P11). Generally, these two middle layers of the Toyota Way model are supported by its foundational layer, the long-term philosophy.
3. The top layer is the problem-solving philosophy, which uses various improvement tools such as *genchi genbutsu* and *kaizen* to achieve continuous improvement. The problem-solving mindset supports what has been discussed in the "process" category of the Toyota Way.

Given the vast differences that exist between the manufacturing and construction industries, each Toyota Way principle within these four broad categories has been translated with care to address its applicability in the construction context (see Chap. 4). Since lean construction practices have been receiving increasing recognition, tools such as the Last Planner System were adopted for linking the Toyota Way process model (see Sect. 4.7.2.1). As a result, Objective 1 was realized with a basket of attributes (91 in total) that were identified in the construction context. The derived attributes later formed the basis of the questionnaire survey.

12.2.2 Assessing Toyota Way-Styled Practices in the Chinese Construction Industry

Objective 2 could only be fulfilled after Objective 1 has been accomplished. Given the problems of poor quality, inefficiency, low profitability, and others, that were haunting the Chinese construction industry, it became apparent that there was a need for a framework that could help the industry achieve improvement. However, the literature relating to lean or the Toyota Way was hardly known in the Chinese construction industry, apart from a few applications in China's manufacturing industry. To fill this knowledge gap, and prior to introducing Toyota Way-styled guidelines to the Chinese construction industry, it was necessary to explore the current state of Toyota Way-styled practices within the Chinese construction firms, as well as the attitudes of Chinese construction professionals towards these Toyota Way attributes. This was what Objective 2 set out to investigate and to achieve. Furthermore, the corresponding last sub-objective was to investigate the relationship between the implementation level of Toyota Way practices and project performance in the Chinese construction industry. All of these needed to be evaluated as far as was possible in a quantitative manner; thus, questionnaire surveys were used. Objective 2 was achieved with the following findings:

In Terms of the Implementation Level

1. Overall, the vast majority of Toyota Way-styled attributes were adopted in some way, but the degree varied. Statistically, of the 91 attributes, 15 attributes were found to be not significantly implemented ($p > 0.05$ or $p > 0.01$), or else insignificant efforts were put into implementing them.
2. Practices under Principle 1 were rated most highly over other principles in terms of the degree of their implementation. Most of these practices were rated over 4, as "moderately implemented".
3. Five of the ten most implemented attributes in the Toyota Way process model were actually from "built-in quality" (Principle 5), which rendered it the most implemented principle among the process-focused principles.
4. The least amount of attention was paid to visual management practices, followed by standardization practices, in the process model.
5. People management-related practices (P10) were notably poorly practised n the Toyota Way people and partner model.
6. There was a much higher level of *genchi genbutsu* practices than of the remaining principles in the problem-solving model.

In Terms of Perceived Importance

1. The respondents from the LCCFs rated the practices as at least "moderately important", with many considering this to be "highly important".
2. In general, the perceived importance of the Toyota Way-styled attributes was rated higher than the extent to which they were implemented.

Gap Between the Implementation and Perceived Importance

1. Statistically significant differences were found between the actual implementation level and the corresponding perceived importance.
2. This implies that the responding firms were aware of the importance of lean or Toyota Way practices, but were not yet fully ready to implement them.

Relationship Between Implementation and Performance

1. All correlation coefficients of the relationships between implementation level and the various performance indicators were positive.
2. 6 out of a possible 70 correlation coefficients between the Toyota Way principles and the performance measures were above 0.5.
3. The Pull *kanban* system (P3) was found to be insignificantly correlated with client satisfaction ($r = .189, p = 0.68$), while *genchi genbutsu* (P12) exhibited an insignificant association with profitability ($r = .15, p = 0.152$).
4. It can be concluded that the introduction of the Toyota Way model to the Chinese construction industry has the potential to contribute to client satisfaction, to reduce costs, to shorten the construction schedule, to improve quality, and yield other benefits.

12.2.2.1 Conclusion

A number of researchers, including Liker (2004), Hines et al. (2004), Paez et al. (2004), among others, have argued that when lean or Toyota Way approaches were adopted as a "popular" recipe by most manufacturers, it was viewed as sort of an "operations improvement" technique—as tools whose focus was exclusively on quick performance improvement. However, in the Chinese context, the philosophy element was most important, while the process-oriented initiatives were among the least important and most poorly implemented attributes for most responding firms.

12.2.3 Understanding the Challenges in the Implementation of the Toyota Way

Following Objective 2, some unresolved issues still remain. For example, the causes for the low level of implementation of some of the Toyota Way-styled attributes in the Chinese construction industry remained unanswered. This became the Objective 3 of this study: to identify the potential factors that could hinder the implementation of the Toyota Way within Chinese construction firms. This was motivated by the fact that understanding the challenges and using appropriate strategies to overcome them should increase the chances of the Toyota Way being successfully implemented in these firms. Objective 3 was met using the mixed-method research approaches (questionnaire survey and interviews) with Chinese building professionals. For a start, some 20 hindrances, having been identified in the literature, were assessed using a Likert scale (see Sect. 8.7). It was discovered that in Chinese construction firms, the most significant barriers that can be ascertained from the ranking results were "lack of a long-term philosophy", "absence of a 'lean' culture in the organization" and "multilayer subcontracting". This phase was followed by semi-structured interviews with the LCCFs (see Chap. 9) to further investigate two sub-objectives, namely:

1. To understand the current practices of LCCFs in relation to the Toyota Way-styled practices, i.e. how would this lead to opportunities for the Toyota Way principles to be introduced?
2. To investigate the constraints present in real-life projects that hinder the successful implementation of Toyota Way principles.

The results from the interviews highlighted that, compared to Toyota Way-styled practices, the gap was enormous, and the interviewees faced considerable challenges in their attempts to adopt Toyota Way practices.

12.2.4 Toyota Way Implementation Framework and Guidelines

The final objective was to establish guidelines for the implementation of the Toyota Way for the LCCFs. This final objective contains three sub-objectives (see Sect. 1.3). Based on a literature review of the state of the art of lean production and lean construction, the conceptual framework of the Toyota Way was developed, assessed, and validated through mixed quantitative and qualitative inquiries in the Chinese construction industry. To fulfil the three sub-objectives, the following steps were taken:

1. Firstly, SWOT analysis was used to summarize the strengths, weaknesses, opportunities and threats to the LCCFs in terms of the implementation of the Toyota Way (see Sect. 11.3). Management would then be in a better position to develop plans for Toyota Way implementation by focusing on weak areas, and thus increasing the likelihood of success in the implementation of the Toyota Way.
2. Secondly, in order to narrow the gaps with Toyota Way-styled practices and to reduce the threats, a set of strategies were proposed and validated by a group of participants. These strategies were aimed at helping the LCCFs obtain the potential to move towards better, more effective, and more efficient implementations of the Toyota Way.
3. Subsequently, all of these contributed to the formulation of guidelines for the Toyota Way implementation framework (see Tables 11.6–11.9). Similarly, for some of the principles which were less implemented because of constraints, necessary modifications are recommended to better suit the Chinese construction context. Such a framework would highlight the various aspects affecting the Toyota way implementation. It would also assist firms in identifying their needs and the current status of the various key factors that affect the Toyota Way implementation. This provides management with effective guidance that contributes to meeting their business objectives.

12.3 Summary of Hypotheses

The research hypotheses set out in Sect. 1.5 are reviewed and summarized in this section. The first two hypotheses relate to the implementation level and perceived importance of Toyota Way principles and attributes among Chinese building professionals. Of the 91 attributes identified (derived from 14 Toyota Way principles), only 15 attributes were found to be not significantly implemented, or implemented with insignificant effort ($p > .05$ or $p > .01$) (see Table 8.8). In other words, the results of these 15 attributes did not support the first hypothesis. Overall, the first hypothesis (H_1) was partially accepted (see Appendix F). With respect to the importance level, the *t*-test results show that, in all instances, the significance level (*p*-value) is <0.05, suggesting that the second hypothesis (H_2) is supported.

Hypothesis 3 predicts the differences between the extent to which respondents perceived Toyota Way attributes to be important and the extent to which they have implemented Toyota Way attributes (see Tables 8.4–8.7). In all instances (except P12.4), the mean scores were higher for the importance scale than for implementation scale. In addition, the Wilcoxon Signed-Rank two-tailed significance levels showed that all (except P12.4 and P1.8) are <0.05, suggesting that there is a statistically significant difference between perceived importance and implementation level for all ranges of the Toyota Way attributes (expect P12.4 and P1.8).

For the two hypotheses concerning the barriers associated with the Toyota Way implementation, as shown in Table 8.11, all the hindrances have a *p*-value <0.05, meaning that $H_{4.1}$ was fully supported, and that there are hindrance when Chinese construction firms implement Toyota Way principles. This refers to a list of 22 hindrances, ranging from lack of a long-term philosophy, through the absence of a lean culture in the organization, to a lack of support from the government. In addition, for hypothesis $H_{4.2}$ to predict whether there are insignificant differences in the mean score on the perceived importance scale for hindrance to Toyota Way implementation between "premier" and "first-grade" Chinese construction firms, the results revealed that this hypothesis was partially supported, given that the perceptions of H5, H14 and H18 were statistically found to be significantly different between the two groups.

For the final hypothesis, the results showed that the level of Toyota Way implementation has a direct correlation with various performance measurements. In other words, the more comprehensive the implementation undertaken, the greater the advantage gained by large Chinese construction firms. There are two exceptions, namely the pull *kanban* system (P3) and client satisfaction, and *genchi genbutsu* (P12) and profitability. Both exhibited insignificant correlation coefficients ($p > 0.05$). This suggests that a company can still improve its performance by focusing on implementing different Toyota Way principles.

12.4 Contributions to Knowledge and Practice

This study makes a number of contributions to knowledge and practice, and particularly to an improved understanding of Toyota Way-styled practices in the context of construction. The contributions listed in the following subsections constitute an original contribution to existing knowledge and to practice.

12.4.1 Contributions to Knowledge

Firstly, this study contributes to the knowledge of lean construction by developing and successfully testing a conceptual framework of the Toyota Way model that embraces various dimensions. This conceptual framework was developed through a synthesis of different literature sources with lean production principles, lean construction practices, tools, and others. Conceptually, the Toyota Way model has its roots in socio-technical systems, given that the model is supported by two pillars—respect for people and continuous improvement. This explains why organizations struggled in their efforts to implement lean or Toyota Way because they failed to understand and appreciate both the social and technical aspects of Toyota Way deployment. Furthermore, this study contributes to a better understanding of the Toyota Way model by linking appropriate theories from the domains of production and management to each sub-model of the Toyota Way (see Chaps. 2 and 5). This includes the following:

1. Drucker's (1994) "theory" of the business was linked to the first principle of the Toyota Way model. The "theory" of the business was built upon an assumptive framework (Drucker 1994), including the assumptions about the environment of the organization, specific mission of the organization, and core competencies of the organization. This, combined with the long-term philosophy of the Toyota Way (Liker 2004) is able to better guide and nurture the organization's activities. Consequently, this goes beyond making profits or gaining market share, but aims to create value for customers, employees, suppliers and the society (see Sect. 5.2).
2. Koskela's (1992, 2000) flow model relating to the TFV production paradigm was linked with the tactical and operational aspects of the Toyota Way—the process sub-model. Arguably, Koskela's (1992, 2000) flow model and its heuristic principles were mainly focused on the reduction of non-value-adding activities by following a set of actions or steps, along with using lean tools and techniques. This study contributes to a better understanding of the flow model by highlighting the significant role of people in process improvement as people are the implementers who identify wastes in the work processes. Consequently, people need to be trained and empowered for this purpose.
3. For the Toyota Way's People and Partners model, the servant leadership theory and a portfolio of motivation theories under the Human Relations paradigm were linked. These were also tested in the Chinese construction industry where it was found that the Chinese building professionals who are in leadership positions do exhibit some of the good qualities and abilities that are in line with the Toyota Way servant leadership features. The functional features of servant leadership such as service to others, appreciation of others and empowerment is essential for the successful implementation of the Toyota Way. Moreover, this study showed that the perception of frontline workers in China's construction industry predominately reflects Theory X of McGregor's (1960) XY theory, as against Theory Y being a precondition in the design of learning organizations such as Toyota. This suggests that the application of Theory Y, along with other motivation theories (e.g. Skinner 1948; Locke 1968) should be considered at this point to develop people and teams when the Chinese construction firms consider Toyota Way implementation.
4. Toyota Way's Problem-solving model: in this sub-model, Deming's (1986) PDCA was linked as this is the heart of the problem-solving approach in Toyota (see Sect. 5.5.1). This study contributes to knowledge of the problem-solving approach with *kaizen* thinking in a project environment.

Thereafter, the Toyota Way model was operationalized in the context of construction and was found to have common grounds with Koskela's (2000) TFV perspective on the construction process (see Sect. 5.3), as well as with the Last Planner System (see Sect. 4.7.2). This framework forms a foundation for research into lean construction and has proven to be helpful in enabling researchers to understand lean in construction through a different perspective. Implementing lean thinking (Egan 1998) in construction is one way, while the Toyota Way model can be seen as an alternative way, for the reasons below:

1. A comparison of several existing lean construction frameworks (Gao and Low 2014) has highlighted that the current lean construction frameworks focus specifically, though not exclusively, on lean practices (tools and techniques) on the shop floor.
2. This study returns to the origin from which the concepts of lean and later of lean construction were derived—the Toyota Way. Using the mother platform of lean eliminates many possibilities of missing out on certain relevant points. For example, the social implications of lean have been ignored in real-life applications (see Green 2002). The choice of the Toyota Way model for construction firms takes full recognition of both the technical and social views, which are currently lacking from lean implementation. It also alerts construction practitioners to the importance of complementing the operational factors with the human factors of the Toyota Way (or of lean approaches) in the construction workplace.
3. The operational measure of the Toyota Way model appears to be more comprehensive than other measures observed in the literature, particularly for the construction industry. Credit should be given to Liker's (2004) Toyota Way model, which broadly considered both technical and social dimensions.

Secondly, this study echoes concerns (Green 2002; Green and May 2005) that a majority of the lean construction frameworks do not consider the implications of lean construction for HRM. As Green (2002, p. 151) put it, "*a considerable body of research that equates the implementation of lean production to regressive policies of human resource management is strangely ignored by lean construction researchers*". This study does consider how such policies would affect the development of human resources (see Berggren 1993; Garrahan and Stewart 1992; Williams et al. 1992), and then follows Toyota's guiding principle—to "respect people"—to develop best practices and strategies for leaders, employees and partners. This contribution is in agreement with Shah and Ward (2007), Liker (2004), and others when they stated that lean or Toyota Way is not a singular concept, and it cannot be equated solely to waste elimination or continuous improvement.

12.4.2 Contributions to Practice

This research is expected to yield practical significance as detailed below.

12.4.2.1 The Development of the Toyota Way Model in Construction

Koskela (1992) called for research that would take the initiative in overcoming barriers and interpreting lean in the construction industry. The primary practical contribution made here is that the interpretation of the Toyota Way in the construction context has been completed. The Toyota Way model was chosen and has been reviewed and evaluated, along with the lean construction tools and techniques, resulting in a checklist of Toyota Way-styled attributes that can easily be used to evaluate the state of a construction company's Toyota Way implementation.

This approach provides construction firms with a framework for their philosophy, for redesigning their process operations, and for focus on human resource issues, as well as on Toyota Way problem-solving practices. Companies in the construction industry that are interested in working with lean construction or Toyota Way-styled practices in a systematic way, but do not know where o how to start, can follow these guidelines based on the discussions and implications of this study.

12.4.2.2 Reporting on Lean in the Chinese Construction Context

The lean philosophy has begun to emerge in construction, although it is still in its infancy. However, the discussion of lean construction practices in developing countries, such as China, is hardly seen in the English-language literature. In filling this gap, the present study may be the first in the lean construction domain to establish the Toyota Way—a model from the manufacturing context—for construction firms in China. To the best knowledge of the researcher, this study is also the first to assess the level of Toyota Way implementation in the Chinese construction industry. For instance, the scale developed here may be used by managers to self-evaluate their progress in implementing Toyota Way-styled practices. The Toyota Way implementation framework developed in this study provides specific, actionable items that can be used in practice to further restructure LCCFs and to provide useful guidelines for solving the current quality, productivity, and health and safety issues of the Chinese construction industry. They can also serve to meet the threat of foreign competition, now that China is a member of the WTO, and can guide China's aggressive entry into the international market.

Moreover, this study has also shed light on the challenges and barriers faced by the LCCFs in their day-to-day operations. It has also identified strategies and proposed guidelines that can be used to facilitate the implementation of the Toyota Way-styled practices for these firms. The findings here will contribute to a better understanding of the applicability of the Toyota Way model in the construction industry and will provide guidelines for its implementation. Finally, during the fieldwork—and especially in the interview stages and case studies—this study created certain level of awareness among the LCCFs about the practices of the Toyota Way and its application within their organizations.

12.5 Limitations

Limitations are inevitable in research. The present study has taken on a challenging task in considering the breadth of the Toyota Way as its subject area. Other challenges included the time constraints on the research and the difficulties that occurred during data collection in China. The following limitations can be identified:

1. *Research participants*: The research was primarily focused on the LCCFs, on the assumption that (1) they may possess sufficient resources to better carry out Toyota Way implementation; and (2) the leadership of these LCCFs may be more open-minded to this new management philosophy. The limitation here is

that small and medium-sized enterprises (SMEs) in China's construction industry were excluded from this research. These SMEs should be taken seriously. Although they are small in terms of their business turnover, they are actively involved with undertaking substantial works that are subcontracted from the LCCFs' projects. In other words, they play a critical role in the construction process, the improvement of which is closely relevant to the Toyota Way process model and its various process-oriented initiatives.

2. *More stakeholders should be engaged*: The second limitation emerged when the research progressed further into the fieldwork stage. Apart from the construction firms, it would also be meaningful to consult the opinions of other project stakeholders, including clients, Design Institutes (DI), supervision firms (*jianli*), suppliers, and other stakeholders in the Chinese construction industry. This would make the data more reliable, given that it would originate from different stakeholders with different interests in the projects. For example, the client has been identified as a critical factor from the perspective of a general contractor (see SWOT analysis), and that would affect the implementation of the Toyota Way. However, there is nowhere to validate alternative perspectives if others are also not included in the research. In addition, this points to a promising direction for future research: to involve more stakeholders in the research process.
3. *No one-size-fits-all guidelines*: a third limitation lies in the assumption that there are "one-size-fits-all" guidelines for individual LCCFs. Given that LCCFs in China vary in terms of the operational strategies, aims, and environments that govern their application of the Toyota Way implementation guidelines, informed judgment must be exercised in determining the suitability of any specific guidelines proposed. This is supported by the contingency approach, which argues that there is no one best way to manage an organization. As shown in the case studies and questionnaire surveys conducted in this research, different companies are in different stages of implementing the Toyota Way. In addition, their motivations and their perceived barriers to implementation also appear to be distinct. Nevertheless, it is hoped that this research will help LCCFs to tailor their initiatives to meet their implementation needs and requirements.
4. *Nature of the measurement*: mean values were computed in the survey stage to assess the state of implementation of the Toyota Way attributes. This measurement, using the Likert scale, is ordinal in nature. Several researchers in the domain of construction management have suggested that ordinal data are more appropriately analysed using procedures that require rank-order information, such as non-parametric methods (see Fellows and Liu 2008; Naoum 2007). It is worth mentioning that the use of a questionnaire survey was intended to partly achieve this aim. It should be borne in mind that mean values are nevertheless adopted throughout to give an overall picture of the status quo of the LCCFs in implementing the Toyota Way-styled practices. To overcome this limitation, a mixed research method was employed, including using face-to-face interviews for further justification.

12.6 Recommendations

The recommendations represented in this section are divided into three areas: (1) LCCFs, (2) the Chinese construction industry and (3) future research.

12.6.1 Recommendations for Large Chinese Construction Firms

12.6.1.1 Systematic Implementation

The LCCFs need to implement Toyota Way-styled practices, and in a holistic manner. Systematic implementation has two meanings here. Firstly, the Toyota Way is not simply about copying a set of tools (Liker 2004; Liker and Meier 2006) and using them in the same way as Toyota does (Lander and Liker 2008), i.e. to remove waste from processes. What should be remembered is that the Toyota Way developed the process-oriented initiatives within the context of a wider socio-technical system, and without that, the process-oriented initiatives will fail or at least underperform. In other words, apart from implementing improvement initiatives derived from the Toyota Way Process model, it is also important to establish a better people system, by developing leadership, truly valuing their employees and cultivating long-term relationships with partners. The pertinent principles and strategies formulated to achieve better people systems have been discussed earlier (see Sect. 11.3). Secondly, within an individual layer, the principles are considered by the researcher to be intrinsically connected—especially in the case of the process model, which embraces a number of principles. For example, the visual management principle is closely associated with principles of pull (*kanban*), built-in quality (*andon* system), etc. These indicate the interrelationships between different process-related principles. Again, for the LCCFs, the point is not to reach for the most applicable or handy tool, but to understand the principles behind the tools and the linkages between the improvement initiatives, and to apply these in a holistic manner. In the event that the Chinese construction firms may not be fully aware of these principles coming together as a whole-system strategy, it is more important to raise the awareness of lean or Toyota Way principles by emphasizing an understanding of the true purpose of each of these principles.

12.6.1.2 Mindset Change

A change in mindset is needed before or when LCCFs embark on the implementation of the Toyota Way practices. This is because a fundamental part of any lean strategy, including the Toyota Way implementation, must stem from a change of mindset and of organizational culture (Liker and Meier 2006). Accordingly, this study found that the current attitudes of Chinese building professionals to inventory management, workers, empowerment, relationships with partners, quality problems, decision-making, etc. (see Chap. 9) would potentially cause problems when attempting to implement the Toyota Way, and have therefore been identified as potential barriers (see Sect. 8.7). The mindset change has been elaborated in among the strategies proposed for LCCFs in various areas (see Sect. 11.3). Compared to the implementation of Toyota Way tools and technique or other improvement initiatives, which are fairly quickly and easily understood, the mindset of people or the related cultural issues are less tangible and require change for implementing the Toyota Way. As Pun (2001, p. 330) observed, "*the alternation of beliefs, assumptions and values that define the behavioral norms and expectations that determine corporate culture is both a difficult and long-term undertaking*". Therefore, those at the very top of the organization have a key role to play here. They must intervene and require people to behave differently, starting with changing their conventional mindset, and allowing them to experience a better set of results as the Toyota Way promises. As this process is repeated, a different set of beliefs and value—a new organizational culture—will eventually evolve. This, however, requires a long-term endeavour and commitment.

12.6.1.3 Understanding Their SWOT

Every LCCF is different and will need to apply a different approach to develop the Toyota Way-styled practices to suit its needs and fill the gaps in the desired Toyota Way implementation. A key to success for any firms starting on the pathway to Toyota Way implementation is planning, and successful planning requires that the people involved have an understanding of their current status and external conditions. SWOT is useful in addressing these needs. However, prior to that, the checklist containing the Toyota Way attributes in the context of construction is a good starting point for understanding the strengths and weaknesses of a firm, as well as the opportunities and threats it faces. This can be determined readily by referring to a series of Tables containing the Toyota Way attributes (see Sect. 11.4), as discussed earlier in Chap. 11. Information discovered during the assessment can provide the foundation for a SWOT analysis. For example, if the assessment, using the checklist of Toyota Way attributes, highlights that attributes such as P5.1 "employees are dedicated to providing built-in quality to every aspect of operations" is given a relatively low score for implementation (i.e. below average), this may suggest that lack of dedication in built-in quality is likely to be identified as a weakness in the SWOT analysis. Hence, corresponding strategies should be formulated to tackle the weaknesses identified, while at the same time maintaining its strengths as these are rated highly. Moreover, it is also important to keep in mind that Toyota Way implementation is not a one time, one off event. Therefore, organizations are recommended to perform the self-assessment and followed-up SWOT analysis on a regular basis to see the improvements they have achieved.

12.6.2 Recommendations for the Chinese Construction Industry

Although this study did not involve stakeholders such as clients, suppliers, or subcontractors as research participants, there are recommendations here for them, as well as for the government.

12.6.2.1 Clients

Clients play a key role in the implementation process. Unlike the government, the SWOT analysis indicates that clients can play a role opposite to that of a facilitator, and can constitute a substantial threat to the general contractors, for example by setting unreasonable deadlines, having traditional mindsets, and others. For these particular threats, the recommendation can include ensuring that clients also understand the Toyota Way principles and the possible benefits that can be derived from them. In addition, it is necessary to change their perceptions on site management, and not to let their traditional perceptions hinder the proper management of projects. It worth mentioning that changing clients' behaviour and their perceptions is not an easy task, given that clients are the ones who finance the project. Moreover, satisfying the client's needs is a common project culture in China's construction industry (Zuo et al. 2009), rather than challenging and educating them. A good strategy is to highlight the benefits claimed from a part of the project, in which Toyota Way principles have been applied.

12.6.2.2 Suppliers and Subcontractors

These partners play crucial roles in assisting the implementation of the Toyota Way by the general contractors. This is because the general contractor, as a team, is present on site mainly in the form of a project manager, while a large portion of the work is subcontracted—and these subcontractors' workers are not part of the general contractor's internal workforce. In this regard, there is a requirement that partners must proactively cooperate and collaborate with the general contractor in implementing the Toyota Way, especially for those who are also directly involved with (such as JIT delivery, collaborative planning, and so on). The ultimate goal is to achieve what Liker (2004) has described as the real extension of the enterprise.

12.6.2.3 Government

The recommendation to the Chinese construction industry, and for that matter to the government, is to be more committed to supporting and promoting the concept of lean and Toyota Way-styled practices, and to learning from the best practices outside of construction. In addition, as far as the earlier SWOT analysis is concerned, many opportunities are linked to the role of the government (e.g. quality auditing, visual management and adoption of reliable technology). Since most LCCFs are SOEs and thus have close relationships with the government in terms of their policies and strategies, the government needs to also work correspondingly on these aspects—for example, to facilitate and mobilize the physical environment in which Toyota Way-styled practices can be better implemented. This includes prefabrication for standardization, design as well as build for lean delivery, migration schools for more skilled workers, and other initiatives.

12.6.3 Future Work

Although this study is a meaningful starting point to garner a better understanding of the Toyota Way model in the Chinese construction industry, several issues remain to be resolved for this research. In view of the research findings and the conclusions, the following areas are recommended for future research.

1. It would be beneficial to perform a longitudinal analysis of the impact of Toyota Way principles on company and project performance. In particular, the causal relationships of the elements within the Toyota Way model, presented in this study, are to some extent undetermined (although correlations have been considered). As a result, all the correlation coefficients are positive and therefore all the elements within the Toyota Way are presented and recommended. This may look comprehensive, but can result in a lack of focus or priority, given the long checklist provided. Future work should focus on the elements which have priority, and which are closely related to a specific type of performance.
2. As the entire Toyota Way model is constructed from a socio-technical point of view (apart from the performance indicators like quality, productivity and profitability), future research should also include measurements of performance pertaining to people and organizational aspects (see Respect for Pe ople Working Group 2004), such as personnel morale, employee satisfaction, staff turnover and others.
3. Future work on improving the Toyota Way implementation guidelines in the context of construction should include an evaluation of the effectiveness of the set of guidelines itself on a periodic basis in order to modify these guidelines according to time-related changes within the construction industry.
4. Research should be carried out on the framework relative to the current tools available for the application of project management. This is to provide more "practical" applications, and also to improve the ease of deployment of the framework.
5. Further research should also be conducted in other categories of the construction industry, such as large infrastructure work, "routine" building projects (Winch 2010), or refurbishment projects (Bryde and Schulmeister 2012)—not least for validation and further generalization. It is expected that the model's characteristics will need to be updated to adapt to different subsectors as the operational environment changes. The comparisons these provide would give greater confidence to the analyses presented and guidelines proposed.

12.7 Closing Remarks

Initially applied to the automotive industry, this study confirms the applicability of the Toyota Way to construction firms in China. This research has made significant contributions to lean construction by identifying key components of the Toyota Way model in the context of construction. In the field to which the research belongs, interests in this subject have led to numerous models, and the application within construction has been documented. This study has addressed several related issues with the aim to enable better implementation of the Toyota Way in the construction industry, such as in the integration of socio-technical views, addressing implications for human resource management, and so on. The study takes the Chinese construction industry as an example, and shows the state of, and constraints deriving from, the implementation of Toyota Way practices within large Chinese construction firms. Implementing the Toyota Way requires a change in both thinking and culture, as well as changes in processes. Management must be committed to the change and be willing to train everyone to match the new culture, to ensure that the organization's culture fosters employee participation, and to monitor the results over time. Most importantly, endeavours have been made to understand the SWOT factors relevant to construction firms prior to their implementation of the Toyota Way. This results in the formulation of strategies for better implementation within the Chinese construction industry. The future of the Toyota Way is bright, but there remains a long road ahead for implementation in the construction industry.

Appendix A: Questionnaire Survey (English Version)

About the Survey

Learning from Toyota and its underlying principles is a novel undertaking in the construction industry. Establishing an implementation framework of the Toyota Way model for the Chinese construction firms is important as this management philosophy has the potential to help solve the problems which plague the Chinese construction industry. The aim of this survey is to investigate the extent/importance which the large Chinese construction firms have attributed to the Toyota Way principles which can be summarized into four categories namely Philosophy, Process, People and Partner, and Problem-solving. Please be assured that the information you give will be kept strictly confidential and will be used for academic purpose only. Thank you for your participation!

Part 1: Introduction to the Toyota Way Principles

1. Base your management decisions on long-term philosophy, even at the expense of short-term financial goals.
2. Create continuous process flow to bring problems to the surface.
3. Use "pull" systems to avoid overproduction.
4. Level out the workload (*Heijunka*).
5. Build a culture of stopping to fix problems, to get quality right the first time all the time.
6. Standardized tasks are the foundation for continuous improvement and employee empowerment.
7. Use visual control so no problems are hidden.
8. Use only reliable, thoroughly tested technology that serves your people and processes.
9. Grow leaders who thoroughly understand the work, live the philosophy and teach it to others.

S. Gao and S.P. Low, *Lean Construction Management*,
DOI 10.1007/978-981-287-014-8, © Springer Science+Business Media Singapore 2014

10. Develop exceptional people and teams who follow your company's philosophy.
11. Respect your extended network of partners and suppliers by challenging them and helping them improve.
12. Go and see for yourself to thoroughly understand the situation (*Genchi Genbutsu*).
13. Make decisions slowly by consensus, thoroughly considering all options; implement decisions rapidly.
14. Become a learning organization through relentless reflection and continuous improvement.

Part 2: General Information

Respondent's name: (optional)
Company name:
Company size: number of employees
Registered capital:
Position held:
Area of expertise:
Highest professional qualification:
Years of industrial experience:
Contact: (Email)
(Telephone)

Company's headquarter:
Service provided:
☐General Contractor ☐Professional Contractor
☐Labour Subcontractor
Company type:
☐State-owned ☐Collective-owned ☐Private
Qualification grade of company:
☐Premier ☐ First ☐ Second ☐ Third and below
Engaged in overseas projects: ☐ Yes ☐ No
ISO 9001/9002 certified: ☐ Yes ☐ No

Part 3: The Toyota Way Principles in China's Construction Firms

1. Please rate to what extent (using a scale of 1–5) the following management philosophy/activities are adopted in your company (1 = not at all; 2 = very little; 3 = somewhat; 4 = moderately; 5 = to a large extent).
2. Please rate how important do you consider the following management philosophy/practices to be in contributing towards organization's performance (1 = not important; 2 = less important; 3 = neutral; 4 = important; 5 = very important).

3.1 Philosophy (Toyota Way Principle 1)

P1: Long term philosophy	Extent of practice					Level of importance				
	1	2	3	4	5	1	2	3	4	5
(1) Sustain a constant purpose (company vision, mission and values)	☐	☐	☐	☐	☐	☐	☐	☐	☐	☐
(2) Have a high purpose or mission which is to generate value towards employees, society and customers	☐	☐	☐	☐	☐	☐	☐	☐	☐	☐
(3) Formulate a plan towards the realization of company's long-term vision	☐	☐	☐	☐	☐	☐	☐	☐	☐	☐
(4) Short-term losses affect decision making, but are less important than pursuing long-term goals	☐	☐	☐	☐	☐	☐	☐	☐	☐	☐
(5) Have a clear view of its core competency and endeavour to become an expert in this area	☐	☐	☐	☐	☐	☐	☐	☐	☐	☐
(6) Be responsible for products, employees and society	☐	☐	☐	☐	☐	☐	☐	☐	☐	☐
(7) Understanding customer's requirement is priority work	☐	☐	☐	☐	☐	☐	☐	☐	☐	☐
(8) Be able to rapid response to meet the changing requirement of the customers (e.g. design change)	☐	☐	☐	☐	☐	☐	☐	☐	☐	☐
(9) Treat employees/suppliers as internal customers	☐	☐	☐	☐	☐	☐	☐	☐	☐	☐

3.2 Process (Toyota Way Principles 2–8)

P2: One-piece flow	Extent of practice					Level of importance				
	1	2	3	4	5	1	2	3	4	5
(1) Employee is concerned with waste[1] elimination in their daily work	☐	☐	☐	☐	☐	☐	☐	☐	☐	☐
(2) Material flow is adhered to consistently throughout the daily work activities	☐	☐	☐	☐	☐	☐	☐	☐	☐	☐
(3) Material, equipment, and other resources are provided in a "just-in-time" manner when needed	☐	☐	☐	☐	☐	☐	☐	☐	☐	☐
(4) Site layout is organized to enhance material flow, employee movement, etc to minimize wastes due to movement, motion, travel, etc.	☐	☐	☐	☐	☐	☐	☐	☐	☐	☐
(5) Strive to cut back to zero the amount of time any work is sitting idle or waiting for someone to work on it	☐	☐	☐	☐	☐	☐	☐	☐	☐	☐
(6) Make flow evident through organizational culture	☐	☐	☐	☐	☐	☐	☐	☐	☐	☐

[1]Waste can mean rework, poor quality, defects, waiting, etc.

P3: Use "pull" system	Extent of practice					Level of importance				
	1	2	3	4	5	1	2	3	4	5
(1) Materials are ordered as close as possible to exact needs	☐	☐	☐	☐	☐	☐	☐	☐	☐	☐
(2) Strive for possible low level of (even stockless) material inventory in construction site	☐	☐	☐	☐	☐	☐	☐	☐	☐	☐
(3) Use simple signals – cards, empty bins, etc to monitor the level of inventory and to order the needed material/component	☐	☐	☐	☐	☐	☐	☐	☐	☐	☐
(4) Monitor the quantity of material/component/equipment that the teams actually take away	☐	☐	☐	☐	☐	☐	☐	☐	☐	☐
(5) Clear job contents, work time, material requirements, among other information are prepared before releasing a work task to a crew	☐	☐	☐	☐	☐	☐	☐	☐	☐	☐

P4: Level out the workload (*Heijunka*)	Extent of practice					Level of importance				
	1	2	3	4	5	1	2	3	4	5
(1) Foremen (Last Planners) make commitments on what the crews will do each week based on what is ready to be done	☐	☐	☐	☐	☐	☐	☐	☐	☐	☐
(2) Project manager plans the work with input from other parties including subcontractors, clients, suppliers, etc.	☐	☐	☐	☐	☐	☐	☐	☐	☐	☐
(3) Daily work activities are planned to balance material availability, manpower, machine availability, and workload between operations	☐	☐	☐	☐	☐	☐	☐	☐	☐	☐
(4) Weekly/Daily work assignments are completed in accordance with the weekly/daily schedule	☐	☐	☐	☐	☐	☐	☐	☐	☐	☐
(5) Levelling the daily work activities without overburdening workers and machinery	☐	☐	☐	☐	☐	☐	☐	☐	☐	☐

P5: Build a culture of stopping to fix problems	Extent of practice					Level of importance				
	1	2	3	4	5	1	2	3	4	5
(1) Employees are dedicated to provide "built-in" quality into every aspect of operations	☐	☐	☐	☐	☐	☐	☐	☐	☐	☐
(2) Preventing defective or "no inspection" assignments from entering the next process	☐	☐	☐	☐	☐	☐	☐	☐	☐	☐
(3) Rejecting defective materials, components and equipment	☐	☐	☐	☐	☐	☐	☐	☐	☐	☐
(4) Employees are encouraged to seek support from their supervisors when something goes wrong at work	☐	☐	☐	☐	☐	☐	☐	☐	☐	☐
(5) Employees are empowered to be responsible for quality	☐	☐	☐	☐	☐	☐	☐	☐	☐	☐
(6) Employees who work in the same team meet on a regular basis to discuss quality problems and lessons learned	☐	☐	☐	☐	☐	☐	☐	☐	☐	☐
(7) Feedback about quality is routinely given by the employees	☐	☐	☐	☐	☐	☐	☐	☐	☐	☐

P6: Standardized tasks	Extent of practice					Level of importance				
	1	2	3	4	5	1	2	3	4	5
(1) Established standard operating procedures (SOPs) (e.g. work processes) are practised by employees for each major operation/process	☐	☐	☐	☐	☐	☐	☐	☐	☐	☐
(2) Employees play a key role in creating the SOPs	☐	☐	☐	☐	☐	☐	☐	☐	☐	☐
(3) Employees are encouraged to improve the existing SOPs based on their own practical experience	☐	☐	☐	☐	☐	☐	☐	☐	☐	☐
(4) Incorporate employee's creative improvement of the standard into new SOPs	☐	☐	☐	☐	☐	☐	☐	☐	☐	☐
(5) Using standardized prefabricated components from offsite shops	☐	☐	☐	☐	☐	☐	☐	☐	☐	☐

P7: Visual control	Extent of practice					Level of importance				
	1	2	3	4	5	1	2	3	4	5
(1) Visual aids are adopted to make wastes, problems, and abnormal conditions readily apparent to employees.	☐	☐	☐	☐	☐	☐	☐	☐	☐	☐
(2) The posted information in terms of job status, schedule, quality, safety, etc is in place that most workers can see it on a daily basis, and it is up-to-date	☐	☐	☐	☐	☐	☐	☐	☐	☐	☐
(3) Appropriate signages are used to identify layouts, traffic, safety concerns, etc.	☐	☐	☐	☐	☐	☐	☐	☐	☐	☐
(4) The construction site is kept clean at all times	☐	☐	☐	☐	☐	☐	☐	☐	☐	☐
(5) Employees take pride in keeping the construction site organized and clean.	☐	☐	☐	☐	☐	☐	☐	☐	☐	☐
(6) The workplace follows the principles of $5\text{-}S^2$ including "sort", "straighten", "shine", "standardize" and "sustain"	☐	☐	☐	☐	☐	☐	☐	☐	☐	☐

Note: $5\text{-}S^2$ comprises a series of activities namely "sort", "straighten", "shine", "standardize" and "sustain" for eliminating wastes.

P8: New technology procurement strategy	Extent of practice					Level of importance				
	1	2	3	4	5	1	2	3	4	5
(1) New technology must support the company's values	☐	☐	☐	☐	☐	☐	☐	☐	☐	☐
(2) New technology must demonstrate its potential to enhance processes	☐	☐	☐	☐	☐	☐	☐	☐	☐	☐
(3) New technology must be specific solution oriented	☐	☐	☐	☐	☐	☐	☐	☐	☐	☐
(4) New technology must be thoroughly tested and proven to provide long-term benefits	☐	☐	☐	☐	☐	☐	☐	☐	☐	☐

3.3 People and Partners (Toyota Way Principles 9–11)

P9: Grow leaders and leadership	Extent of practice					Level of importance				
	1	2	3	4	5	1	2	3	4	5
(1) Leaders are motivated to inspire people to achieve goals	☐	☐	☐	☐	☐	☐	☐	☐	☐	☐
(2) Leaders must have in-depth job knowledge	☐	☐	☐	☐	☐	☐	☐	☐	☐	☐
(3) Leaders possess teaching ability and are able to pass their knowledge on to others	☐	☐	☐	☐	☐	☐	☐	☐	☐	☐
(4) Leaders must support the employees doing their work	☐	☐	☐	☐	☐	☐	☐	☐	☐	☐
(5) Leaders will take time to understand problems and root causes before acting	☐	☐	☐	☐	☐	☐	☐	☐	☐	☐
(6) Leaders strongly encourage employees to develop "continuous improvement" in thinking and action	☐	☐	☐	☐	☐	☐	☐	☐	☐	☐
(7) Leaders must understand the company policy and procedures, and communicate these to their team members	☐	☐	☐	☐	☐	☐	☐	☐	☐	☐

P10: Develop people and promote teamwork	Extent of practice					Level of importance				
	1	2	3	4	5	1	2	3	4	5
(1) Select the best person for a given job	☐	☐	☐	☐	☐	☐	☐	☐	☐	☐
(2) Training is provided to equip the employees with the required skills before they are assigned to work	☐	☐	☐	☐	☐	☐	☐	☐	☐	☐
(3) On-the-job-training is provided to further develop employee's exceptional skills	☐	☐	☐	☐	☐	☐	☐	☐	☐	☐
(4) Employees are cross-trained to perform additional functions	☐	☐	☐	☐	☐	☐	☐	☐	☐	☐
(5) Training materials are standardized	☐	☐	☐	☐	☐	☐	☐	☐	☐	☐
(6) Employees are encouraged to cooperate with others to complete the whole task	☐	☐	☐	☐	☐	☐	☐	☐	☐	☐
(7) Daily work activities are organized into team function	☐	☐	☐	☐	☐	☐	☐	☐	☐	☐
(8) Internal motivation methods (e.g. safe working conditions, good pay, job rotation, etc) are used to motivate employees	☐	☐	☐	☐	☐	☐	☐	☐	☐	☐
(9) External motivation methods (e.g. rewards, commendation, and measurement towards goals) are used to motivate employees	☐	☐	☐	☐	☐	☐	☐	☐	☐	☐

P11: Respect the extended network of partners and suppliers	Extent of practice					Level of importance				
	1	2	3	4	5	1	2	3	4	5
(1) Respect partners' capabilities	☐	☐	☐	☐	☐	☐	☐	☐	☐	☐
(2) Challenge the partners by setting collaborative targets	☐	☐	☐	☐	☐	☐	☐	☐	☐	☐
(3) Take part in partners' production process	☐	☐	☐	☐	☐	☐	☐	☐	☐	☐
(4) Work with the partners to improve project effectiveness	☐	☐	☐	☐	☐	☐	☐	☐	☐	☐

(5) Work with the partners in various areas to develop their technical capabilities	☐	☐	☐	☐	☐	☐	☐	☐	☐	☐
(6) Share information with partners in a structured manner	☐	☐	☐	☐	☐	☐	☐	☐	☐	☐
(7) Conduct joint improvement activities with partners to solve problems	☐	☐	☐	☐	☐	☐	☐	☐	☐	☐
(8) Strive to establish a long-term relationship with reliable partners	☐	☐	☐	☐	☐	☐	☐	☐	☐	☐
(9) Limit the number of suppliers	☐	☐	☐	☐	☐	☐	☐	☐	☐	☐

3.4 Problem-Solving (Toyota Way Principles 12–14)

P12: Practice of *Genchi Genbutsu*[3]	Extent of practice					Level of importance				
	1	2	3	4	5	1	2	3	4	5
(1) Solve problem by going to the places (e.g. construction site) where problems are discovered	☐	☐	☐	☐	☐	☐	☐	☐	☐	☐
(2) Analyzing and thoroughly understand the situation before making decisions	☐	☐	☐	☐	☐	☐	☐	☐	☐	☐
(3) Making decisions based on the verified data	☐	☐	☐	☐	☐	☐	☐	☐	☐	☐
(4) Making decisions based on management team's past experiences	☐	☐	☐	☐	☐	☐	☐	☐	☐	☐
(5) *Genchi Genbutsu* has become part of the company culture	☐	☐	☐	☐	☐	☐	☐	☐	☐	☐

Note[3]: *Genchi Genbutsu* is a Japanese term, which refers to going to the source where value is being created to witness and to understand the real situation.

P13: Decision making strategy	Extent of practice					Level of importance				
	1	2	3	4	5	1	2	3	4	5
(1) Using appropriate problem-solving methodologies (e.g. 5 Whys) to determine the root causes of problems	☐	☐	☐	☐	☐	☐	☐	☐	☐	☐
(2) Possible experiments are conducted to test the potential cause of a problem	☐	☐	☐	☐	☐	☐	☐	☐	☐	☐
(3) Broadly consider alternative solutions	☐	☐	☐	☐	☐	☐	☐	☐	☐	☐
(4) Valuing the process through which the decision was reached	☐	☐	☐	☐	☐	☐	☐	☐	☐	☐
(5) Building consensus within the team, including employees and outside partners	☐	☐	☐	☐	☐	☐	☐	☐	☐	☐
(6) Addressing the root causes of problems via effective communication vehicle	☐	☐	☐	☐	☐	☐	☐	☐	☐	☐

P14: Practice of *Hansei*[4] and *Kaizen*[5]	Extent of practice					Level of importance				
	1	2	3	4	5	1	2	3	4	5
(1) Reflection on mistakes (e.g. defects, rework, safety issues, etc) on a regular basis	☐	☐	☐	☐	☐	☐	☐	☐	☐	☐
(2) Management treats problems as development opportunities for employees	☐	☐	☐	☐	☐	☐	☐	☐	☐	☐
(3) *Kaizen* activities are conducted in your workplace	☐	☐	☐	☐	☐	☐	☐	☐	☐	☐
(4) Management supports the *kaizen* activities	☐	☐	☐	☐	☐	☐	☐	☐	☐	☐
(5) The improvement will be codified into documents and/or policies used by organization	☐	☐	☐	☐	☐	☐	☐	☐	☐	☐
(6) Each hierarchy of the organization develops measurable objectives as well as actions to support the executive-level goals	☐	☐	☐	☐	☐	☐	☐	☐	☐	☐
(7) Managers are keen on measuring the objectives and give feedback	☐	☐	☐	☐	☐	☐	☐	☐	☐	☐
(8) PDCA[6] methodology is used to solve problems	☐	☐	☐	☐	☐	☐	☐	☐	☐	☐

Note: *Hansei* means reflection in Japanese. *Kaizen* is known as continuous improvement in the West.
PDCA means Plan, Do, Check, Act

Part 4: Perspectives on Relationship Between the Toyota Way Principles and Organization's Performance

1. Please indicate the level of your company's operational performance (using a scale of 1–5), as compared with your competitors, described by the following items (1 = very poor; 2 = poor; 3 = average; 4 = good; 5 = excellent).
2. Please rate the extent to which you expect your organization's performance to improve when the Toyota Way principles are fully implemented (1 = no improvement; 2 = little improvement; 3 = some improvement; 4 = good improvement; 5 = very good improvement).

Items	Level of performance					Level of improvement				
	1	2	3	4	5	1	2	3	4	5
(1) Financial performance (profitability)	☐	☐	☐	☐	☐	☐	☐	☐	☐	☐
(2) Productivity	☐	☐	☐	☐	☐	☐	☐	☐	☐	☐
(3) Quality	☐	☐	☐	☐	☐	☐	☐	☐	☐	☐
(4) Eliminate waste on site	☐	☐	☐	☐	☐	☐	☐	☐	☐	☐
(5) Deliver the project on time	☐	☐	☐	☐	☐	☐	☐	☐	☐	☐
(6) Health and safety records	☐	☐	☐	☐	☐	☐	☐	☐	☐	☐
(7) Client satisfaction	☐	☐	☐	☐	☐	☐	☐	☐	☐	☐

3. If the Toyota Way philosophy proves to be a feasible system for improving the performance of construction firms in the above-mentioned areas, would your

organization be willing to implement the Toyota Way philosophy as an investment for a better future?

☐ Yes, we will ☐ we will consider ☐ No, we will not

Part 5: Hindrances of the Toyota Way Principles When Implemented in China's Construction Firms

Please rate to what extent do you consider the following items to be the hindrances that may hamper the implementation of the Toyota Way principles in large China's construction firms (1 = no influence; 2 = little influence; 3 = some influence; 4 = strong influence; 5 = very strong influence).

Items	Level of hindrance				
	1	2	3	4	5
(1) Lack of a long term philosophy	☐	☐	☐	☐	☐
(2) Absence of a "lean" culture in the organization	☐	☐	☐	☐	☐
(3) Limited use of design and build procurement mode	☐	☐	☐	☐	☐
(4) Construction firm's limited involvement in the design stage	☐	☐	☐	☐	☐
(5) Foremen's (last planner) limited involvement in the planning stage	☐	☐	☐	☐	☐
(6) Multi-layers subcontracting	☐	☐	☐	☐	☐
(7) Limited use of off-site construction techniques (e.g. prefabrication)	☐	☐	☐	☐	☐
(8) Lack of project management skills (e.g. leadership skills, problem solving skills, etc.)	☐	☐	☐	☐	☐
(9) Lack of support from the top management	☐	☐	☐	☐	☐
(10)High workforce turnover	☐	☐	☐	☐	☐
(11)Insufficient training	☐	☐	☐	☐	☐
(12)Employee's resistance to change	☐	☐	☐	☐	☐
(13)Management's resistance to change	☐	☐	☐	☐	☐
(14) Employee's tolerance for an untidy or disorganized workplace	☐	☐	☐	☐	☐
(15) Absence of a "lean" culture in the extended network of partners	☐	☐	☐	☐	☐
(16)Unhealthy competition among suppliers	☐	☐	☐	☐	☐
(17)Inadequate delivery performance	☐	☐	☐	☐	☐
(18)Hierarchies in the organizational structure	☐	☐	☐	☐	☐
(19)Financial constraints	☐	☐	☐	☐	☐
(20)Less personal empowerment	☐	☐	☐	☐	☐
(21)Avoid making decisions and take responsibility	☐	☐	☐	☐	☐
(22)Using "guan xi" or relationships to conceal mistakes/errors	☐	☐	☐	☐	☐
(23)Stringent requirements and approvals	☐	☐	☐	☐	☐
(24)Lack of support from the government	☐	☐	☐	☐	☐
(25)Others ____ (Please specify)	☐	☐	☐	☐	☐
(26)Others ____ (Please specify)	☐	☐	☐	☐	☐

Thank you for taking the time to complete this questionnaire

If you have any questions about the survey, please feel free to contact Mr. Gao Shang, National University of Singapore, Department of Building. Tel: +65 82008140 (SINGAPORE) +86 139 6883 7511 (CHINA) or Email: gaoshang@nus.edu.sg

Appendix B: Questionnaire Survey (Chinese Version)

中国大型建筑施工企业建立丰田模式的应用型框架

问卷调查

首先衷心感谢您对本次问卷调查的支持！丰田模式是丰田汽车取得成功的根本。基于对丰田模式的研究,建筑业已经出现类似精益生产，精益建造等新名词。研究丰田模式在中国大型建筑施工企业对于提升企业的**核心竞争力**，提高工程建设项目的质量，杜绝现场的浪费等都具有重要意义。本次调研的目的就是为了了解中国大型建筑施工企业在企业的长期理念、施工管理、员工与合作伙伴、以及解决问题的方法论上是否接近丰田模式的 14 个管理理念。本项调查采用匿名方式，您提供的信息将绝对保密，仅将用于研究所用。

第一部分: 丰田模式概况

丰田生产系统（丰田模式的14个原则）:
原则1：管理决策以长期理念为基础，即使因此牺牲短期财务目标也在所不惜
原则2：建立无间断的操作流程使问题浮现
原则3：实施“拉动式”生产制度以避免生产过剩 (“拉动式”意指在生产流程下游的顾客需求的时候供应给他们正确数量的东西。材料的补充应该由消费量决定)
原则4：使工作负荷水准稳定(生产均衡化)
原则5：建立立即暂停制度以解决问题, 从一开始就重视质量管理的文化
原则6：工作的标准化是持续改进与授权员工的基础
原则7：运用可视化管理使问题无处隐藏
原则8：使用可靠的、已经过充分测试的技术以协助员工及生产流程
原则9：把彻底了解且拥护公司理念的员工培养成为领导者，使他们能教导其他员工
原则10：培养与发展信奉公司理念的杰出人才与团队
原则11：重视事业伙伴与供货商网络，激励并助其改进
原则12：亲临现场查看以彻底了解情况 (现地现物)
原则13：不急于作决策，以共识为基础，彻底考虑所有可能的选择，并快速执行决策
原则14：通过不断省思（日语是*Hansei*，意指“反省”）与持续改进以变成一个学习型组织

姓名:	____(可不填)	公司注册资金：	____万元
公司:	____	公司性质：	□施工总承包 □专业承包 □劳务分包
公司所在地区：	____	公司类别：	□国有 □集体 □私营
公司规模：	____(名员工)	涉足领域：	□房建 □公路/铁路 □港口 □石化
职位：	____		□电力 □机电安装 □市政 □水利
专长领域：	____		□其他，请指明____

S. Gao and S.P. Low, *Lean Construction Management*,
DOI 10.1007/978-981-287-014-8, © Springer Science+Business Media Singapore 2014

最高学历：	▭	公司资质：□特级 □一级 □二级 □三级	
工作年限：	▭	是否有海外项目：	□是 □否
电子邮件：	▭	ISO9001/9002 认证：	□是 □否
联系电话：	▭		

第三部分：丰田模式和中国大型建筑施工企业

(1) 请针对下列各项实践在您所在公司的**实行情况**和该实践对公司绩效的**重要程度**进行评分
（将您的回答用打"√"的方法，填在空格"□"内）

(2) 本次问卷采用 **5 分制**打分的形式

实行程度：①"实行程度非常低" ②"实行程度比较低" ③"实行程度一般水平" ④"实行程度比较高" ⑤"实行程度非常高"

重要程度：①"非常不重要" ②"比较不重要" ③"一般水平" ④"比较重要" ⑤"非常重要"

丰田模式原则 1：管理决策以长期理念为基础

原则 1： 长期理念	**实行程度**					**重要程度**				
	①	②	③	④	⑤	①	②	③	④	⑤
(1) 拥有一个坚定的目标（企业愿景、理念和价值）	□	□	□	□	□	□	□	□	□	□
(2) 把为员工、业主以及整个社会创造价值作为企业的出发点	□	□	□	□	□	□	□	□	□	□
(3) 从企业发展的全局制定出一个长远目标并制定周密实施计划	□	□	□	□	□	□	□	□	□	□
(4) 始终认为短期财务损失可能会影响决策，但远不如长远目标重要	□	□	□	□	□	□	□	□	□	□
(5) 明确企业的核心竞争力并努力保持在该领域的领先地位	□	□	□	□	□	□	□	□	□	□
(6) 对企业产品、员工和社会有强烈的使命感和责任感	□	□	□	□	□	□	□	□	□	□
(7) 优先考虑业主的需求和要求	□	□	□	□	□	□	□	□	□	□
(8) 能对业主提出的要求（比如设计变更等）进行快速的反应，满足业主的要求	□	□	□	□	□	□	□	□	□	□
(9) 把员工和合作伙伴视为内部顾客	□	□	□	□	□	□	□	□	□	□

丰田模式原则 2-8：正确的的生产流程

原则 2: 无间断的生产流程	实行程度					重要程度				
	①	②	③	④	⑤	①	②	③	④	⑤
(1) 员工在工作中会尽量避免浪费【注】	□	□	□	□	□	□	□	□	□	□
(2) 根据工作量明确每天所应消耗材料的需求计划	□	□	□	□	□	□	□	□	□	□

(3) 材料，设备和其他资源在需要时能及时到位	☐	☐	☐	☐	☐	☐	☐	☐	☐	☐
(4) 合理布置施工现场，从而增强物流，人流，工序流等从而减少不必要的运输，移动，和搬运	☐	☐	☐	☐	☐	☐	☐	☐	☐	☐
(5) 对各个工序，各个施工环节制定无缝衔接的准时制施工管理规定，尽力将闲置或等候的时间减少到零	☐	☐	☐	☐	☐	☐	☐	☐	☐	☐
(6) 重视施工工序流程以成为企业文化的一部分	☐	☐	☐	☐	☐	☐	☐	☐	☐	☐

【注】浪费这里是指生产过剩，不必要的运输，过度处理，存货过剩，不必要的移动，搬运，返工等。

原则 3: “拉动式”生产制度	实行程度					重要程度				
	①	②	③	④	⑤	①	②	③	④	⑤
(1) 材料的采购量尽可能接近实际的需求量	☐	☐	☐	☐	☐	☐	☐	☐	☐	☐
(2) 在施工现场建立最小的材料存货，甚至“零”库存	☐	☐	☐	☐	☐	☐	☐	☐	☐	☐
(3) 设立卡片、空箱等简单工具对库存情况进行监测，以便对材料和构件进行即时补充供应	☐	☐	☐	☐	☐	☐	☐	☐	☐	☐
(4) 对施工队实际领取材料/构件/设备的数量进行严格考核,控制损耗	☐	☐	☐	☐	☐	☐	☐	☐	☐	☐
(5) 确保每一场工的准备工作都准备就绪才开工。其中包括确定每道工序的工作内容，工作时间，材料需要量，提料时间等	☐	☐	☐	☐	☐	☐	☐	☐	☐	☐

原则 4: 工作负荷水准稳定 (生产均衡化)	实行程度					重要程度				
	①	②	③	④	⑤	①	②	③	④	⑤
(1) 制定施工进度计划时会尽可能的考虑到包括业主，分包商，供应商等利益相关者提供的信息	☐	☐	☐	☐	☐	☐	☐	☐	☐	☐
(2) 制定施工进度计划时会尽量考虑到材料，设备资源和人员的均衡化	☐	☐	☐	☐	☐	☐	☐	☐	☐	☐
(3) 各个施工队(班组)的督工根据实际中可执行的工作承诺他们每周的工作计划	☐	☐	☐	☐	☐	☐	☐	☐	☐	☐
(4) 各个施工队能完成周计划/日计划规定的施工任务	☐	☐	☐	☐	☐	☐	☐	☐	☐	☐
(5) 均衡每日的施工任务，不使员工和机械设备的负荷过重	☐	☐	☐	☐	☐	☐	☐	☐	☐	☐

原则 5: 追求质量的企业文化	实行程度					重要程度				
	①	②	③	④	⑤	①	②	③	④	⑤
(1) 员工注重质量，将质量“内置”于每一项施工环节	☐	☐	☐	☐	☐	☐	☐	☐	☐	☐
(2) 确保施工过程中上一道工序质量未经检验或者检	☐	☐	☐	☐	☐	☐	☐	☐	☐	☐

验不合格，不得进入下一道工序										
(3) 确保不合格的施工材料，设备不会被用于施工	□	□	□	□	□	□	□	□	□	□
(4) 员工在工作中遇到问题时及时反映给项目经理/工程师，并寻求解决问题的方法	□	□	□	□	□	□	□	□	□	□
(5) 员工被授权并肩负确保工程质量的责任	□	□	□	□	□	□	□	□	□	□
(6) 员工会在施工队的小组会议中定期讨论工程质量问题，总结质量通病的经验教训	□	□	□	□	□	□	□	□	□	□
(7) 员工会向项目经理或工程师对质量问题提出反馈意见	□	□	□	□	□	□	□	□	□	□

原则 6: 工作标准化	**实行程度**					**重要程度**				
	①	②	③	④	⑤	①	②	③	④	⑤
(1) 员工按照标准化的施工程序和操作方法来施工	□	□	□	□	□	□	□	□	□	□
(2) 员工是企业制定标准化施工程序和操作方法的主要贡献者	□	□	□	□	□	□	□	□	□	□
(3) 鼓励员工根据自身的实践经验提出施工程序和操作方法中需要改进的地方	□	□	□	□	□	□	□	□	□	□
(4) 把员工提出的有创意的改进意见纳入新的施工程序和操作方法手册中	□	□	□	□	□	□	□	□	□	□
(5) 企业定制并采用标准化的预置构件	□	□	□	□	□	□	□	□	□	□

原则 7: 可视化管理	**实行程度**					**重要程度**				
	①	②	③	④	⑤	①	②	③	④	⑤
(1) 可视化工具在施工中被采用，让浪费、异常等情况曝光，方便员工改进	□	□	□	□	□	□	□	□	□	□
(2) 员工在施工现场能阅读及时更新的关于施工进度，质量，安全等信息	□	□	□	□	□	□	□	□	□	□
(3) 施工现场有足够的图示标明施工现场布置，交通指示，安全标语等	□	□	□	□	□	□	□	□	□	□
(4) 施工现场一直保持清洁有序	□	□	□	□	□	□	□	□	□	□
(5) 员工会自觉维持清洁有序的施工现场	□	□	□	□	□	□	□	□	□	□
(6) 5S 方案【注】在施工现场中被采用	□	□	□	□	□	□	□	□	□	□

【注】5S 方案由一系列活动包括“整理“，”条理”，”整洁”，”标准化”，以及”维持”组成，目的是为了去除造成错误，瑕疵的浪费情形

原则 8: 新技术引进策略	实行程度					重要程度				
	①	②	③	④	⑤	①	②	③	④	⑤
(1) 引进的新技术必须支持企业价值	□	□	□	□	□	□	□	□	□	□
(2) 引进的新技术能表现出改进施工流程的潜力	□	□	□	□	□	□	□	□	□	□
(3) 以解决具体问题为导向引进新技术，不盲目追求高科技	□	□	□	□	□	□	□	□	□	□
(4) 引进新技术之前经过全面的评估，并证实其具有长期效益	□	□	□	□	□	□	□	□	□	□

丰田模式原则 9-11：员工与合作伙伴

原则 9: 培养领导者和领导力	实行程度					重要程度				
	①	②	③	④	⑤	①	②	③	④	⑤
(1) 项目经理懂得如何自我激励和激励员工	☐	☐	☐	☐	☐	☐	☐	☐	☐	☐
(2) 项目经理具备扎实的专业技术知识	☐	☐	☐	☐	☐	☐	☐	☐	☐	☐
(3) 项目经理有能力把他们具备的专业知识教授给其他员工	☐	☐	☐	☐	☐	☐	☐	☐	☐	☐
(4) 项目经理经常到实际工作现场支持员工的工作执行	☐	☐	☐	☐	☐	☐	☐	☐	☐	☐
(5) 项目经理会在行动前会花时间去了解问题和其发生的根源	☐	☐	☐	☐	☐	☐	☐	☐	☐	☐
(6) 项目经理会鼓励员工去培养精益求精的工作态度和“持续改进”的执行能力	☐	☐	☐	☐	☐	☐	☐	☐	☐	☐
(7) 项目经理熟知公司政策和工作程序，并能和员工及时沟通该方面的信息	☐	☐	☐	☐	☐	☐	☐	☐	☐	☐

原则 10: 培养与发展杰出人才和团队	实行程度					重要程度				
	①	②	③	④	⑤	①	②	③	④	⑤
(1) 企业总是会挑选最胜任的员工	☐	☐	☐	☐	☐	☐	☐	☐	☐	☐
(2) 企业会提供必要的上岗前的技能培训	☐	☐	☐	☐	☐	☐	☐	☐	☐	☐
(3) 企业会提供员工在职培训的机会，发展其出色的专业技能	☐	☐	☐	☐	☐	☐	☐	☐	☐	☐
(4) 企业为员工提供跨部门培训使其成为多面手	☐	☐	☐	☐	☐	☐	☐	☐	☐	☐
(5) 企业有标准化的员工培训手册	☐	☐	☐	☐	☐	☐	☐	☐	☐	☐
(6) 企业鼓励员工之间进行合作	☐	☐	☐	☐	☐	☐	☐	☐	☐	☐
(7) 员工以团队的方式开展现场作业	☐	☐	☐	☐	☐	☐	☐	☐	☐	☐
(8) 企业会运用侧重内在因素的激励机制（比如提供良好的工作环境，好的待遇，职务轮调等）来激励员工	☐	☐	☐	☐	☐	☐	☐	☐	☐	☐
(9) 企业会运用侧重外在因素的激励机制（比如设立全勤奖，绩效评估，设立具有挑战性的目标等）来激励员工	☐	☐	☐	☐	☐	☐	☐	☐	☐	☐

原则 11: 重视合作伙伴（比如材料供应商，分包商）	**实行程度**					**重要程度**				
	①	②	③	④	⑤	①	②	③	④	⑤
(1) 尊重合作伙伴各方面的能力	□	□	□	□	□	□	□	□	□	□
(2) 通过给合作伙伴设置目标，来对他们提出新的挑战	□	□	□	□	□	□	□	□	□	□
(3) 与合作伙伴密切合作，提高项目效益	□	□	□	□	□	□	□	□	□	□
(4) 参与合作伙伴的生产管理过程中去	□	□	□	□	□	□	□	□	□	□
(5) 与合作伙伴在不同的领域展开合作，提升其技术能力	□	□	□	□	□	□	□	□	□	□
(6) 与合作伙伴资源信息共享	□	□	□	□	□	□	□	□	□	□
(7) 与合作伙伴共同实施持续改善计划，共同解决工作中出现的问题	□	□	□	□	□	□	□	□	□	□
(8) 拥有一支稳定可靠和长期合作的供应商队伍	□	□	□	□	□	□	□	□	□	□
(9) 限制供应商的数量，不经常变换供应商	□	□	□	□	□	□	□	□	□	□

丰田模式原则 12-14：解决根本问题

原则 12: 亲临现场查看以彻底了解情况（现地现物【注】）	**实行程度**					**重要程度**				
	①	②	③	④	⑤	①	②	③	④	⑤
(1) 项目经理到实际作业现场（比如施工现场）去查看问题并解决问题	□	□	□	□	□	□	□	□	□	□
(2) 在做决定前务必确认事实并做非常全面的分析	□	□	□	□	□	□	□	□	□	□
(3) 根据亲临现场得到的资料来做决策	□	□	□	□	□	□	□	□	□	□
(4) 做决策时主要要依赖领导的经验	□	□	□	□	□	□	□	□	□	□
(5) 亲临现场查看以彻底了解情况（现地现物）已成为一种企业文化	□	□	□	□	□	□	□	□	□	□

【注】现地现物是在日文里译为亲临现场查看以彻底了解情况.

原则 13: 决策策略	**实行程度**					**重要程度**				
	①	②	③	④	⑤	①	②	③	④	⑤
(1) 采用适当的解决问题的方法（比如问上 5 次"为什么"直到找出问题的根本原因）	□	□	□	□	□	□	□	□	□	□
(2) 通过实验来测试一个问题的可能原因	□	□	□	□	□	□	□	□	□	□
(3) 广泛考虑到各种可能的解决方案	□	□	□	□	□	□	□	□	□	□
(4) 重视实现决策的过程	□	□	□	□	□	□	□	□	□	□
(5) 决策时能和员工以及其他部门达成共识	□	□	□	□	□	□	□	□	□	□
(6) 能通过有效的沟通机制来发掘问题产生的原因	□	□	□	□	□	□	□	□	□	□

原则 14: 不断反思与持续改进，努力成为学习型组织	实行程度					重要程度				
	①	②	③	④	⑤	①	②	③	④	⑤
(1) 定期对诸如质量问题，返工，安全问题等工作中存在的不足进行自我反省	□	□	□	□	□	□	□	□	□	□
(2) 把问题或者错误当做学习的机会	□	□	□	□	□	□	□	□	□	□
(3) 员工在工作中执行“持续改进”方案	□	□	□	□	□	□	□	□	□	□
(4) 领导大力支持“持续改进”方案	□	□	□	□	□	□	□	□	□	□
(5) 任何取得的改进结果都会被记录并得到相关部门的使用	□	□	□	□	□	□	□	□	□	□
(6) 企业对诸如质量，进度等各项目标进行逐级分解，明确各级人员的责任	□	□	□	□	□	□	□	□	□	□
(7) 定期考核目标的进展情况并提出反馈意见	□	□	□	□	□	□	□	□	□	□
(8) 运用 PDCA (规划-执行-检查-行动)的学习循环来解决问题并取得持续改进	□	□	□	□	□	□	□	□	□	□

第四部分: 丰田模式能给中国大型建筑施工企业带来的优势

1．请评价与竞争对手相比，您所在的公司目前在下列几个方面的**表现**如何（将您的回答用打“√”的方法，填在空格“□”内）

表现程度: ① “表现非常差” ② “表现比较差” ③ “表现一般水平” ④ “表现比较好” ⑤ “表现非常好”

2．请评价在何种程度上，丰田模式的推行能帮助您所在的公司在下列几个方面**获得改善**（请将您的回答用打“√”的方法，填在空格“□”内）

改善程度: ① “没有改善” ② “较少改善” ③ “一些改善” ④ “较大改善” ⑤ “显著改善”

	表现程度					改善程度				
	①	②	③	④	⑤	①	②	③	④	⑤
(1) 财务表现	□	□	□	□	□	□	□	□	□	□
(2) 生产效率	□	□	□	□	□	□	□	□	□	□
(3) 质量	□	□	□	□	□	□	□	□	□	□
(4) 杜绝浪费	□	□	□	□	□	□	□	□	□	□
(5) 项目进度	□	□	□	□	□	□	□	□	□	□
(6) 健康与安全问题	□	□	□	□	□	□	□	□	□	□
(7) 业主满意度	□	□	□	□	□	□	□	□	□	□

3. 如果丰田模式理念被证明其对提高施工企业在上述领域的表现是可行的，您所在企业是否考虑将实施丰田模式作为一项投资来改善企业的绩效？

□ 是, 愿意 □ 会考虑 □ 否，不愿意

第五部分: 实施丰田模式的障碍

1. 请对下列因素对实施丰田模式的**影响程度**进行评分（将用打"√"的方法，填在空格"□"内）

影响程度: ① "几乎没有影响" ② "有较小影响" ③ "有一定影响" ④ "有较大影响" ⑤ 有非常大影响"

因素	影响程度				
	①	②	③	④	⑤
(1) 缺乏长期理念	□	□	□	□	□
(2) 缺乏一个"精益"的组织文化	□	□	□	□	□
(3) 设计施工一体化的承包模式非常有限	□	□	□	□	□
(4) 建筑企业在设计阶段的参与非常有限	□	□	□	□	□
(5) 各个施工队负责人没有充分参与项目的进度计划安排	□	□	□	□	□
(6) 多层分包/转包	□	□	□	□	□
(7) 很少采用诸如预置等非现场施工技术	□	□	□	□	□
(8) 缺乏项目管理技能（如领导能力，解决问题的能力等）	□	□	□	□	□
(9) 缺乏领导的支持	□	□	□	□	□
(10) 员工离职率高，流动性大	□	□	□	□	□
(11) 没有充分的培训	□	□	□	□	□
(12) 员工对变革的抵制	□	□	□	□	□
(13) 管理层对变革的抵制	□	□	□	□	□
(14) 员工习惯于"脏"、"乱"、"差"的工地	□	□	□	□	□
(15) 诸如供应商，分包商等事业伙伴缺乏对"精益"的了解	□	□	□	□	□
(16) 供应商之间的恶性竞争	□	□	□	□	□
(17) 供应商的交付能力差	□	□	□	□	□
(18) 多层级的组织结构	□	□	□	□	□
(19) 没有充分的财政支持	□	□	□	□	□
(20) 个人授权太少	□	□	□	□	□
(21) 避免单独决策，不想承担责任	□	□	□	□	□
(22) 找"关系"	□	□	□	□	□
(23) 过于严格繁冗的政府规章	□	□	□	□	□
(24) 缺乏政府部门对的支持	□	□	□	□	□
(25) 其他_________（请注明）	□	□	□	□	□

2. 若您对本研究有其他建议，请在下面方框内提出：

感谢您宝贵时间完成这份问卷

如果您对调查问卷有任何疑问，请随时联系高尚先生, 新加坡国立大学(房屋建筑系)

联系方式: (+86)139 6883 7511(**中国**)(+65)82008140(**新加坡**)**电子邮件**: gaoshang@nus.edu.sg

Appendix C: Correlations Results

	Profitability	Productivity	Quality	Delivery time	Client satisfaction
Toyota Way Philosophy model					
P1 Long-term philosophy	**.320****	**.434****	**.496****	**.322****	**.485****
P1.1 Sustain a constant purpose (company vision, mission and values)	.369** (.000)	.345** (.001)	.294** (.004)	.103 (.322)	.258* (.012)
P1.2 Have a high purpose or mission which is to generate value towards employees, society and customers	.420** (.000)	.301** (.003)	.325** (.001)	.160 (.123)	.225* (.029)
P1.3 Formulate a plan towards the realization of company's long-term vision	.274** (.008)	.302** (.003)	.443** (.000)	.325** (.001)	.429** (.000)
P1.4 Short-term losses affect decision-making, but are less important than pursing long-term goals	.150 (.154)	.287** (.005)	.269** (.009)	.194 (.063)	.195 (.061)
P1.5 Have a clear view of its core competency and endeavour to become an expert in this area	.261* (.012)	.330** (.001)	.326** (.001)	.090 (.391)	.399** (.000)
P1.6 Be responsible for products, employees and society	.323** (.002)	.357** (.000)	.492** (.000)	.411** (.000)	.510** (.000)
P1.7 Understanding customer's requirement is priority work	.123 (.239)	.347** (.001)	.326** (.001)	.310** (.002)	.384** (.000)
P1.8 Be able to rapid response to meet the changing requirement of the customers (e.g. design change)	.221* (.033)	.310** (.002)	.276** (.007)	.356** (.000)	.415** (.000)
P1.9 Treat employees/ suppliers as internal customers	.059 (.571)	.232* (.024)	.401** (.000)	.297** (.004)	.332** (.001)
Toyota Way Process model					

(continued)

DOI 10.1007/978-981-287-014-8, © Springer Science+Business Media Singapore 2014

	Profitability	Productivity	Quality	Delivery time	Client satisfaction
P2 One-piece flow	**.314****	**.451****	**.482****	**.398****	**.365****
P2.1 Employee is concerned with waste elimination	.190 (.068)	.330** (.001)	.354** (.000)	.246* (.017)	.279** (.006)
P2.2 Material flow is adhered to consistently throughout the daily work activities	.294** (.004)	.315** (.002)	.288** (.005)	.369** (.000)	.185 (.074)
P2.3 Material, equipment and other resources are provided in a "just-in-time" manner when needed	.238* (.023)	.263* (.011)	.328** (.001)	.347** (.001)	.303** (.003)
P2.4 Site layout is organized to enhance material flow, employee movement, etc. to minimize wastes due to movement, motion, travel, etc.	.271** (.009)	.422** (.000)	.378** (.000)	.270** (.008)	.193 (.062)
P2.5 Strive to cut back to zero the amount of time any work is sitting idle or waiting for someone to work on it	.165 (.115)	.336** (.001)	.365** (.000)	.267** (.009)	.280** (.006)
P2.6 Make flow evident through organizational culture	.256* (.013)	.316** (.002)	.473** (.000)	.360** (.000)	.369** (.000)
P3 Pull "*kanban*" system	**.211***	**.317****	**.361****	**.238***	**.189 (.068)**
P3.1 Materials are ordered as close as possible to exact needs	.274** (.008)	.141 (.177)	.180 (.083)	.158 (.129)	.057 (.583)
P3.2 Strive for possible low level of (even stockless) material inventory in construction site	.128 (.221)	.117 (.263)	.185 (.074)	.123 (.236)	.060 (.564)
P3.3 Use simple signals—cards, empty bins, etc. to monitor the level of inventory and to order the needed material/component	.092 (.383)	.250* (.015)	.318** (.002)	.179 (.085)	.059 (.570)
P3.4 Monitor the quantity of material/component/ equipment that the teams actually take away	.079 (.450)	.352** (.001)	.335** (.001)	.214* (.039)	.285** (.005)
P3.5 Clear job contents, work time, material requirements, among other information are prepared before releasing a work task to a crew	.200 (.055)	.318** (.002)	.315** (.002)	.282** (.006)	.222* (.032)
P4 Level out the workloads (*Heijunka*)	**.287****	**.350****	**.499****	**.376****	**.422****
P4.1 Project manager plans the work with input from other parties including subcontractors, clients, suppliers, etc.	.282** (.006)	.378** (.000)	.444** (.000)	.283** (.006)	.371** (.000)

(continued)

	Profitability	Productivity	Quality	Delivery time	Client satisfaction
P4.2 Daily work activities are planned to balance material availability, manpower, machine availability, and workload between operations	.258* (.013)	.195 (.060)	.472** (.000)	.302** (.003)	.323** (.001)
P4.3 Foremen (Last Planners) make commitments on what the crews will do each week based on what is ready to be done	.191 (.067)	.252* (.014)	.360** (.000)	.260* (.011)	.339** (.001)
P4.4 Weekly/daily work assignments are completed in accordance with the weekly/daily schedule	.318** (.002)	.251* (.015)	.278** (.007)	.326** (.001)	.335** (.001)
P4.5 Levelling the daily work activities without overburdening workers and machinery	.056 (.592)	.198 (.056)	.302** (.003)	.163 (.117)	.142 (.173)
P5 Built-in quality	**.295****	**.296****	**.529****	**.337****	**.321****
P5.1 Employees are dedicated to provide "built-in" quality into every aspect of operations	.156 (.136)	.214* (.038)	.497** (.000)	.267** (.009)	.292** (.004)
P5.2 Preventing defective or "no inspection" assignments from entering the next process	.276** (.008)	.323** (.002)	.450** (.000)	.388** (.000)	.274** (.007)
P5.3 Rejecting defective materials, components and equipment	.235* (.024)	.314** (.002)	.504** (.000)	.400** (.000)	.348** (.001)
P5.4 Employees are encouraged to seek support from their supervisors when something goes wrong at work	.263* (.011)	.229* (.028)	.371** (.000)	.128 (.222)	.174 (.095)
P5.5 Employees are empowered to be responsible for quality	.217* (.037)	.204* (.049)	.344** (.001)	.223* (.031)	.239* (.021)
P5.6 Employees who work in the same team meet on a regular basis to discuss quality problems and lessons learned	.251* (.015)	.237* (.022)	.323** (.001)	.195 (.060)	.197 (.057)
P5.7 Feedback about quality is routinely given by the employees	.263* (.011)	.171 (.100)	.317** (.002)	.172 (.098)	.184 (.076)
P6 Standardized work	**.385****	**.338****	**.474****	**.357****	**.367****
P6.1 Established standard operating procedures (SOPs) (e.g. work processes) are practised by employees for each major operation/process	.339** (.001)	.317** (.002)	.425** (.000)	.292** (.004)	.330** (.001)

(continued)

	Profitability	Productivity	Quality	Delivery time	Client satisfaction
P6.2 Employees play a key role in creating the SOPs	.285** (.006)	.208* (.045)	.388** (.000)	.269** (.009)	.251* (.015)
P6.3 Employees are encouraged to improve the existing SOPs based on their own practical experience	.368** (.000)	.307** (.003)	.393** (.000)	.330** (.001)	.348** (.001)
P6.4 Incorporate employee's creative improvement of the standard into new SOPs	.258* (.012)	.243* (.018)	.389** (.000)	.277** (.007)	.255* (.013)
P6.5 Using standardized prefabricated components from offsite shops	.295** (.004)	.334** (.001)	.327** (.001)	.329** (.001)	.361** (.000)
P7 Visual Management	**.371****	**.403****	**.583****	**.387****	**.483****
P7.1 Visual aids are adopted to make wastes, problems and abnormal conditions readily apparent to employees	.290** (.005)	.253* (.014)	.451** (.000)	.312** (.002)	.370** (.000)
P7.2 The posted information in terms of job status, schedule, quality, safety, etc. is in place that most workers can see it on a daily basis, and it is up-to-date	.335** (.001)	.257* (.012)	.475** (.000)	.304** (.003)	.426** (.000)
P7.3 Appropriate signages are used to identify layouts, traffic, safety concerns, etc.	.335** (.001)	.282** (.006)	.392** (.000)	.242* (.019)	.392** (.000)
P7.4 The construction site is kept clean at all times	.261* (.012)	.398** (.000)	.406** (.000)	.282** (.006)	.325** (.001)
P7.5 Employees take pride in keeping the construction site organized and clean	.320** (.002)	.352** (.001)	.538** (.000)	.332** (.001)	.396** (.000)
P7.6 The workplace follows the principles of 5-S	.252* (.015)	.366** (.000)	.502** (.000)	.388** (.000)	.384** (.000)
P8 The use of reliable technology	**.285****	**.360****	**.472****	**.224***	**.262***
P8.1 New technology must support the company's values	.218* (.036)	.258* (.012)	.418** (.000)	.185 (.074)	.264* (.010)
P8.2 New technology must demonstrate its potential to enhance processes	.250* (.016)	.300** (.003)	.348** (.001)	.149 (.152)	.175 (.092)
P8.3 New technology must be specific solution oriented	.293** (.004)	.356** (.000)	.430** (.000)	.211* (.041)	.296** (.004)
P8.4 New technology must be thoroughly tested and proven to provide long-term benefits	.256* (.013)	.382** (.000)	.460** (.000)	.227* (.028)	.213* (.039)

(continued)

	Profitability	Productivity	Quality	Delivery time	Client satisfaction
Toyota Way People and Partner model					
P9 Leaders and Leadership	**.296****	**.464****	**.509****	**.342****	**.297***
P9.1 Leaders are motivated to inspire people to achieve goals	.250* (.016)	.479** (.000)	.356** (.000)	.240* (.020)	.191 (.065)
P9.2 Leaders must have in-depth job knowledge	.267** (.010)	.216* (.037)	.287** (.005)	.130 (.210)	.118 (.257)
P9.3 Leaders possess teaching ability and are able to pass their knowledge on to others	.273** (.008)	.358** (.000)	.361** (.000)	.347** (.001)	.221* (.033)
P9.4 Leaders must support the employees doing their work	.218* (.036)	.271** (.008)	.345** (.001)	.111 (.285)	.223* (.030)
P9.5 Leaders will take time to understand problems and root causes before acting	.203 (.052)	.338** (.001)	.277** (.007)	.200 (.054)	.216* (.037)
P9.6 Leaders strongly encourage employees to develop "continuous improvement" in thinking and action	.235* (.023)	.373** (.000)	.558** (.000)	.327** (.001)	.330** (.001)
P9.7 Leaders must understand the company policy and procedures, and communicate these to their team	.196 (.060)	.433** (.000)	.478** (.000)	.376** (.000)	.253* (.014)
P10 People management	**.310****	**.396****	**.527****	**.333****	**.357****
P10.1 Select the best person for a given job	.271** (.008)	.347** (.001)	.483** (.000)	.222* (.031)	.337** (.001)
P10.2 Training is provided to equip the employees with the required skills before they are assigned to work	.197 (.059)	.351** (.001)	.466** (.000)	.339** (.001)	.318** (.002)
P10.3 On-the-job-training is provided to further develop employee's exceptional skills	.190 (.068)	.318** (.002)	.399** (.000)	.189 (.068)	.298** (.004)
P10.4 Employees are cross-trained to perform additional functions	.169 (.106)	.331** (.001)	.353** (.000)	.322** (.002)	.222* (.032)
P10.5 Training materials are standardized	.168 (.107)	.172 (.098)	.313** (.002)	.119 (.254)	.200 (.053)
P10.6 Employees are encouraged to cooperate with others to complete the whole task	.323** (.002)	.296** (.004)	.351** (.001)	.238* (.021)	.258* (.012)
P10.7 Daily work activities are organized into team function	.199 (.056)	.259* (.012)	.482** (.000)	.384** (.000)	.325** (.001)
P10.8 Internal motivation methods	.270** (.009)	.376** (.000)	.480** (.000)	.293** (.004)	.354** (.000)
P10.9 External motivation methods	.379** (.000)	.375** (.000)	.439** (.000)	.204* (.049)	.313** (.002)

(continued)

	Profitability	Productivity	Quality	Delivery time	Client satisfaction
P11 Partners relationships	**.337****	**.396****	**.584****	**.320****	**.376****
P11.1 Respect partners' capabilities	.218* (.035)	.308** (.003)	.510** (.000)	.204* (.049)	.267** (.009)
P11.2 Challenge the partners by setting collaborative targets	.227* (.028)	.397** (.000)	.568** (.000)	.376** (.000)	.346** (.001)
P11.3 Take part in partners' production process	.333** (.001)	.389** (.000)	.560** (.000)	.248* (.016)	.317** (.002)
P11.4 Work with the partners to improve project effectiveness	.301** (.003)	.355** (.000)	.591** (.000)	.409** (.000)	.402** (.000)
P11.5 Work with the partners in various areas to develop their technical capabilities	.225* (.030)	.290** (.005)	.423** (.000)	.296** (.004)	.324** (.001)
P11.6 Share information with partners in a structured manner	.303** (.003)	.280** (.006)	.422** (.000)	.311** (.002)	.314** (.002)
P11.7 Conduct joint improvement activities with partners to solve problems	.302** (.003)	.301** (.003)	.386** (.000)	.201 (.052)	.280** (.006)
P11.8 Strive to establish a long-term relationship with reliable partners	.373** (.000)	.314** (.002)	.415** (.000)	.149 (.152)	.240* (.020)
P11.9 Limit the number of suppliers	.249* (.017)	.260* (.012)	.398** (.000)	.279** (.007)	.357** (.000)
Toyota Way Problem-Solving model					
P12 *Genchi Genbutsu*	**.150**	**.377****	**.437****	**.296****	**.238***
P12.1 Solve problem by going to the places (e.g. construction site) where problems are discovered	.096 (.360)	.296** (.004)	.362** (.000)	.275** (.007)	.274** (.008)
P12.2 Analysing and thoroughly understand the situation before making decisions	.228* (.028)	.374** (.000)	.382** (.000)	.257* (.012)	.204* (.048)
P12.3 Making decisions based on the verified data	.168 (.107)	.357** (.000)	.482** (.000)	.299** (.003)	.240* (.020)
P12.4 Making decisions based on management team's past experiences	−.043 (.684)	.138 (.186)	.082 (.432)	.095 (.365)	−.039 (.708)
P12.5 *Genchi Genbutsu* has become part of the company culture	.204* (.050)	.327** (.001)	.344** (.001)	.219* (.034)	.177 (.088)
P13 Decision-making	**.395****	**.388****	**.494****	**.350****	**.385****
P13.1 Using appropriate problem-solving methodologies (e.g. 5 Whys) to determine the root causes of problems	.433** (.000)	.332** (.001)	.392** (.000)	.327** (.001)	.379** (.000)

(continued)

	Profitability	Productivity	Quality	Delivery time	Client satisfaction
P13.2 Possible experiments are conducted to test the potential cause of a problem	.297** (.004)	.346** (.001)	.463** (.000)	.439** (.000)	.424** (.000)
P13.3 Broadly consider alternative solutions	.311** (.002)	.187 (.071)	.359** (.000)	.142 (.172)	.234* (.023)
P13.4 Valuing the process through which the decision was reached	.363** (.000)	.279** (.006)	.334** (.001)	.225* (.029)	.269** (.009)
P13.5 Building consensus within the team, including employees and outside partners	.244* (.018)	.290** (.005)	.389** (.000)	.326** (.001)	.255* (.013)
P13.6 Addressing the root causes of problems via effective communication vehicle	.255* (.013)	.375** (.000)	.367** (.000)	.184 (.075)	.244* (.018)
P14 *Kaizen* or continuous improvement	**.365****	**.452****	**.501****	**.349****	**.360****
P14.1 Reflection on mistakes (e.g. defects, rework, safety issues, etc.) on a regular basis	.327** (.001)	.322** (.002)	.478** (.000)	.290** (.005)	.311** (.002)
P14.2 Management treats problems as development opportunities for employees	.325** (.001)	.365** (.000)	.344** (.001)	.189 (.068)	.154 (.138)
P14.3 *Kaizen* activities are conducted in your workplace	.360** (.000)	.392** (.000)	.535** (.000)	.322** (.002)	.364** (.000)
P14.4 Management supports the *kaizen* activities	.291** (.005)	.374** (.000)	.386** (.000)	.284** (.006)	.399** (.000)
P14.5 The improvement will be codified into documents and/or policies used by organization	.261* (.012)	.342** (.001)	.362** (.000)	.389** (.000)	.295** (.004)
P14.6 Each hierarchy of the organization develops measurable objectives as well as actions to support the executive-level goals	.284** (.006)	.381** (.000)	.368** (.000)	.194 (.061)	.234* (.023)
P14.7 Managers are keen on measuring the objectives and give feedback	.275** (.008)	.391** (.000)	.499** (.000)	.296** (.004)	.317** (.002)
P14.8 PDCA methodology is used to solve problems	.265* (.010)	.422** (.000)	.311** (.002)	.363** (.000)	.319** (.002)

**denotes the correlation is significant at the 0.01 level;
*denotes the correlation is significant at the 0.05 level;

Appendix D: Interview Protocol

Institution: National University of Singapore (Department of Building)

Programme: PhD in Project Management

Research Topic: The Toyota Way Model: An Implementation Framework for Large Chinese Construction firms

To attain a deeper understanding of how the principles of the Toyota Way can be implemented by the Chinese construction firms in their work, semi-structured interviews will be conducted, in addition to observation, with a representative sample of interviewees from site engineers, to project managers, and managers in large Chinese construction firms. Interviews comprise a core set of questions which are listed below.

A: Interviewee Profile

Firm:

Position:

Working experience:

B: Awareness of Lean or Toyota Way in the Construction Context

1. What is your knowledge about lean principles in construction or lean construction?
2. Do you currently practise any form of the Toyota Way or lean in your daily work?

C: Toyota Way Philosophy Model

[P1: Base your management decisions on a long-term philosophy, even at the expense of short-term financial goals]

1. How the long-term philosophy has been adhered to, and implemented within your firm?

DOI 10.1007/978-981-287-014-8,

2. Are you and your colleagues aware of the constant purpose of your firm?
3. Has the management attempted to maximize the value to the clients?
4. Generally, are you aware of your firm's core competencies?

D: Toyota Way Process Model

[P2: Create a continuous process flow to bring problems to the surface]
1. Are the efforts made to achieving a one-piece or uninterrupted work-flow in your project?
2. Do you see any construction trades conflicts in the flow?
3. What is your strategy to eliminate the non-value-adding activities?

[P3: Use "pull" systems to avoid overproduction]
1. What material planning systems are you currently adopting?
2. What is the inventory level of the commonly used materials?
3. If it is high, what are the challenges you normally encounter in pursuing the low level of inventory at site?

[P4: Level out the workload (*heijunka*)]
1. What are the levels of project planning adopted at the project level?
2. Who is the last planner in the project (e.g. general foreman, engineer, or project manager)?
3. How is the reliability of the project planning (e.g. weekly plan)?

[P5: Build a culture of stopping to fix problems, to get quality right the first time]
1. How are the non-conforming parts identified in operations?
2. What are the challenges you normally encounter in implementing built-in quality in the daily operations?
3. How are the quality circles (QCs) conducted (e.g. at project level)?

[P6: Standardized tasks and processes are the foundation for continuous improvement and employee empowerment]
1. Generally, have you practised standardization in terms of work processes, or perhaps in other areas pertaining to your daily operations?
2. Are standard operations procedures (SOPs) available on the site?
3. Please state the role of workers in improving the current process or SOPs?
4. Any awards for such improvement?

[P7: Use visual control so no problems are hidden]
1. Generally, how do you implement the visual control tools in your daily work? (e.g. in the areas of material, health and safety, and so on.)
2. Are you aware of the 5-S practice in the field of construction? If so, do you and your colleagues practise any form of the 5-S in your project?

[P8: Use only reliable, thoroughly tested technology that serves your people and processes]

1. In addition to the "*Ten new emerging construction technologies in China's construction industry*" promoted by the Chinese Construction Bureau, did your firm pursue any new technology and apply it in your project?
2. What are the guiding principles set by your firm in terms of the new technology adoption? (provide an example if any)

D: Toyota Way People and Partner Model

[P9: Grow leaders who thoroughly understand the work, live the philosophy and teach it to others]

1. Did your firm frequently identify the leaders from outside or grow leaders within the firm?
2. What is the level of technical knowledge of your leaders or supervisors?
3. Did management attempt to support the employees doing the work?

[P10: Develop exceptional people and teams who follow your company's philosophy]

1. How would you rate the employees and the sub-contracted workforces?
2. What kind of trainings are available for different levels of employees? (e.g. frontline workers, site personnel, and management)
3. How would you rate your workforce at the project level in terms of teamwork ability?
4. Please state the motivation method adopted for your firm.

[P11: Respect your extended network of partners and suppliers by challenging them and helping them improve]

1. How are the suppliers/subcontractors selected?
2. Did your firm use single- or multi-sourcing?
3. What is the size of your main suppliers/sub-contractor base?
4. What kind of delivery systems are adopted in your firm?
5. How is collaboration conducted between your firm and the partners (e.g. suppliers, subcontractors, client and so on)
6. How is your relationship with your clients?

E: Toyota Way Problem-Solving Model

[P12: Go and see for yourself to thoroughly understand the situation (*genchi genbutsu*)]

1. Did management attempt to go and see what actually happens on site?

2. Generally, what are the steps for problem-solving at the project level?

[P13: Make decisions slowly by consensus, thoroughly considering all options; implement decisions rapidly (*nemawashi*)]
1. How are the decision(s) for problem-solving made at the project level?
2. Did your project team practise "consensus" to reach final agreement?
3. What is the role of the employees in decision-making?

[P14: Become a learning organization through relentless reflection (*hansei*) and continuous improvement (*kaizen*)]
1. What is the general attitude of management towards problems?
2. Have you heard of continuous improvement activities or *kaizen* being conducted in your firm or project?
3. If so, what types of *kaizen* activities are implemented?
4. Do you and your colleagues practise reflection or *hansei* of the progress, quality, objectives, etc. of the project periodically?

Appendix E: Validation Results

Strategies for better implementation of the Toyota Way-styled practices in China's construction industry	Average
1.1 Aiming for generating value for clients, employees and society at large. While, the bottom line is not act irresponsibly in the pursuit of short-term profits.	4.17
1.2 Educate all the employees to understand and accept the company vision, mission and values.	4.00
1.3 Acknowledge people are the most important asset of a firm; respect them, develop and grow with them.	4.33
1.4 Establish the firm's culture of "clients are the priority", and help the employees to understand who their internal clients are.	4.67
1.5 Improve the project process by employing Toyota Way process-oriented initiatives to create more value for clients.	4.33
2.1 Enhance project leaders' understanding of site issues; more *genchi genbutsu* practice is encouraged to identify the constraints for better improvement of work-flow, material flow and manpower flow.	4.00
2.2 An awareness programme on the theme of "eliminating *muda*" should be introduced on-site.	4.17
2.3 Involve subcontractors/suppliers in the planning process, so as to improve the availability of manpower and material resources.	4.33
3.1 For those material that are subject to fluctuations in market prices, purchasing partnerships with trusted working partners should be established for material purchase to reap mutual benefits.	4.50
3.2 For those materials that cannot be stored on-site, and which are vulnerable to burglary, or which have to meet with customs requirements, a pull system should be encouraged.	4.17
3.3 Develop foremen's ability in effective preparing work for the coming weekly plan, e.g. tracking material usage, understand the constraints, and others.	4.50
4.1 Update weekly plans and look-ahead (monthly) plans in a timely manner if unexpected events occur, and communicate these updates and changes with the frontline workers in order for them to understand where they are and what the follow-up actions are.	4.50
4.2 From a long-term perspective, trust should be established between foremen, supervisors and contractor's employees for them to be included in the planning process.	4.17
4.3 Track the reliability of the weekly performance is of importance; root causes should be revealed and lessons should be learnt.	5.00

(continued)

S. Gao and S.P. Low, *Lean Construction Management*,
DOI 10.1007/978-981-287-014-8, © Springer Science+Business Media Singapore 2014

Strategies for better implementation of the Toyota Way-styled practices in China's construction industry	Average
4.4 Adopt appropriate motivational strategies to maintaining the core workforce at least until the project is accomplished.	4.17
5.1 Improve the firm's culture relating to quality, i.e. promote "zero tolerance" for substandard quality.	3.83
5.2 Apart from the difficult part of quality control, employees should also be encouraged to give feedback on quality issues, engage in quality improvement dialogues, provide teamwork in problem-solving, etc.	4.00
5.3 QCs should be done more frequently. The management must not ignore the importance of recognizing and rewarding QC achievements.	4.33
6.1 The essence of the construction methods should be made easily accessible for convenience, and frequently shared with workers in order for them to truly understand the procedures.	4.00
6.2 Recognition for innovative ideas should be encouraged to improve the conventional approach, even for small improvement.	4.00
6.3 For appropriate projects in which standardized prefabrication components can be applied, the use of prefabrication should be promoted.	3.50
7.1 Highlighting the visual management components in the government's "beautifying the construction site" campaign to all the employees.	4.50
7.2 Prior to the implementation of "5-S", this concept should be introduced through training to all employees.	4.50
7.3 Efforts should be made to change the clients'—as well as the contractors'—mindset of the site environment. The goal of working towards a tidy, organized, clean site should be set.	4.33
8.1 The adoption of emerging technologies, as promoted by the government, should be based on the characteristics of the project, the actual situation of project location, and other factor considerations.	4.50
8.2 If resistance to adoption is encountered, long-term interests should be taken into consideration, as it may be helpful to the employees, their work, and the firm at large.	4.50
9.1 Building the company's culture in a way that its leaders are encouraged to *genchi genbutsu* on the projects, to understand the projects and their employees' work, as well as their own capacity.	4.50
9.2 The management should develop a long-term plan for leadership development, with a focus on developing their teaching skills.	4.67
10.1 Build up the workforce into different layers based on their abilities and skill sets—that is, to maintain the appropriate levels of skilled and semiskilled workers.	4.67
10.2 Make sure that adequate resources are available to carry out a variety of on-site programmes for executive and workforce training. The breadth, types, and variety of training should be considered and enhanced. Training topics such as "identifying the non-value-adding activities in work", "possible improvement in current working procedures" and so on are worth incorporating into current training programmes.	4.67
10.3 Possible forms of teamwork should be encouraged include QC teams, *kaizen* teams, 5-S teams, and multidisciplinary teams to tackle a particular problem.	4.17
10.4 Improve the presently used motivation strategies.	4.00
11.1 Carefully review the existing portal that integrates the information of suppliers and subcontractors. Efforts should be made to remove working partners with unsatisfactory records and performances.	4.83

(continued)

Strategies for better implementation of the Toyota Way-styled practices in China's construction industry	Average
11.2 Eliminating the illegal multi-subcontracting practice. The partners' portal should be used as a reference for choosing reliable firms for material delivery and subcontracting work.	4.33
11.3 Create more opportunities for team-working with their partners to improve potential areas and work problems of the project. They should capture the lessons learnt and share best practices with other partners.	4.17
11.4 Establish training to raise awareness of the Toyota Way practice (or lean practice) for partners, and to periodically evaluate their compliance with the Toyota Way principles, so as to maintain an ongoing dialogue with the contractor to foster continuous improvement.	4.67
12.1 A *genchi genbutsu* firm culture should be established. The attitude of *genchi genbutsu* should be strengthened, even going so far as to write this into the core value or guiding principles of the companies.	4.00
12.2 Decision-making should not only be based on experience, but also needs to incorporate an understanding of the condition.	4.83
12.3 *Genchi genbutsu* should be set as a priority to show the determination of the leadership, and to boost the morale of employees.	4.33
13.1 Establish a "no blame and no complaints" work environment. All employees should be encouraged to voice their opinions and their suggestions for decisions.	4.17
13.2 Encourage the use of decision-making tools in a systematic way. Decision-making tools, such as 5 whys, cause and effects, and others should be encouraged and introduced in the form of workshops or classroom training.	4.33
14.1 Establish the "continuous improvement" firm culture. Employees are motivated to engage in continuous improvement activities on a regular basis.	4.33
14.2 The firm's culture should be shifted to allow seeing problems as opportunities. Employees are encouraged to expose problems.	4.17
14.3 Management should play the role of champions in taking the initiative to facilitate continuous improvement in the lower level of the firm. Management should also introduce recognition rewards for any *kaizen* improvement achieved on the project.	4.50
14.4 Efforts are also required to document the improvements made, especially the valid solutions that are generated.	4.83

Note: A five-point Likert scale is used from 1 (strongly disagree) to 5 (strongly agree) for measuring their level of agreement on the strategies listed above

Appendix F: Summary of the Hypotheses Testing

Hypotheses	Results	Remarks
H_1: Large Chinese construction firms have implemented implementation of Toyota Way principles	Partially supported	15 attributes (from principles) were found to be not significantly implemented ($p > 0.05$ or $p > 0.01$)
$H_{1.1}$: Large Chinese construction firms have implemented Toyota Way Principle 1 (long-term philosophy)	Supported	The significance level (p-value) of all attributes in Toyota Way principle 1 is <0.05
$H_{1.2}$: Large Chinese construction firms have implemented Toyota Way Principle 2 (one-piece flow)	Supported	The significance level (p-value) of all attributes in Toyota Way Principle 2 is <0.05
$H_{1.3}$: Large Chinese construction firms have implemented Toyota Way Principle 3 (pull *kanban* system)	Partially supported	P3.2 ($p > 0.01$) and P3.3 ($p > 0.05$) were found to be not significantly implemented
$H_{1.4}$: Large Chinese construction firms have implemented Toyota Way Principle 4 (level out workload)	Partially supported	P4.5 ($p > 0.01$) was found to be not significantly implemented
$H_{1.5}$: Large Chinese construction firms have implemented Toyota Way Principle 5 (built-in quality)	Supported	The significance level (p-value) of all attributes in Toyota Way Principle 5 is <0.05
$H_{1.6}$: Large Chinese construction firms have implemented Toyota Way Principle 6 (standardized work)	Partially supported	P6.4 ($p > 0.01$) and P6.5 ($p > 0.05$) were found to be not significantly implemented
$H_{1.7}$: Large Chinese construction firms have implemented Toyota Way Principle 7 (visual management)	Partially supported	P7.1 ($p > 0.05$), P7.2 ($p > 0.01$), P7.5 ($p > 0.05$) and P7.6 ($p > 0.05$) were found to be not significantly implemented
$H_{1.8}$: Large Chinese construction firms have implemented Toyota Way Principle 8 (adoption of reliable technology)	Supported	The significance level (p-value) of all attributes in Toyota Way Principle 8 is <0.05
$H_{1.9}$: Large Chinese construction firms have implemented Toyota Way Principle 9 (leaders and leadership)	Supported	The significance level (p-value) of all attributes in Toyota Way Principle 9 is <0.05
$H_{1.10}$: Large Chinese construction firms have implemented Toyota Way Principle 10 (partners relationship)	Partially supported	P10.4 ($p > 0.05$) was found to be not significantly implemented
$H_{1.11}$: Large Chinese construction firms have implemented Toyota Way Principle 11 (long-term partnership)	Partially supported	P11.4 ($p > 0.05$), P11.5 ($p > 0.05$) and P11.6 ($p > 0.05$) were found to be not significantly implemented

(continued)

S. Gao and S.P. Low, *Lean Construction Management*,
DOI 10.1007/978-981-287-014-8, © Springer Science+Business Media Singapore 2014

Hypotheses	Results	Remarks
$H_{1.12}$: Large Chinese construction firms have implemented Toyota Way Principle 12 (*genchi genbutsu*)	Supported	The significance level (p-value) of all attributes in Toyota Way Principle 12 is <0.05
$H_{1.13}$: Large Chinese construction firms have implemented Toyota Way Principle 13 (consensus decision-making)	Partially supported	P13.2 ($p > 0.05$) was found to be not significantly implemented
$H_{1.14}$: Large Chinese construction firms have implemented Toyota Way Principle 14 (continuous improvement)	Partially supported	P14.5 ($p > 0.05$) was found to be not significantly implemented
H_2: Large Chinese construction firms perceive the 14 Toyota Way principles as important factors in firm performance	Supported	The significance level (p-value) of all attributes is <0.01
H_3: There is a difference between the extent to which respondents perceived Toyota Way attributes as important and the extent to which they implemented the Toyota Way attributes	Partially supported	There was a statistically significant difference between perceived importance and implementation level for all ranges of Toyota Way attributes, expect P12.4 ($p = 0.928 > 0.05$) and P1.8 ($p = 0.299 > 0.05$)
$H_{4.1}$: There are hindrances when Chinese construction firms implement Toyota Way principles	Supported	All hindrance are statistically significant barriers at the significance level of <0.05
$H_{4.2}$: There are no significant differences in the mean scores on the perceived importance scale for hindrances to Toyota Way implementation between "premier" and "first graded" Chinese construction firms	Partially supported	There were statistically insignificant differences between the two groups in terms of the perceptions of hindrance to the Toyota Way implementation, except for the perception of H5 "foremen's insufficient knowledge on project planning", H14 "employee's tolerance for an untidy or disorganized workplace" and H18 "hierarchies in the organizational structure", which were statistically significant
H_5: The level of Toyota Way implementation has a direct correlation with the performance measurement	–	–
$H_{5.1}$: The level of Toyota Way implementation has a direct correlation with profitability	Partially supported	The correlation matrix (see Table 8.10) shows that of the 14 principles, the implementation level of *genchi genbutsu* (P3) is insignificantly correlated with profitability ($r = .150$, $p = 0.152$)
$H_{5.2}$: The level of Toyota Way implementation has a direct correlation with productivity	Supported	The correlation matrix (see Table 8.10) shows that all the Toyota Way principles are significantly correlated with productivity measures ($p > 0.05$)
$H_{5.3}$: The level of Toyota Way implementation has a direct correlation with quality	Supported	The correlation matrix (see Table 8.10) shows that all the Toyota Way principles are significantly correlated with quality measures ($p > 0.05$)

(continued)

Hypotheses	Results	Remarks
$H_{5.4}$: The level of Toyota Way implementation has a direct correlation with project delivery time	Supported	The correlation matrix (see Table 8.10) shows that all the Toyota Way principles are significantly correlated with delivery time ($p > 0.05$)
$H_{5.5}$: The level of Toyota Way implementation has a direct correlation with client satisfaction	Partially supported	The correlation matrix (see Table 8.10) shows that of the 14 principles, the implementation level of pull *kanban* system (P3) is insignificantly correlated with client satisfaction ($r = .189$, $p = 0.068$)

References

Abdel-Wahab, M. S., Dainty, A. R. J., Ison, S. G., Bowen, P., & Hazlehurst, G. (2008). Trends of skills and productivity in the UK construction industry. *Engineering, Construction and Architectural Management, 15*(4), 372–382.

Abduh, M., & Roza, H. A. (2006). *Indonesian contractors' readiness towards lean construction.* Proceedings of the 14th Annual Conference of the International Group for Lean Construction (IGLC-14), Santiago.

Abramowitz, I. (1967). *Production management: Concepts and analysis for operation and control.* New York, NY: Ronald Press.

Adam, E. E. J. R. (1983). Towards a typology of production and operations management system. *Academy of Management Review, 8*(3), 365–375.

Adler, P. S. (1999). Building better bureaucracies. *The Academy of Management Executive, 13*(4), 36–47.

Adler, P. S., & Borys, B. (1996). Two types of bureaucracy: Enabling and coercive. *Administrative Science Quarterly, 41*(1), 61–89.

Agbulos, A., Mohamed, Y., Al-hussein, M., Abourizk, S., & Roesch, J. (2006). Application of lean concepts and simulation analysis to improve efficiency of drainage operations maintenance crews. *Journal of Construction Engineering and Management, 132*(3), 291–299.

Aggarwal, R. L. (1985). MRP, JIT, OPT, FMS? *Harvard Business Review, 85*(5), 8–16.

Akintoye, A. (1995). Just-in-time application and implementation for building material management. *Construction Management and Economics, 13*(2), 105–113.

Ala-Risku, T., & Kärkkäinen, M. (2006). Material delivery problems in construction projects: A possible solution. *International Journal of Production Economics, 104*(1), 19–29.

Al-Sudairi, A. A. (2007). Evaluating the effect of construction process characteristics to the applicability of lean principles. *Construction Innovation: Information, Process, Management, 7*(1), 99 121.

Alex, D. P., Fernando, S., Al-Hussein, M., & Kung, D. (2008). Application of lean thinking to improve the productivity of water and sewer service installations. *Canadian Journal of Civil Engineering, 35*(4), 418–430.

Alinaitwe, H. M. (2009). Prioritizing lean construction barriers in Uganda's construction industry. *Journal of Construction in Developing Countries, 14*(1), 15–30.

Aminpour, S., & Woetzel, J. R. (2006). Applying lean manufacturing in China. *The McKinsey Quarterly, 2006*(special edition), 106–115.

Anderson, J. C., Rangtusanthan, M., & Schroeder, R. G. (1994). A theory of quality management underlying the Deming management method. *Academy of Management Review, 19*(3), 472–509.

Andrew, C. G., & Johnson, G. A. (1982). The crucial importance of production and operations management. *Academy of Management Review, 7*(1), 143–147.

Ang, Y. K., & Ofori, G. (2001). Chinese culture and successful implementation of partnering in Singapore's construction industry. *Construction Management and Economics, 19*(6), 619–632.

S. Gao and S.P. Low, *Lean Construction Management*,
DOI 10.1007/978-981-287-014-8, © Springer Science+Business Media Singapore 2014

Aoki, K. (2008). Transferring Japanese kaizen activities to overseas plants in China. *International Journal of Operations and Production Management, 28*(6), 518–539.

Arbulu, R., Ballard, G., & Harper, N. (2003). *Kanban in construction*. Proceedings of the 11th Annual Conference of the International Group for Lean Construction (IGLC-11), Blacksburg, VA, USA, 22–24 July.

Arif, M., & Egbu, C. (2010). Making a case for offsite construction in China. *Engineering, Construction and Architectural Management, 17*(6), 536–548.

Atkinson, P. (2010). Lean is a cultural issue. *Management Services, 54*(2), 35–41.

Bajaj, D., & Zhang, R. (2003). Managing construction industry development in China. *AACE International Transaction INT, 04*, 1–6.

Balfour Beatty. (2011). *Balfour Beatty updates, issue two 2011*. Available at: http://www.balfourbeatty.com/files/publications/2011/BB_update_issue2_2011.pdf. Accessed December 27, 2012.

Ball, M. (1988). *Rebuilding construction: Economic change and the British construction industry*. London: Routledge.

Ballard, G. (1999). *Improving work flow reliability* (pp. 275–286). Proceeding of the 7th Annual Conference of the International Group for Lean Construction (IGLC-7), University of California, Berkeley, CA, 26–28 July.

Ballard, G. (2000). *The last planner system of production control* (p. 198) (unpublished PhD thesis). Department of Civil Engineering, University of Birmingham, Birmingham.

Ballard, G., & Howell, G. (1994). *Implementing lean construction: Stabilizing work flow*. Proceeding of the 2nd Annual Conference of International Group for Lean Construction (IGLC-2), Santiago, Chile, 28–30 September.

Ballard, G., & Howell, G. (1998). Shielding production: An essential step in production control. *Journal of Construction Engineering in Management, 124*(1), 18–24.

Ballard, G., & Howell, G. (1998b). *What kind of production is construction?* Proceedings of the 6th Conference of the International Group for Lean Construction (IGLC-6), Guaruja, Brazil, 13–15 August.

Ballard, G., & Howell, G. (2003). *An update on last planner*. Proceedings of the 11th Annual Conference of the International Group for Lean Construction (LGIC-11), Blacksburg, VA, USA, July, 13 pp.

Ballard, G., & Zabelle, T. R. (2000). *Lean design: Process, tools, and techniques*. Lean Construction Institute White Paper No. 10, Lean Construction Institute, Ketchum, ID.

BAOYE. (2011). *Annual report 2011*. Available at: http://www.baoyegroup.com/NetGroup/Upload/201204/20120423203658051.pdf. Accessed April 25, 2012.

Barker, T. S., & Smith, H. W. (1997). Strategic planning: Evolution of a model. *Innovative Higher Education, 21*(4), 287–306.

Barrett, P., & Sexton, M. (2006). Innovation in small, project-based construction firms. *British Journal of Management, 17*(4), 331–346.

Basin, S., & Burcher, P. (2006). Lean viewed as a lean philosophy. *Journal of Manufacturing Technology Management, 17*(1), 56–72.

Basu, S. (1999). *Corporate purpose: Why it matters more than strategy*. New York: Garland Publishing.

Beach, R., Webster, M., & Campbell, K. M. (2005). An evaluation of partnership development in the construction industry. *International Journal of Project Management, 23*(8), 611–621.

Berggren, C. (1993). Lean production-the end of history? *Work, Employment and Society, 7(2)*, 163–188.

Bertalanffy, L. V. (1973). *General system theory: Foundations, development, applications*. New York, NY: George Braziller.

Bertelsen, S. (2004). Lean construction: Where are we and how to proceed? *Lean Construction Journal, 1*(1), 46–70.

Bhasin, S. (2012). Performance of Lean in large organisations. *Journal of Manufacturing Systems, 31*(2012), 349–357.

Black, T. R. (1999). *Doing quantitative research in the social sciences – an integrated approach to the research design: Measurement and statistics*. London: Sage.
Blythe, B. J., & Tripodi, T. T. (1989). *Measurement in direct social work practice*. Newbury Park, CA: Sage.
Bossink, B. A. G., & Brouwers, H. J. H. (1996). Construction waste: Quantification and source evaluation. *Journal of Construction Engineering and Management, 122*(1), 55–60.
Bresnen, M., & Marshall, N. (2001). Understanding the diffusion and application of new management ideas in construction. *Engineering, Construction and Architectural Management, 8*(5/6), 335–345.
Broad Group. (2012). *30-Story building built in 15 days – construction time lapse*. Available at: http://youtu.be/Hdpf-MQM9vY. Accessed January 25, 2012.
Brunet, A. P., & New, S. (2003). Kaizen in Japan: An empirical study. *International Journal of Operations and Production Management, 23*(12), 1426–1446.
Bryde, D. J., & Schulmeister, R. (2012). Applying Lean principles to a building refurbishment project: Experiences of key stakeholders. *Construction Management and Economics, 30*(9), 777–794.
Bryman, A. (2004). *Social research methods*. Oxford: Oxford University Press.
Bryman, A., & Bell, E. (2003). *Business research methods*. Oxford: Oxford University Press.
Burrill, C. W., & Ledolter, J. (1999). *Achieving quality through continual improvement*. New York, NY: John Wiley.
Cederblom, J. B., & Paulsen, D. W. (2001). *Critical reasoning: Understanding and criticizing arguments and theories*. Belmont, CA: Thomson Wadsworth.
Chan, A., & Fan, L. (1999). Quality management in construction. In S. K. M. Ho (Ed.), *Proceedings of the 4th international conference on ISO 9000 and total quality management* (pp. 167–187). Hong Kong: Hong Kong Baptist University.
Chan, W. K. L., Wong, F. K. W., & Scott, D. (1999). Managing construction projects in China – the transitional period in the Millennium. *International Journal of Project Management, 17*(4), 257–263.
Chase, R. B., & Aquilano, N. J. (1992). *Production and operations management: A life cycle approach* (6th ed.). Homewood, IL: IRWIN.
Cheah, C. Y. J., & Chew, D. A. S. (2005). Dynamics of strategic management in the Chinese construction industry. *Management Decision, 43*(4), 551–567.
Cheah, C. Y. J., Kang, J., & Chew, D. A. S. (2007). Strategic analysis of large local construction firm in China. *Construction Management and Economics, 25*(1), 25–38.
Chen, J. J. (1998). The characteristics and current status of China's construction industry. *Construction Management and Economics, 16*(6), 711–719.
Chen, Y. Q., Lu, H., Lu, W., & Zhang, N. (2010). Analysis of project delivery systems in Chinese construction industry with data envelopment analysis (DEA). *Engineering, Construction and Architectural Management, 17*(6), 598–614.
Chen, P., & Partington, D. (2004). An interpretive comparison of Chinese and Western conceptions of relationships in construction project management work. *International Journal of Project Management, 22*(5), 397–406.
Chen, I. J., & Paulraj, A. (2004). Towards a theory of supply chain management: The constructs and measurements. *Journal of Operations Management, 22*(2), 119–150.
Chen, P., Qiang, M. S., & Wang, J. N. (2009). Project management in the Chinese construction industry: Six-case study. *Journal of Construction Engineering and Management, 135*(10), 1016–1026.
Cheng, E. W. L., Li, H., Fang, D. P., & Xie, F. (2004). Construction safety management: An exploratory study from China. *Construction Innovation: Information, Process, Management, 4*(4), 229–241.
China Construction Association. (2009). *Guo Yunchong's (Vice Minister of Ministry of Housing and Urban–rural Development) speech at the 3rd Construction Quality Forum and Luban Award Presentation Ceremony*. Available at: http://www.zgjzy.org/guild/sites/ccia/detail.asp?i=LDJH&id=22548. Accessed December 2, 2010.

China Daily. (2010a). *Short-lived buildings create huge waste*. Available at: http://www.chinadaily.com.cn/cndy/2010-04/06/content_9687627.htm. Accessed April 14, 2010.
China Daily. (2010b). *Expo pavilion built in 14 hours*. Available at: http://www.chinadaily.com.cn/cndy/2010-03/15/content_9588476.htm. Accessed June 7, 2010.
China International Construction Association. (2010). *China international contractors association press conference successfully held*. Available at: http://www.chinca.org/templates/en/index.aspx?nodeid=127&page=ContentPage&contentid=10371. Accessed April 12, 2010.
Chini, A. R., & Valdez, H. E. (2003). ISO 9000 and the U.S. construction industry. *Journal of Management in Engineering, 19*(2), 69–77.
Cho, S., & Ballard, G. (2011). Last planner and integrated project delivery. *Lean Construction Journal, 2011*(issue), 67–78.
CIOB. (2009). *Building modern China: The men and women behind the biggest construction boom in history*. Berkshire: Chartered Institute of Building.
Clarke, L., & Wall, C. (2000). Craft versus industry: The division of labour in European housing. *Construction Management and Economics, 18*(6), 689–698.
Coleman, B. J., & Vaghefi, M. R. (1994). *Heijunka*: A key to the Toyota production system. *Journal of Production and Inventory Management, 35*(4), 31–35.
Collins, K. F., & Muthusamy, S. K. (2007). Applying the Toyota Production System to a healthcare organization: A case study on a rural community healthcare provider. *Quality Management Journal, 14*(4), 41–50.
Construction 21 Steering Committee. (1999). *Re-inventing construction*. Singapore: Ministry of Manpower and Ministry of National Development.
Conti, R., Angelis, J., Cooper, C., Faragher, B., & Gill, C. (2006). The effects of lean production on worker job stress. *International Journal of Operations and Production Management, 26*(9), 1013–1038.
Convis, G. (2001). Role of management in a lean manufacturing environment. *Automotive Manufacturing and Production, 7*, 1–7.
Cooke, B., & Williams, P. (2009). *Construction planning, programming and control* (3rd ed.). Ames, IA: Blackwell.
Coombs, R., Saviotti, P., & Walsh, V. (1987). *Economics and technological change*. Basingstoke: Macmillan Education.
Crawford, P., & Vogl, B. (2006). Measuring productivity in the construction industry. *Building Research and Information, 24*(3), 208–219.
Creswell, J. W. (1994). *Research design: Qualitative and quantitative approaches*. London: Sage.
Creswell, J. W. (2003). *Research design: Qualitative, quantitative and mixed methods approaches*. London: Sage.
Crowly, A. (1998). Construction as a manufacturing process: Lessons from the automotive industry. *Computers and Structures, 67*(5), 389–400.
Crowther, D., & Green, M. (2004). *Organizational theory*. London: Chartered Institute of Personnel and Development.
Cusumano, M. A. (1994). The limits of lean. *Sloan Management Review, 35*(4), 27–32.
Datamonitor. (2011). *China communication construction company limited*. Available at: http://www.datamonitor.com. Accessed December 2, 2011.
Deming, W. E. (1986). *Out of the crisis*. Cambridge, MA: Massachusetts Institute of Technology, Center for Advanced Engineering Study.
Dentz, J., Nahmens, I., & Mullens, M. (2009). Applying lean production in factory homebuilding. *A Journal of Policy Development and Research, 11*(1), 81–104.
Department of Industry, Science and Resources. (1999). *Building for growth: An analysis of the Australian building and construction industries (competitive Australia)*. Canberra: Commonwealth of Australia.
Diekmann, J. E., Krewedl, M., Balonick, J., Stewart, T., & Won, S. (2004). *Application of lean manufacturing principles to construction*. Austin, TX: The University of Texas at Austin.

Doolen, T., & Hacker, M. (2005). A review of Lean assessment in organizations: An exploratory study of lean practices by electronic manufacturers. *Journal of Manufacturing Systems, 24*(1), 55–67.

Drucker, P. F. (1986). *The practices of management*. New York, NY: Harper & Row Perennial Library.

Drucker, P. F. (1990). The emerging theory of manufacturing. *Harvard Business Review, 68*(3), 94–102.

Drucker, P. F. (1994). The theory of the business. *Harvard Business Review, 72*(5), 95–104.

Druker, J., White, G., Hegewisch, A., & Mayne, L. (1996). Between hard and soft HRM: Human resource management in the construction industry. *Construction Management and Economics, 14*(5), 405–416.

Dubrin, A. J. (2008). *Essentials of management* (8th ed.). Mason, OH: South-Western Cengage Learning.

Dulaimi, M. F., & Tanamas, C. (2001). *The principles and applications of lean construction in Singapore*. Proceedings of the 9th Annual Conference of the International Group for Lean Construction (IGLC-9), National University of Singapore, Singapore, 6–8 August.

Dyer, J. H., & Nobeoka, K. (2000). Creating and managing a high-performance knowledge-sharing network: The Toyota case. *Strategic Management Journal, 21*(3), 345–367.

Edum-Fotwe, F. T., & McCaffer, R. (2000). Developing project management competency: Perspectives from the construction industry. *International Journal of Project Management, 18*(2), 111–124.

Egan, J. (1998). *Rethinking construction* (the report of the construction task force). London: Department of Environment, Transport and Regions.

Egbu, C. (2006). Knowledge production and capabilities – their importance and challenges for construction organisations in China. *Journal of Technology Management in China, 1*(3), 304–321.

Eisenhardt, K. M. (1989). Building theories from case study research. *Academy of Management Review, 14*(4), 532–550.

Emiliani, M. L., & Stec, D. J. (2005). Leaders lost in transformation. *Leadership and Organization Development Journal, 26*(5), 370–387.

Emiliani, M. (2006). Origins of lean management in America: The role of Connecticut businesses. *Journal of Management History, 12*(2), 167–184.

Emmitt, S. (2007). *Design management for architects*. Oxford: Blackwell.

Engineering News-Record (ENR). (2012a). *The top 225 international contractors, ENR.com*. Available at: http://enr.construction.com/toplists/Top-International-Contractors/001-100.asp. Accessed December 10, 2012.

Engineering News-Record (ENR). (2012b). *The top 225 global contractors, ENR.com*. Available at: http://enr.construction.com/toplists/Top-Global-Contractors/001-100.asp. Accessed December 10, 2012.

Erhardt Construction. (2012). *Lean project delivery*. Available at: http://www.erhardtcc.com/about.php. Accessed December 28, 2012.

Evans, J. R., & Lindsay, W. M. (2008). *The management and control of quality* (7th ed.). Mason, OH: Thomson/South-Western.

Everett, R. J., & Sohal, A. S. (1991). Individual involvement and intervention in quality improvement programmes: Using the *andon* system. *International Journal of Quality and Reliability Management, 8*(2), 21–34.

Fan, Y. (2000). A classification of Chinese culture. *Cross Cultural Management: An International Journal, 7*(2), 3–10.

Fang, D. P., Huang, J. X., & Anderson, J. (2008). Expansion of Chinese economy and construction safety. *ICE Proceedings 'Municipal Engineer', 161*(1), 51–57. a special "Municipal Engineering in China" edition.

Fang, D. P., Li, M. G., Fong, P. S., & Shen, L. Y. (2004). Risks in Chinese construction market-contractors' perspective. *Journal of Construction Engineering and Management, 130*(6), 853–861.

Fellows, R., & Liu, A. M. M. (2008). *Research methods for construction* (3rd ed.). Oxford: Blackwell.

Filippini, R. (1997). Operations management research: Some reflections on evolution, models and empirical studies in OM. *International Journal of Operations and Production Management, 17* (7), 655–670.

Finnimore, B. (1989). *Houses from the factory: System building and the welfare state 1942–74.* London: Rivers Oram Press.

Flanagan, R., & Li, S. R. (1997). *International construction: A perspective on China.* Ascot, Berkshire: Chartered Institute of Building.

Forbes, L. H., Ahmed, S. M., & Barcala, M. (2002). *Adapting lean construction theory for practical application in developing countries* (pp. 219–227). Proceedings of the 1st CIB W107 International Conference: Creating a Sustainable Construction Industry in Developing Countries (Eds. Division of Building Technology, CSIR), Stellenbosch, South Africa, 11–13 November.

Forza, C. (1996). Work organization in lean production and traditional plants what are the differences? *International Journal of Operations and Production Management, 16*(2), 42–62.

Frisch, R. (1965). *Theory of production.* Dordrecht: D. Reidel.

Fullerton, R. R., & McWatters, C. S. (2001). The production performance benefits from JIT implementation. *Journal of Operations Management, 19*(1), 81–96.

Gaither, N., & Frazier, G. (1999). *Production and operations management* (8th ed.). Cincinnati, OH: South-Western College Publishing.

Galsworth, G. D. (2005). *Visual workplace, visual thinking: Creating enterprise excellence through the technologies of the visual workplace.* Portland, OR: Visual-Lean Enterprise Press.

Gann, D. M. (1996). Construction as a manufacturing process? Similarities and differences between industrialized housing and car production in Japan. *Construction Management and Economics, 14*(5), 437–450.

Gao, S., & Low, S. P. (2014). The Toyota way model: An alternative framework for lean construction. *Total Quality Management & Business Excellence, 25*(5–6), 664–682.

Gao, S., Low, S. P., Hwang, B. G., & Ofori, G. (2012). *Lean construction in large Chinese construction firms: A SWOT analysis.* Proceedings of the CIOB World Construction Conference 2012, Colombo, Sri Lanka, 28–30 June.

Garnett, N., Jones, D. T., & Myrray, S. (1998). *Strategic application of Lean thinking.* Proceedings of the 6th Annual Conference of the International Group for Lean Construction (IGLC-6), Guarujá, Brazil, 13–15 August.

Garrahan, P., & Stewart, P. (1992). *The Nissan enigma: Flexibility at work in a local economy.* London: Mansell.

Garvin, D. A. (1993). Building a learning organization. *Harvard Business Review, 71*(4), 78–91.

Gibb, A. G. F., & Isack, F. (2001). Client drivers for construction projects: Implications for standardization. *Engineering, Construction and Architectural Management, 8*(1), 46–58.

Goetsch, D. L., & Davis, S. B. (2009). *Quality management for organizational excellence: Introduction to total quality* (6th ed.). Upper Saddle River, NJ: Pearson.

Green, S. D. (1999). The missing arguments of lean construction. *Construction Management and Economic, 17*, 133–137.

Green, S. D. (2002). The human resource management implications of lean construction: Critical perspectives and conceptual chasms. *Journal of Construction Research, 3*(1), 147–166.

Green, S. D., & May, S. C. (2005). Lean construction: Arenas of enactment, models of diffusion and the meaning of "leanness". *Building Research and Information, 33*(6), 498–511.

Greenleaf, R. K. (1977). *Servant leadership: A journey into the nature of legitimate power and greatness.* New York, NY: Paulist Press.

Gupta, M. C., & Boyd, L. H. (2008). Theory of constraints: A theory for operations management. *International Journal of Operations and Production Management, 28*(10), 991–1012.

Haghirian, P. (2010). *Understanding Japanese management practices*. New York, NY: Business Expert.
Hall, R. W. (1983). *Zero inventories*. Homewood, IL: Dow Jones-Irwin.
Hall, A. (1995). A structure for organizational learning. *The Journal of Technology Transfer, 20*(3/4), 11–19.
Hall, M. A. (2001). 'Root' cause analysis: A tool for closer supply chain integration in construction. In A. Akintoye (Ed.), *17th Annual ARCOM Conference, 5–7 September 2001, University of Salford* (Vol. 1, pp. 929–938). Salford: Association of Researchers in Construction Management.
Hampson, I. (1999). Lean production and the Toyota Production System or, the case of the forgotten production concepts. *Economic and Industrial Democracy, 20*(3), 369–391.
Han, Y., Kakabadse, N. K., & Kakabadse, A. (2010). Servant leadership in the People's Republic of China: A case study of the public sector. *Journal of Management Development, 29*(3), 265–281.
Han, S. S., & Ofori, G. (2001). Construction industry in China's regional economy, 1990–1998. *Construction Management and Economics, 19*(2), 189–205.
Hao, F. F. (2010, March 6). Oversea investment became the Chinese State owned construction company's new focus (in Chinese). *China Construction News*, p 3. Available at: http://press.idoican.com.cn/detail/articles/20100306066031/. Accessed April 13, 2010.
Harber, D., Samson, D. A., Sohal, A. S., & Wirth, A. (1990). Just-in-time: The issue of implementation. *International Journal of Operations and Production Management, 10*(1), 21–30.
Hari, S., Egbu, C., & Kumar, B. (2005). A knowledge capture awareness tool: An empirical study on small and medium enterprises in the construction industry. *Engineering, Construction and Architectural Management, 12*(6), 533–567.
Hari, S., Egbu, C., & Kumar, B. (2006). A knowledge capture awareness tool. *Engineering, Construction and Architectural Management, 12*(6), 533–567.
Herzberg, F. (1959). *The motivation to work*. New York, NY: Wiley.
Hines, P., Holweg, M., & Rich, N. (2004). Learning to evolve: A review of contemporary lean thinking. *International Journal of Operations and Production Management, 24*(10), 994–1011.
Hirano, H. (1995). *5 pillars of the visual workplace: The sourcebook for 5-S implementation*. New York, NY: Productivity Press.
Ho, S. K. M. (1999). *Operations and quality management*. London: International Thomson Business.
Hofstede, G., & Bond, M. H. (1988). The Confucious connection: From cultural roots to economic growth. *Organizational Dynamics, 16*(Spring), 4–21.
Hofstede, G., & Hofstede, G. J. (2005). *Culture and organizations: Software of the mind* (2nd ed.). New York, NY: McGraw-Hill.
Höök, M., & Stehn, L. (2008). Applicability of lean principles and practices in industrialized housing production. *Construction Management and Economics, 26*(10), 1091–1100.
Hopp, W. J., & Spearman, M. L. (1996). *Factory physics: Foundations of manufacturing management*. Chicago, IL: Irwin.
Hopp, W. J., & Spearman, M. L. (2000). *Factory physics: Foundations of manufacturing management*. Boston, MA: McGraw-Hill.
Hormozi, A. M. (2001). Agile manufacturing: The next logical step. *Benchmarking: An International Journal, 8*(2), 132–143.
Howell, G. (1999). *What is lean construction?* (pp. 1–10). Proceeding of the 7th Annual Conference of the International Group for Lean Construction (IGLC-7), University of California, Berkeley, CA, 26–28 July.
Hu, D. Y. (2003). Status quo and prospect for the development of project management and general contracting in China (in Chinese). *Journal of Chinese Engineering Consultant, 2*, 10–18.
Huang, Y. L., Lan, Y., & Bai, Y. (2012). The development of the Chinese construction industry after the cultural revolution: Administration framework, economic growth, and market structure. *Journal of Architectural Engineering, 19*(1), 41–50.

Huda, F., & Preston, D. (1992). *Kaizen*: The applicability of Japanese techniques to IT. *Software Quality Journal, 1*(1), 9–26.

Hunt, S., & Morgan, R. (1994). Organizational commitment: One of many commitments or key mediating construct? *Academy of Management Journal, 37*(6), 1568–1587.

Imai, M. (1986). *Kaizen (Ky'zen): The key to Japan's competitive success.* New York, NY: McGraw Hill.

Imai, M. (1997). *Gemba Kaizen: A commonsense low-cost approach to management.* New York, NY: McGraw Hill.

Ishikawa, K. (1990). *Introduction to quality control.* Tokyo: 3A Corp.

J.D. Powers & Associates. (2013). *2013 U.S. vehicle dependability study.* Available at: http://autos.jdpower.com/ratings/dependability-press-release.htm. Accessed June 4, 2013.

Jang, J. W., & Kim, Y. W. (2007). *Using the Kanban for construction production and safety control.* Proceeding of the 15th Annual Conference of the International Group for Lean Construction (IGLC-15), University of Michigan, Michigan, 18–20 July.

Jankowicz, A. D. (2000). *Business research project* (3rd ed.). London: Chapman and Hall.

Johansen, E., & Walter, L. (2007). Lean construction: Prospects for the German construction industry. *Lean Construction Journal, 3*(1), 19–32.

Johnson, P., & Harris, D. (2002). Qualitative and quantitative issues in research design. In D. Partington (Ed.), *Essential skills for management research* (pp. 99–116). London: Sage Publications.

Jørgensen, B., & Emmitt, S. (2008). Lost in transition: The transfer of lean manufacturing to construction. *Engineering, Construction and Architectural Management, 15*(4), 383–398.

Kang, J. (2006). *Strategic analysis of large local construction companies in China.* Unpublished PhD thesis, Nanyang Technological University, Singapore.

Karlsson, C., & Åhlström, P. (1996). Assessing changes towards lean production. *International Journal of Operations and Production Management, 16*(2), 24–41.

Katayama, H., & Bennett, D. (1996). Lean production in a changing competitive world: A Japanese perspective. *International Journal of Operations and Production Management, 16*(2), 8–23.

Kenley, R., & Seppänen, O. (2010). *Location-based management for construction: Planning, scheduling and control.* London: Spon.

Keys, J. B., & Miller, T. R. (1984). The Japanese management theory jungle. *Academy of Management Review, 9*(2), 342–353.

Khalfan, M. M. A., McDermott, P., Oyegoke, A. S., Dickinson, M. T., & Li, X. (2008). *Application of kanban in the UK construction industry by public sector clients.* Proceeding of the 16th Annual Conference of the International Group for Lean Construction (IGLC-16), Manchester, UK, 16–18 July, pp. 347–357.

Khaw, B. W. (2012, February 29). Record breaking. *Housing Matters.* Available at: http://mndsingapore.wordpress.com/tag/construction-productivity/. Accessed May 27, 2013.

Kim, D. Y. (2002). *Exploratory study of lean consruction: Assessment of lean implementation.* Unpublished PhD thesis, The University of Texas at Austin, USA.

Kim, Y. W., & Jang, J. W. (2005). *Case study: An application of last planner to heavy civil construction in Korea.* Proceeding of the 13th Annual Conference of the International Group for Lean Construction (IGLC-13), Sydney, Australia.

Kim, J. B., & Michell, P. (1999). Relationship marketing in Japan: The buyer-supplier relationships of four automakers. *Journal of Business and Industrial Marketing, 14*(2), 118–129.

Kim, D. Y., & Park, H. S. (2006). Innovative construction management method: Assessment of lean construction implementation. *KSCE Journal of Civil Engineering, 10*(6), 381–388.

Klenke, K. (2008). *Qualitative research in the study of leadership.* Bingley: Emerald Group Publishing.

Koskela, L. (1992). *Application of the new production philosophy to construction* (CIFE Technical Report: 72). Stanford University, 75 p.

Koskela, L. (1999). *Management of production in construction: A theoretical view* (pp. 241–252). Proceeding of the 7th Annual Conference of the International Group for Lean Construction (IGLC-7), University of California, Berkeley, CA, 26–28 July.
Koskela, L. (2000). *An exploration towards a production theory and its application to construction*. Espoo, Finland: Technical Research Centre of Finland.
Koskela, L. (2004). Moving-on – beyond lean thinking. *Lean Construction Journal, 1*(1), 24–37.
Koskela, L., Howell, G., Ballard, G., & Tommelein, I. (2002). The foundations of lean construction. In B. Hellingsworth, R. Best, & G. de Valence (Eds.), *Design and construction: Building in value* (pp. 211–226). Amsterdam: Elsevier.
Koskela, L., & Vrijhoef, R. (2001). Is the current theory of construction a hindrance to innovation? *Building Research and Information, 29*(3), 197–207.
Kreitner, R. (2007). *Management* (10th ed.). Boston, MA: Houghton Mifflin Company.
Kululanga, G. K., Edum-Fotwe, F., & McCaffer, R. (2001). Measuring construction contractors organizational learning. *Building Research and Information, 29*(1), 21–29.
Kumar, J. (2005). *Research methodology: A step-by-step guide for beginners* (2nd ed.). London: Sage.
Lam, Y. T., & Cheng, Z. G. (2004). The development of the construction legal system in China. *Construction Management and Economics, 22*(4), 347–356.
Lan, P. (1999). Management in the Chinese construction industry. *Asia Pacific Business Review, 5* (3–4), 94–118.
Lan, P., & Jackson, J. T. (2002). Current characteristics of the main stakeholders in the Chinese industry. *International Journal for Construction Marketing, 1*(1). Available at: www.brookes.ac.uk/other/conmark/IJCM/issue_01/lan.html. Accessed May 5, 2009.
Lander, E., & Liker, J. K. (2008). The Toyota production system and art: Making highly customized and creative products the Toyota way. *International Journal of Production Research, 45*(16), 3681–3698.
Latham, M. (1994). *Constructing the team*. London: HMSO.
Lee, R. G., & Dale, B. G. (1998). Policy deployment: An examination of the theory. *International Journal of Quality and Reliability Management, 15*(5), 520–540.
Leonard, D. (2006). Building quality at Veridian Homes. *Quality Progress, 39*(10), 49–54.
Levinson, W. A., & Rerick, R. A. (2002). *Lean enterprise: A synergistic approach to minimizing waste*. Milwaukee, WI: ASQ Quality.
Levy, S. M. (1990). *Japanese construction: An American perspective*. New York, NY: Van Nostrand Reinhold.
Lewis, M. A. (2000). Lean production and sustainable competitive advantage. *International Journal of Operations and Production Management, 20*(8), 959–978.
Li, F. F. (2009). The largest Chinese firms continue to grow. *ENR: Engineering News-Record*, November 23, 2009, 263, p. 1-1.
Li, L. D. (2012). *Discussion on Lean management based on the profitability issues of SOE construction firms in China (in Chinese)*. Available at: http://www.zgjzy.org/NewsShow.aspx?id=2833. Accessed May 24, 2013.
Li, J. H., Anderson, A. R., & Harrison, R. T. (2003). Total quality management principles and practices in China. *International Journal of Quality and Reliability Management, 20*(9), 1026–1050.
Li, H., & Love, P. E. D. (1998). Developing a theory of construction problem solving. *Construction Management and Economics, 16*(6), 721–727.
Li, B., & Peng, H. (2006). *The social protection of rural workers in the construction industry in Urban China, CASE Paper 113*. London: LSE.
Liao, G. R. (2004). Discussion on honest and credit in construction market (in Chinese). *Optimization of Capital Construction, 25*(1), 20–21.
Li-Hua, R., & Khalil, T. M. (2006). Technology management in China: A global perspective and challenging issues. *Journal of Technology Management in China, 1*(1), 9–26.

Liker, J. K. (2004). *The Toyota way: 14 management principles from the world's greatest manufacturer*. New York, NY: McGraw-Hill.

Liker, J. K. (2010, January 28). Toyota's lost its quality edge? Not so fast. *Business Week*, Viewpoint.

Liker, J. K., & Choi, T. Y. (2004). Building deep supplier relationships. *Harvard Business Review, 82*(12), 104–113.

Liker, J. K., & Hoseus, M. (2008). *Toyota culture: The heart and soul of the Toyota way*. New York, NY: McGraw-Hill.

Liker, J. K., & Meier, D. (2006). *The Toyota way field book: A practical guide for implementing Toyota's 4Ps*. New York, NY: McGraw-Hill.

Lillrank, P. (1995). The transfer of management innovations from Japan. *Organization studies, 16* (6), 971–989.

Lincoln, Y. S., & Guba, E. G. (2000). Paradigmatic controversies, contradictions, and emerging confluences. In N. K. Denzin & Y. S. Lincoln (Eds.), *Handbook of qualitative research* (2nd ed., pp. 163–188). Thousand Oaks, CA: Sage.

Ling, F. Y. Y., Ibbs, C. W., & Kumaraswamy, M. M. (2005). Enablers that help foreign architectural, engineering, and construction firms win construction contracts in China. *Journal of Management in Engineering, 21*(2), 63–69.

Ling, F. Y. Y., Hartmann, A., Kumaraswamy, M., & Dulaimi, M. (2006). Influences on innovation benefits during implementation: Client's perspective. *Journal of Construction Engineering and Management, 133*(4), 306–315.

Ling, F. Y. Y., Ang, A. M. H., & Lim, S. S. Y. (2007). Encounters between foreigners and Chinese: Perception and management of cultural differences. *Engineering, Construction and Architectural Management, 14*(6), 501–518.

Ling, F. Y. Y., Ibbs, C., & Chew, E. (2008). Strategies adopted by international architectural, engineering, and construction firms in Southeast Asia. *Journal of Professional Issues in Engineering Education and Practice, 134*(3), 248–256.

Ling, F. Y. Y., Vu, M. C. P., & To, P. H. (2009). Strengths, weaknesses, opportunities, and threats for architectural, engineering, and construction firms: Case study of Vietnam. *Journal of Construction Engineering and Management, 135*(10), 1105–1113.

Liu, M., Ballard, G., & Ibbs, W. (2011). Work flow variation and labour productivity: Case study. *Journal of Management in Engineering, 27*(4), 236–242.

Liu, J. Y., & Low, S. P. (2007). Enhancing buildability in China's construction industry using Singapore's buildable design appraisal system. *Journal of Technology Management in China, 2*(3), 264–278.

Liu, G. W., Shen, Q. P., Li, H., & Shen, L. Y. (2004). Factors constraining the development of professional project management in China's construction industry. *International Journal of Project Management, 22*(3), 203–211.

Liu, Y. S., Zhao, X. F., & Liao, Y. P. (2013). Market structure, ownership structure and performance of China's construction industry. *Journal of Construction Engineering and Management, 139*(7), 852–857.

Locke, E. A. (1968). Toward a theory of task motivation and incentives. *Organizational Behavior and Human Performance, 3*(2), 157–189.

Locke, E. A., & Latham, G. P. (2002). Building a practically useful theory of goal setting and task motivation: A 35-year odyssey. *American Psychologist, 57*(9), 705–717.

Lockyer, K. G. (1984). Production management – the art of the possible. *International Journal of Operations and Production Management, 4*(4), 28–36.

Long, K. (2006). Unit of analysis. In M. S. Lewis-Beck, A. Bryman, & T. F. Liao (Eds.), *The SAGE encyclopedia of social science research methods*. Thousand Oaks, CA: Sage.

Loosemore, M., Dainty, A., & Lingard, H. (2003). *Human resource management in construction projects: Strategic and operational approaches*. London: Spon.

Love, P. E. D., Huang, J. C., Edwards, D. J., & Irani, Z. (2004). Nurturing a learning organization in construction: A focus on strategic shift, organizational transformation, customer orientation and quality centered learning. *Construction Innovation: Information, Process, Management, 4*(2), 113–126.

Love, P. E. D., & Irani, Z. (2004). An exploratory study of information technology evaluation and benefits management practices of SMEs in the construction industry. *Information and Management, 42*(1), 227–242.
Love, P. E. D., Irani, Z., & Edwards, D. J. (2003). Learning to reduce rework in projects: Analysis of firms learning and quality practices. *Project Management Journal, 34*(3), 13–25.
Love, P. E. D., & Li, H. (2000). Overcoming the problems associated with quality certification. *Construction Management and Economics, 18*(2), 139–149.
Low, S. P., & Ang, G. K. (2003). Integrating JIT and 5-S concepts for construction site management: A case study. *International Journal of Construction Management, 3*, 31–47.
Low, S. P., & Chan, Y. M. (1996). The application of Just-in-time principles to process layout for precast concrete production. *Singapore Management Review, 18*, 23–39.
Low, S. P., & Chan, Y. M. (1997). *Managing productivity in construction, JIT operations and measurements*. Brookfield, VT: Ashgate.
Low, S. P., & Choong, J. C. (2001). Just-in-time management in precast concrete construction: A survey of the readiness of main contractors in Singapore. *Integrated Manufacturing system, 12*(6), 416–429.
Low, S. P., & Jiang, H. B. (2003). Internationalization of Chinese construction enterprises. *Journal of Construction Engineering and Management, 129(6)*, 589–598.
Low, S. P., & Gao, S. (2011a). Bridging Western management theories and Japanese management practices: Case of the Toyota Way model. *Emerald Emerging Markets Case Studies, 1*(1), 1–20.
Low, S. P., & Gao, S. (2011b). The application of the Just-in-Time philosophy in the Chinese construction industry. *Journal of Construction in Developing Countries, 16*(1), 91–111.
Low, S. P., & Mok, S. H. (1999). The application of JIT philosophy to construction: A case study in site layout. *Construction Management and Economics, 17*(1), 37–48.
Low, S. P., & Teo, J. A. (2004). Implementing total quality management in construction firms. *Journal of Management in Engineering, 20*(1), 8–15.
Lu, X. (2002). Review of booming up and recession of precast building elements industry in China and redevelopment proposal (in Chinese). *Architecture Technology, 32*(2), 82–84.
Lu, Y. J., & Fox, P. W. (2001). *The construction industry in the 21st century: Its image, employment prospects and skill requirements: Case study from China.* Unpublished report for the ILO.
Lu, W. S., Li, H., Shen, L. Y., & Huang, T. (2009). Strengths, weaknesses, opportunities, and threats analysis of Chinese construction companies in the global market. *Journal of Management in Engineering, 25*(4), 166–176.
Lu, W. S., Shen, L. Y., & Yam, M. C. H. (2008). Critical success factors for competitiveness of contractors: China study. *Journal of Construction Engineering and Management, 132*(12), 972–982.
Lu, S., & Yan, H. (2007). An empirical study on incentives of strategic partnering in China: Views from construction companies. *International Journal of Project Management, 25*(3), 241–249.
Lutz, J. D., & Hijazi, A. (1992). Planning repetitive construction: Current practice. *Construction Management and Economics, 11*(2), 99–110.
Lyons, P., & Doueck, H. J. (2010). *The dissertation: From beginning to end.* New York, NY: Oxford University Press.
MacDuffie, J. P., & Helper, S. (1997). Creating lean suppliers. *California Management Review, 39*(4), 118–151.
Marksberry, P. (2011). The Toyota Way – an quantitative approach. *International Journal of Lean Six Sigma, 2*(2), 132–150.
Martinsons, M. G. (1994). *Why Chinese managers will not use MIS*. Vancouver: Pacific Rim Institute for Studies of Management.
Maslow, A. H. (1954). *Motivation and personality*. New York, NY: Harper.
Maturana, S., Alarcón, L., Gazmuri, P., & Vrsalovic, M. (2007). On-site subcontractor evaluation method based on lean principles and partnering practices. *Journal of Management in Engineering, 23*(2), 67–74.

McCormick, R. E. (1993). *Managerial economics*. Englewood Cliffs, NJ: Prentice Hall.

McGeorge, D., & Palmer, A. (2002). *Construction management: New directions*. Oxford: Blackwell Science.

McGill, M. E., & Slocum, J. W. (1993). Unlearning the organization. *Organizational Dynamics, 22*(2), 67–79.

McGregor, D. (1960). *The human side of enterprise*. New York, NY: McGraw-Hill.

Miller, C. J. M., Packham, G. A., & Thomas, B. C. (2002). Harmonization between main contractors and subcontractors: A prerequisite for lean construction. *Journal of Construction Research, 3*(1), 67–82.

Milosevic, I. N. (2010). Application of SWOT analysis in the management of a construction project. *Leadership and Management in Engineering, 10*(2), 78–86.

Miltenburg, J. (2001). U-shaped production lines: A review of theory and practice. *International Journal of Production Economics, 70*(3), 201–214.

Ministry of Commerce. (2009). *A brief statistics on revenue from projects outside China between 2000–2009*. People's Republic of China: Department of Outward Investment and Economic Cooperation.

Ministry of Housing and Urban-Rural Development. (2008). *Reform and development: A report on construction industry and market in China (in Chinese)*. Beijing: China Architecture & Building.

Ministry of Housing and Urban-Rural Development. (2009a). *Reform and development: A report on construction industry and market in China (in Chinese)*. Beijing: China Architecture & Building.

Ministry of Housing and Urban-Rural Development. (2009b). *A briefing on national quality supervision and enforcement inspection in 2009 (in Chinese)*. Available at: http://www.cin.gov.cn/zcfg/jswj/gczl/200912/t20091229_198788.htm. Accessed January 2, 2010.

Ministry of Housing and Urban-Rural Development. (2010a). *A notice on continually carry out the State Council General Office on "safety year" activities (in Chinese)*. Available at: http://www.cin.gov.cn/zcfg/gwywj/201002/t20100222_199792.htm. Accessed March 2, 2010.

Ministry of Housing and Urban-Rural Development. (2010b). *Reform and development: A report on construction industry and market in China (in Chinese)*. Beijing: China Architecture & Building.

Monden, Y. (1983). *Toyota production system: Practical approach to production management*. Norcross, GA: Engineering and Management.

Monden, Y. (1998). *Toyota production system* (3rd ed.). Norcross, GA: Engineering and Management.

Moore, R. (2007). *Selecting the right manufacturing improvement tools: What tool? When?* Amsterdam: Elsevier Butterworth-Heinemann.

Morgan, J. M., & Liker, J. K. (2006). *The Toyota product development system: Integrating people, process, and technology*. New York, NY: Productivity Press.

Morris, T., & Lancaster, Z. (2005). Translating management ideas. *Organization Studies, 27*(2), 207–233.

Mullins, L. J. (2006). *Essentials of organizational behavior*. Harlow: Financial Times.

Nahmens, I., Ikuma, L. H., & Khot, D. (2012). Lean and job satisfaction: A case study in industrialized homebuilding. *Lean Construction Journal, 2012*, 91–104.

Nakagawa, Y. (2006). Importance of standard operating procedure documents and visualisation to implement lean construction. In R. Kenley (Ed.), Proceedings of the 13th International Group for Lean Construction (IGLC-13), Sydney, Australia, July 2005, pp. 207–215.

Naoum, S. G. (2007). *Dissertation research and writing for construction students* (2nd ed.). Oxford: Butterworth-Heinemann.

National Audit Office of China. (2012). *No. 3 of 2012 (General Serial No.110): Follow-up audit results of construction project for Beijing-Shanghai high-speed railway in 2011*. Available at: http://www.cnao.gov.cn/main/articleshow_ArtID_1209.htm. Accessed May 17, 2013.

National Bureau of Statistics of China (NBSC). (2003). *China statistic yearbook 2003*. Beijing: China Statistic. Available at: http://www.stats.gov.cn/tjsj/ndsj/. Accessed March 2, 2009.

National Bureau of Statistics of China (NBSC). (2008). *China statistic yearbook 2008*. Beijing: China Statistic. Available at: http://www.stats.gov.cn/tjsj/ndsj/. Accessed March 2, 2009.

National Bureau of Statistics of China (NBSC). (2011). *China statistic yearbook 2011*. Beijing: China Statistic. Available at: http://www.stats.gov.cn/tjsj/ndsj/. Accessed December 23, 2012.

Nave, D. (2002). How to compare six sigma, lean and the theory of constraints: A framework for choosing what's best for your organization. *Quality Progress, 35*(3), 73–78.

Neely, A. D. (1991). Production management: A two-dimensional function? *International Journal of Operations and Production Management, 11*(7), 49–54.

Nesan, L. J., & Holt, G. D. (1999). *Empowerment in construction: The way forward for performance improvement*. Baldock: Research Studies.

O'Connor, P. D. T. (1994). *The practice of engineering management: A new approach*. Chichester: Wiley.

O'Grady, P. J. (1988). *Putting the just-in-time philosophy into practice*. New York, NY: Nichols.

O'Keefe, R. M., Balci, O., & Smith, E. P. (1986). *Validation of expert system performance*. Working paper (TR 86–37), Virginia Tech.

Ogawa, E. (1984). *Modern production management: A Japanese experience*. Tokyo: Asian Productivity Organization.

Oglesby, C. H., Parker, H. W., & Howell, G. A. (1989). *Productivity improvement in construction*. New York, NY: McGraw-Hill.

Ohno, T. (1988). *Toyota production system: Beyond large-scale production*. Cambridge, MA: Productivity Press.

Olatunji, J. O. (2008). *Lean in Nigerian construction: State, barriers, strategies and "go-to-gemba" approach*. Proceeding of the 16nd Annual Conference of the International Group for Lean construction (IGLC-16), Manchester, United Kingdom, 16–18 July.

Olivella, J., Cuatrecasas, L., & Gavilan, N. (2008). Work organization practices for lean production. *Journal of Manufacturing Technology Management, 19*(7), 798–811.

Oliver, N., & Wilkinson, B. (1992). *The Japanization of British industry*. Oxford: Blackwell Business.

Orr, C. (2005) *Lean leadership in construction*. Proceeding of the 13th Annual Conference of the International Group for Lean Construction (IGLC-13), University of Sydney, Australia, 19–21 July.

Osono, E., Shimizu, N., & Takeuchi, H. (2008). *Exteme Toyota: Radical contradictions that drive success at the world's best manufacturer*. New York: Wiley.

Ouchi, W. C. (1981). *Theory Z: How American business can meet the Japanese challenge*. Reading, MA: Addison-Wesley.

Paez, O., Dewees, J., Genaidy, A., Tuncel, S., Karwowski, W., & Zurada, J. (2004). The lean manufacturing enterprise: An emerging socio-technological system integration. *Human Factors and Ergonomics in Manufacturing, 14*(3), 285–306.

Paez, O., Salem, S., Solomon, J., & Genaidy, A. (2005). Moving from lean manufacturing to lean construction: Toward a common socio-technological framework. *Human Factors and Ergonomics in Manufacturing, 15*(2), 233–245.

Panizzolo, R. (1998). Applying the lessons learned from 27 lean manufacturers: The relevance of relationships management. *International Journal Production Economics, 55*(3), 223–240.

Paolini, A., Leu, B., & Chinn, R. (2005). Exporting lean to China: Know before you go. *PRTM Insight, 17*(2), 133–139.

Pappas, M. (1999). *Evaluating innovative construction management methods through the assessment of intermediate impacts*. An unpublished thesis for a Master of Science in Engineering, University of Texas at Austin.

Pearson, G. J. (2009). *The rise and fall of management: A brief history of practice, theory and context*. Burlington, VT: Gower.

Perloff, J. M. (2001). *Microeconomics*. Boston, MA: Addison-Wesley Longman.

Picchi, F. A., & Granja, A. D. (2004). *Construction sites: Using lean principles to seek broader implementations.* Proceedings of the 12th Annual Conference of the International Group for Lean Construction (IGLC-12), Helsingør, Denmark, 3–6 August.

Pil, F. K., & Fujimoto, T. (2007). Lean and reflective production: The dynamic nature of production models. *International Journal of Production Research, 45*(16), 3741–3761.

Polat, G., & Arditi, D. (2005). The JIT materials management system in developing countries. *Construction Management and Economics, 23*(7), 697–712.

Polat, G., & Ballard, G. (2004). *Waste in Turkish construction: Need for lean construction.* Proceedings of the 12th Annual Conference of the International Group for Lean Construction (IGLC-12), Helsingør, Denmark, 3–6 August.

Pun, K. F. (2001). Cultural influences on total quality management adoption in Chinese enterprises: An empirical study. *Total Quality Management, 12*(3), 323–342.

Pun, N., & Lu, H. L. (2010). A culture of violence: The labour subcontracting system and collective action by construction workers in post-socialist China. *The China Journal, 64* (July), 143–158.

Ramaswamy, K. P., & Kalidindi, S. N. (2009). *Waste in Indian building construction projects.* Proceedings of the 17th Annual Conference of the International Group for Lean Construction (IGLC-17), Taiwan, 15–17 July.

Respect for People Working Group. (2004). *Respect for people: A framework for action.* London: Constructing Excellence.

Riley, M. J., & Clare-Brown, D. (2001). Comparison of cultures in construction and manufacturing industries. *Journal of management in Engineering, 17*(3), 149–158.

Russell, R. F., & Stone, A. G. (2002). A review of servant leadership attributes: Developing a practical model. *The Leadership and Organization Development Journal, 23*(3), 145–157.

Sacks, R., & Goldin, M. (2007). Lean management model for construction of high-rise apartment buildings. *Journal of construction engineering and Management, 133*(5), 374–384.

Sacks, R., Treckmann, M., & Rozenfeld, O. (2009). Computer-aided visualization of work flow to support lean construction. *Journal of Construction Engineering and Management, 135*(12), 1307–1315.

Salem, O., Solomon, J., Genaidy, A., & Mihkarah, I. (2006). Lean construction: From theory to implementation. *Journal of Management in Engineering, 22*(4), 168–175.

Sanchez, A. M., & Perez, M. P. (2001). Lean indicators and manufacturing strategies. *International Journal of Operations and Production Management, 21*(11), 1433–1451.

Santos, A. (1999). *Application of fow principles in the production management of construction sites.* Unpublished PhD thesis, University of Salford, Salford, UK

Santos, A., Formoso, C. T., & Tookey, J. E. (2002a). Expanding the meaning of standardization within construction processes. *The TQM Magazine, 14*(1), 25–33.

Santos, A., Powell, J. A., & Sarshar, M. (2002b). Evolution of management theory: The case of production management in construction. *Management Decision, 40*(8), 788–796.

Scazzieri, R. (1993). *A theory of production: Tasks, processes, and technical practices.* Oxford: Clarendon.

Schmenner, R. W., & Swink, M. L. (1998). On theory in operations management. *Journal of Operations Management, 17*(1), 97–113.

Senaratne, S., & Wijesiri, D. (2008). Lean construction as a strategic option: Testing its suitability and acceptability in Sri Lanka. *Lean Construction Journal, 2008*, 34–48.

Senge, P. M. (1990). *The fifth discipline: The art and practice of the learning organization.* New York, NY: Doubleday/Currency.

Sepheri, M. (1986). How *kanban* system is used in an American Toyota motor facility. *Industrial Engineering, 17*(2), 50–56.

Shah, R., & Ward, P. T. (2007). Defining and developing measures of lean production. *Journal of Operations Management, 25*(4), 785–805.

Shelderake, J. (1996). *Management theory: From taylorism to Japanization.* London: International Thomson Business.

Shen, Z. G., Jensen, W., Berryman, C., & Zhu, Y. M. (2011). Comparative study of activity-based construction labor productivity in the United States and China. *Journal of Management in Engineering, 27*(2), 116–124.

Shewhart, W. A. (1931). *Economic control of quality of manufactured products*. New York, NY: D. Van Nostrand.

Shingo, S. (1988). *Non-stock production: The Shingo system for continuous improvement*. Cambridge, MA: Productivity Press.

Shook, J. Y. (1998). Bringing the Toyota production system to the United States: A personal perspective. In J. K. Liker (Ed.), *Becoming lean* (pp. 41–69). Portland: Productivity Press.

Shook, J. Y. (2009). Toyota's secret: The A3 report. *MIT Sloan Management Review, 50*(4), 30–33.

Sim, K. L., & Rogers, J. W. (2009). Implementing lean production systems: Barriers to change. *Management Research News, 32*(1), 37–49.

Singh, N., Zhao, H. X., & Hu, X. R. (2003). Cultural adaptation on the web: A study of American companies' domestic and Chinese websites. *Journal of Global Information Management, 11*(3), 63–80.

Skanska. (2009). *Skanska worldwide*. Available at: http://group.skanska.com/Global/Media/Worldwide/2009/Worldwide-2-2009.pdf. Accessed December 28, 2012.

Skinner, B. F. (1948). *The behavior of organisms: An experimental analysis*. New York, NY: D. Appleton-Century.

Slattery, D. K., & Sumner, M. R. (2011). Leadership characteristics of rising stars in construction project management. *International Journal of Construction Education and Research, 7*(3), 159–174.

Smith, D. C. (1992). From lean to nimble: Is agile production the next paradigm? *Ward's Auto World, September*, 7.

Sobek, D. K., & Smalley, A. (2008). *Understanding A3 thinking: A critical component of Toyota's PDCA management system*. Boca Raton, FL: CRC.

Song, L. G., & Liang, D. A. (2011). Lean construction implementation and its implication on sustainability: A contractors' case study. *Canadian Journal of Civil Engineering, 38*(1), 1–10.

South China Morning Post. (2012). *China must improve quality of construction or face scepticism*. Available at: http://www.scmp.com/comment/insight-opinion/article/1024730/china-must-improve-quality-construction-or-face-scepticism. Accessed October 2, 2012.

Sowards, D. (2007). Lean construction: New tools and old combine to reduce waste. *Quality Digest, 27*(100), 32–36.

Spear, S., & Bowen, H. K. (1999). The DNA of the Toyota production system. *Harvard Business Review, 77*(5), 97–106.

Sugimori, Y., Kusunoki, K., Cho, F., & Uchikawa, S. (1977). Toyota production system and *Kanban* system materialization of just in time and respect for human system. *International Journal of Production Research, 15*(6), 553–564

Taj, S. (2005). Applying lean assessment tools in Chinese hi-tech industries. *Management Decision, 43*(4), 628–643.

Tang, S. L., Ahmed, S. M., Aoieong, R. T., & Poon, S. W. (2005). *Construction quality management*. Hong Kong: Hong Kong University Press.

Tapping, D., Luyster, T., & Shuker, T. (2002). *Value stream management: Eight steps to planning, mapping, and sustaining lean improvement*. New York, NY: Productivity Press.

Taylor, F. W. (1934). *The principles of scientific management*. New York, NY: Harper.

Tener, R. K. (1993). Empowering high-performing people to promote project quality. *Journal of Management in Engineering, 9*(4), 321–328.

Tezel, A., Koskela, L., & Tzortzopoulos, P. (2010). *Visual management in construction: Study report on Brazilian cases*. Salford: Salford Centre for Research and Innovation in the Built and Human Environment (SCRI), University of Salford.

Thomas, H., Horman, M., Minchin, R., & Chen, D. (2003). Improving labor flow reliability for better productivity as lean construction principle. *Journal of Construction Engineering Management, 129*(3), 251–261.

Thomas, H., Riley, D., & Messner, J. (2005). Fundamental principles of site material management. *Journal of Construction Engineering Management, 131*(7), 808–815.

Thomassen, M. A., Sander, D., Barnes, K. A., & Nielsen, A. (2003). *Experience and results from implementing lean construction in a large Danish contracting firms*. Proceedings of the 11th Annual Conference of the International Group for Lean Construction (IGLC-11), Blacksburg, VA, USA, 22–24 July.

Tommelein, I. D. (1998). Pull-driven scheduling for pipe-spool installation: Simulation of a lean construction technique. *Journal of Construction Engineering Management, 124*(4), 279–288.

Toyota Motor Corporation. (2003). *Environment and social report 2003*. Tokyo: Toyota Motor Corporation. 84 pp.

Toyota Motor Corporation. (2009). *Financial summary-FY 2009*. Tokyo: Toyota Motor Corporation.

Tsuji, M. (2003). Transformation of the Japanese system towards a network economy. In E. Giovannetti, M. Kagami, & M. Tsuji (Eds.), *The internet revolution: A global perspective* (pp. 7–20). Cambridge: Cambridge University Press.

Tsutsui, W. M. (1998). *Manufacturing ideology: Scientific management in twentieth-century Japan*. Princeton, NJ: Princeton University Press.

Tucker, A. L., Edmondson, A. C., & Spear, S. (2002). When problem solving prevents organizational learning. *Journal of Organizational Change Management, 15*(2), 122–137.

Turner Construction Company. (2012). *Benefits of lean construction*. Available at: http://www.turnerconstruction.com/experience/lean/lean-read-more. Accessed December 27, 2012.

Valence, G. (2010). Theory and construction economics. In G. Valence (Ed.), *Modern construction economics: Theory and application*. Abingdon, Oxon: Spon Press.

Valmohammadi, C. (2011). The impact of TQM implementation on the organizational performance of Iranian manufacturing SMEs. *The TQM Journal, 23*(5), 496–509.

Vogel, E. F. (1975). *Modern Japanese organization and decision-making*. Berkeley, CA: University of California Press.

Voss, C., Tsikriktsis, N., & Frohlich, M. (2002). Case research in operations management. *International Journal of Operations and Production Management, 22*(2), 195–219.

Wang, P. (2012, August 13). Shrinking profits in Shenzhen's construction sector in the first half of year 2012. *China Construction News*, P1. Available at: http://www.chinajsb.cn/bz/content/2012-08/13/content_67545.htm. Accessed June 7, 2013.

Wang, D. S., Hadavi, A., & Krizek, R. J. (2006). Chinese construction firms in reform. *Construction Management and Economics, 24*(5), 509–519.

Wang, X., & Huang, J. (2006). The relationships between key stakeholders' project performance and project success: Perceptions of Chinese construction supervising engineers. *International Journal of Project Management, 24*(3), 253–260.

Wang, J. Y., Liu, J., Liao, Z. G., & Tang, P. (2009). Identification of key liability risks of supervision engineers in China. *Construction Management and Economics, 27*(12), 1157–1173.

White, B. (2000). *Dissertation skills for business and management students*. London: Cassell.

White, R. E., Pearson, J. N., & Wilson, J. R. (1999). JIT manufacturing: A survey of implementation in small and large U.S. manufacturers. *Management Science, 45*(1), 1–15.

Whitelaw, R. R. P. (1969). *Marketing and economics: An introduction to the use of economic indicators*. University of Minnesota: Pergamon Press.

Williams, K., Haslam, C., Williams, J., Clutler, T., Adcroft, A., & Willaiams, S. (1992). Against lean production. *Economy and Society, 21*(3), 321–354.

Winch, G. M. (2003). Models of manufacturing and the construction process: The genesis of re-engineering construction. *Building Research and Information, 31*(2), 107–118.

Winch, G. M. (2006). Towards a theory of construction as production by projects. *Building Research and Information, 34*(2), 154–163.

Winch, G. M. (2010). *Managing construction projects: An information processing approach* (2nd ed.). Oxford: Blackwell Science.

Womack, J. P., & Jones, D. T. (1996). *Lean thinking: Banish waste and create wealth in your corporation*. New York, NY: Simon & Schuster.

Womack, J. P., Jones, D. T., & Roos, D. (1990). *The machine that changed the world*. New York, NY: Rawson Associates.
Wren, D. A., & Bedeian, A. G. (2009). *The evolution of management thought*. Hoboken, NJ: Wiley.
Wyman, O. (2008). *The Harbour Report™ North America 2008*. Detroit: Oliver Wyman.
Xiao, H., & Proverbs, D. (2002). Construction time performance: An evaluation of contractors from Japan, the UK and the US. *Engineering Construction and Architectural Management, 9*(2), 81–89.
Xinhua. (2008). *Subway tunnel collapse leaves one dead, 18 missing*. Accessed October 19, 2011, from http://news.xinhuanet.com/english/2008-11/15/content_10363256.htm
Xinhua. (2009a). *China to be largest construction market by 2018*. Available at: http://news.xinhuanet.com/english/2009-11/13/content_12445518.htm. Accessed January 29, 2010.
Xinhua. (2009b). *Skills training key to future for China's jobless migrants*. Available at: http://news.xinhuanet.com/english/2009-03/13/content_11007727.htm. Accessed March 17, 2009.
Xinhua. (2012). *China regulates contracting to improve construction quality*. Available at: http://www.chinadaily.com.cn/business/2012-11/23/content_15952681.htm. Accessed May 16, 2013.
Xiong, L. (2007). A reflection on the status quo of the construction project profits in China (in Chinese). *Enterprise Economy, 322*(6), 105–108.
Xu, T., & Greenwood, D. (2006). Using design-and build as an entry strategy to the Chinese construction market. *International Journal of Project Management, 24*(5), 438–445.
Xu, T., Tiong, R. L. K., Chew, D. A. S., & Smith, N. J. (2005). Development model for competitive construction industry in the People's Republic of China. *Journal of Construction Engineering and Management, 131*(7), 844–853.
Xue, X. L., Shen, Q. P., Wang, Y. W., & Lu, J. F. (2008). Measuring the productivity of the construction industry in China by using DEA-based malmquist productivity indices. *Journal of Construction Engineering and Management, 134*(1), 64–71.
Yin, R. K. (1994). *Case study research: Design and methods* (2nd ed.). Beverly Hills, CA: Sage.
Yin, R. K. (2006). Case study methods. In L. G. Judith, C. Gregory, & B. E. Patricia (Eds.), *Handbook of complementary methods in education research*. Washington, DC: American Educational Research Association.
Yin, R. K. (2008). *Case study research: Design and methods* (4th ed.). Sage: Thousand Oaks, CA.
Yu, H. T., Al-Hussein, M., Al-Jibouri, S., & Telyas, A. (2013). Lean transformation in a modular building company: A case for implementation. *Journal of Management in Engineering, 29*(1), 103–111.
Yu, H. T., Tweed, T., Al-Hussein, M., & Nasseri, R. (2009). Development of lean model for house construction using value stream mapping. *Journal of Construction Engineering and Management, 135*(8), 782–790.
Yung, P. (2009). Institutional arrangements and construction safety in China: An empirical examination. *Construction Management and Economics, 27*(5), 439–450.
Yung, P., & Yip, B. (2009). Construction quality in China during transition: A review of literature and empirical examination. *International Journal of Project Management, 28*(1), 79–91.
Yusuf, Y., Gunasekaran, A., & Dan, G. (2007). Implementation of TQM in China and organizational performance: An empirical investigation. *Total Quality Management, 18*(5), 509–530.
Zangwill, W. I., & Kantor, P. B. (1998). Toward a theory of continuous improvement and the learning curve. *Management Science, 44*(7), 910–920.
Zeng, S. X., Chen, H. M., & Tam, C. M. (2005). Market structure of the construction industry of China. *Architectural Science Review, 48*(4), 367–376.
Zeng, S. X., Tam, C. M., Deng, Z. M., & Wang, H. C. (2003). The rise and fall of total quality management in the Chinese construction industry. *Architectural Science Review, 46*(2), 105–114.
Zhang, Z. H., Waszink, A., & Wijngaard, J. (2000). An instrument for measuring TQM implementation for Chinese manufacturing companies. *International Journal of Quality and Reliability, 17*(7), 730–755.
Zhao, K. Q. (2007). Is lean applicable in China? (in Chinese). *PKU Business Review, 35*(6), 50–52.

Zhao, Z. Y., & Shen, L. Y. (2008). Are Chinese contractors competitive in international markets? *Construction Management and Economics, 26*(3), 225–236.

Zheng, Y. J. (2008). *Zheng Yi-jun's speech at the conference on "National Scientific and Technological Progress and Technology Innovations in the Construction Industry"*. Available at: http://www.cpmchina.com/book/story.php?id=11. Accessed March 11, 2010.

Zou, P. X. W., Fang, D. P., Wang, S. Q., & Loosemore, M. (2007). An overview of the Chinese construction market and construction management practice. *Journal of Technology Management in China, 2*(2), 163–176.

Zuo, J., Zillante, G., & Coffey, V. (2009). Project culture in the Chinese construction industry: Perceptions of contractors. *Australasian Journal of Construction Economics and Building, 9*(2), 17–28.

MIX
Papier aus verantwortungsvollen Quellen
Paper from responsible sources
FSC® C105338

If you have any concerns about our products,
you can contact us on
ProductSafety@springernature.com

In case Publisher is established outside the EU,
the EU authorized representative is:
Springer Nature Customer Service Center GmbH
Europaplatz 3, 69115 Heidelberg, Germany

Printed by Libri Plureos GmbH
in Hamburg, Germany